Statis
Second Edition

Frank Owen
Ron Jones

LONGMAN
London and New York

First Published 1977
Second Edition 1982
Fourth Impression 1985

All rights reserved. No part of this publication may be reproduced, stored in a retrieval system, or transmitted in any form or by any means, electronic, mechanical, photocopying, recording or otherwise, without the prior permission of the copyright owner.

© 1977, 1982 Copyright Polytech Publishers Limited

ISBN 0 85505 063 2

Produced by Longman Group (FE) Ltd
Printed in Hong Kong

Contents

		Preface	v
Chapter	1	The Organisation of Data	1
Chapter	2	The Presentation of Data I	18
Chapter	3	The Presentation of Data II	40
Chapter	4	Approximation and Error	59
Chapter	5	The Averaging of Data	71
Chapter	6	The Time Series I	98
Chapter	7	The Time Series II	106
Chapter	8	Index Numbers	126
Chapter	9	Dispersion	142
Chapter	10	The Normal Distribution	161
Chapter	11	Probability	171
Chapter	12	Decision Criteria	196
Chapter	13	Probability Distributions	211
Chapter	14	Sample Design	227
Chapter	15	Planning a Sample Survey	235
Chapter	16	Sampling Theory	250
Chapter	17	Quality Control	267
Chapter	18	Estimation	280
Chapter	19	Hypothesis Testing	298
Chapter	20	Statistical Significance	309
Chapter	21	The Chi Square Test	330
Chapter	22	Regression and Correlation	347
Chapter	23	Statistical Sources	379
		Answers to Questions	396
		Tables	407
		Index	414

Preface to First Edition

Recent years have seen many changes in both the content and style of examinations of the standard which used to be called 'Introduction to Statistics'. These changes reflect the growing belief that it is not enough to be able to perform statistical calculations. Important as it s to have some degree of calculative ability, this alone is no longer sufficient to ensure a pass in the examination room. More and more it is becoming necessary for the student to understand, not only what he is doing, but also the meaning of the results he obtains.

In surveying the literature, it seems to us that although there are many excellent books on statistical method there is a marked deficiency of textbooks designed to cover the present first-year syllabus of the major professional bodies. It is this gap that this book is designed to fill. We hope that it will prove to give full coverage of the Ordinary National Certificate Statistics examinations in both Business Studies and in Public Administration, the examination of the Institute of Certified Accountants, and the Institute of Secretaries and Administrators. Additionally it covers the content of the Statistics section of the Institute of Cost and Management Accountants paper in Mathematics and Statistics, and should prove useful for large parts of the Quantitative Methods paper of the Institute of Public Finance and Accountancy.

Every author knows that his book is never entirely his own work. We owe a great deal to the comments and criticisms of our colleagues at Liverpool Polytechnic; we believe that all students using this book will have so willingly allowed us to use their past examination questions. We are deeply appreciative of the encouragement and practical help of our publishers without which this work might never have seen the light of day.

The debt we owe to previous writers is beyond measure. If we have not acknowledged every single one it is because their ideas are so much a part of our own that it is impossible to identify with certainty what is theirs. We hope that each one of them will accept our acknowledgement of their contribution to the existing state of knowledge.

Such merits as this book may have are direct results of the help we have received from these and from many others. But the final manuscript is ours and the responsibility for undetected errors that remain must be ours alone.

Above all our gratitude is expressed to our wives, who have endured many hours of loneliness during the writing of this book. Their forbearance and encouragement have contributed in no small part to the completion of this manuscript. To them it is dedicated.

Frank Owen
Ronald H. Jones

Liverpool Polytechnic

Preface to Second Edition

Opportunity has been taken to revise completely the content of this book and to incorporate a new chapter on Statistical Decisions which have become increasingly important since the book was first published.

We would like to thank the many people who have made suggestions for improvement, most of which we have incorporated in the revised text. We would like to thank too the many students using this book whose encouragement we have found highly rewarding.

Frank Owen
Ronald H. Jones

Acknowledgements

We would like to express our thanks to the following examinations bodies who have allowed us to make such a liberal use of their past examination questions:

The Association of Certified Accountants
The Chartered Institute of Secretaries and Administrators
The Institute of Cost and Management Accountants
The Association of International Accountants
The Chartered Institute of Public Finance and Accountancy

Chapter One

The Organisation of Data

The modern business world has a great hunger for facts and data. Well organised data improves our understanding of problems, and helps us to take decisions wisely. Badly organised data is little better than worthless. Unfortunately, you will all-too-often come across data that is not organised — most firms have filing cabinets full of data that someone intends to organise 'one day'. In this chapter we will suggest methods of how data can be organised for meaningful analysis — we will take our first steps in the rewarding (though often confusing) world of statistics.

Most people are vaguely aware that Statistics is concerned with figures in one way or another. Equally, we think, most people are rather distrustful of the statistics that they see quoted in the press or on television. We must admit that we ourselves have some sympathy for the housewife who is told on the news one evening that the cost of living has gone up by only 2% this month, and then finds in the shops next morning that everything she buys has, in fact, risen in price by between 5% and 10%. When this sort of thing happens it is no wonder that people get the impression that statistics can be made to prove anything. And yet — if our figures are accurate and the information is presented properly — how can this be so? We would like you to believe right from the start that no genuine statistician will ever deliberately misrepresent information or use it to mislead people. It can be done of course. In life many people are unscrupulous, and later in this course we will tell you how they misrepresent information, with the strict warning that *you* must never do it.

The great weakness of Statistics, is that to the man in the street who has never studied it, the methods used by statisticians are a closed book. We hope that as you work through this course your own personal book will be opened and that you will understand the dilemma in which our housewife finds herself.

But before we begin to think of the techniques you will use and the calculations you will perform, let us stop for a minute to consider the raw material you will be dealing with.

Suppose that the student union in your school or college wishes to obtain information about its members — their age, sex, home area, whether they live in a flat, or at home, or in lodgings and so on. How would the union secretary go about collecting this information? The most obvious way is for each student to be issued with a questionnaire, posing the relevant questions, and asking for it to be returned to the secretary's office. No doubt some of the forms will be incorrectly completed: some students may genuinely misunderstand the questions: some may refuse to answer certain questions which they regard as personal: doubtless some, in the fashion of the great petitions of the nineteenth century, will be signed by Queen Victoria or Karl Marx. Yet, with all its faults, this mass of completed questionnaires is the basic raw material for the statistical report that the union secretary wishes to produce.

Raw material such as this, collected at first hand, in response to specific questions is known as *primary data*; its characteristics are that it is obtained directly for the purpose of the survey which is being undertaken, and is, as yet, unanalysed.

Now, if your union secretary is lucky, he may also be able to obtain a great deal of information from the College administration, who, using enrolment forms as their primary data, may already have produced for their own purposes a fair amount of statistical information about students. Such information will, of course, have been produced for college purposes and may not be exactly what the union wants: but it is often useful additional information. Such data, which has already been collected for another, and different purpose, we know as *secondary data*. Usually it is of less use than primary data since it has already been processed and the original questionnaire is unlikely to have asked all the questions you would like to have asked. But whether it is primary or secondary, there can be very little statistical information which was not at one time to be found only in a pile of completed forms or questionnaires. The main task of any writer on statistics is to explain what the statistician does with his raw data between collecting it and presenting his report. So let us go back to your union secretary.

It is obvious that no-one would sit down and write a report in the form of 'Mary Smith is 17, lives in Durham, and is in lodgings here; Susan Yeung comes from Singapore and is in lodgings here ...'

We might as well hand over the completed forms to anyone who is interested since all that this type of report does is detail the information which is already given in detail on the questionnaire.

We can get a clue about the next stage of the analysis if we ask ourselves what it is that the union really wants to know. Surely the sort of information that is really wanted is how many students are 16, how many are 17 and so on; what percentage of students live at home; what proportion of students come from overseas. It is not the individual we are interested in so much as total numbers in given categories. The categories in this investigation may be age, sex, type of residence, number of hours a week spent on study and so on. Within each category students will vary. Some are 16; others are 17; some live at home, others in lodgings. We call each of these categories a *variable* because within each category students will vary. So we may now say that we are interested in a number of variables such as age, and more specifically in the value we can assign to each student within the range of values over which the variable extends. We may find, for example, that when we consider the variable 'age', 267 students are aged 17, 164 are aged 18 and so on up to the eldest student. The numbers of students whom we can place at each value of the variable we will call the *frequency*, because it tells us how often we will come across a student with this particular characteristic (that is, aged 18, or doing 27 hours a week private study, or travelling more than 15 miles to college). Thus the first step we must take is to decide what aspects of student life we are interested in and count up how many students are found within each of these categories. In so doing we are simplifying our data — reducing it to a more manageable form. In the process some detail is lost. We no longer know how old Brenda Jones is; but if we are interested we still have her completed questionnaire. On the other hand we do know that 267 students are aged 17 as well as much other general information.

Once we have reached this stage we are in a position to summarise our results in the form of a table and our work begins to look more like that of a statistician. Probably as a first tentative step we would produce a simple table dealing with only one variable. It might appear like this:

Age of Students attending ABC College

Age (the Variable)	Number of Students (the Frequency)
17	267
18	164
19	96
20	74
21 and over	23
	624

There is nothing wrong with our producing 15 or 20 tables like this, each concerned with a single variable, but surely it is better for presentation purposes if we could produce a small number of compound tables each showing several variables at once. Thus we could construct a double table showing the two variables, age and sex of students at the same time.

We have constructed this table by listing one of our variables vertically (age) and the other horizontally (sex). There is no golden rule, but it generally looks better if we tabulate the variable with the greater number of values vertically and that with the smaller number of values horizontally. Notice too that we have totalled both the vertical and the horizontal columns and that this adds to our information. We not only have the age distribution of male students and of female students but also the age distribution of the entire student population, and the total number of male and female students.

Age	Number of Students		
	Male	Female	Total
17	151	116	267
18	98	66	164
19	70	26	96
20	52	22	74
21 +	18	5	23
Total	389	235	624

You may of course still argue that the table is still concerned with only one variable, age, and that all we have is two age distributions. Let us then extend our table to consider three variables, age, sex and type of accommodation. Obviously now we must further subdivide either the horizontal or the vertical columns. Again it is a good general guide to say that we believe it better to subdivide the horizontal rather than the vertical columns. But in doing this the variable in the vertical column tends to become the more important. So we must consider which is the most important variable, and this often depends on what we are trying to show. Let us suppose that in this case we are aiming to show that the type of accommodation a student occupies depends on his or her age. In this case we will list the ages vertically and subdivide the horizontal columns. Our table may now appear like this:

Age	Number of Students						
	At Home		In Lodgings		In Flat		Total
	M	F	M	F	M	F	
17	112	92	16	20	23	4	267
18	64	42	24	16	10	8	164
19	31	12	28	7	11	7	96
20	8	4	16	10	28	8	74
21+	2	3	3	1	13	1	23
Total	217	153	87	54	85	28	—
	370		141		113		624

You will readily appreciate what a vast amount of information a table such as this can give us: the number of students who live at home, subdivided into male and female and classified according to age, as well as the same information for those who are living in lodgings or living in a flat. You can understand too how much more information could be incorporated if we subdivided further the horizontal axis as well as some subdivision of the vertical axis such as the area of origin followed in each case by the age range.

There is one problem — the more we subdivide, the more complicated our table becomes, and there comes a time when it is so difficult to read it and understand it that we find that clarity has been lost rather than gained. It is true that one treble table, such as the one above, is better than three single tables. It is equally true that if we are considering eight or nine variables, three treble tables are better than one very complex one. And if you are wondering why clarity is so important think again what we have been doing. We have collected primary data, simplified it and classified it, and are now trying to present it to our union executive in a readily digestible form. How much notice do you think the executive will take of us if they cannot understand what our tables are all about?

Just in case you are ever in the position of having to construct tables to present the raw material you have collected, there are several points you should bear in mind. Let us call them the 'Principles of Good Tabulation'.

(a) Every table should have a short explanatory title at the head. At the end you should put a note of the source of the information you have used, whether it is based on your own survey or secondary data.

(b) The unit of measurement should be clearly stated, and if necessary defined in a footnote. Not many people, for example, would know offhand what a 'Long Ton' is. In addition the heading to every column should be clearly shown.

(c) Use different rulings to break up a larger table — double lines or thicker lines add a great deal to the ease with which a table is understood.

(d) Whenever you feel it useful insert both column and row totals.

(e) If the volume of data is large, two or three simple tables are better than one cumbersome one.

(f) Before you start to draft a table be quite sure what you want it to show. Remember that although most people read from left to right, most people find it easier to absorb figures which are in columns rather than rows.

As with most things practice is the best way of learning, and these principles will soon become second nature after you have drafted a few tables for yourself.

You might well ask at this stage whether this is all there is in the subject of Statistics. It if were you would all end up with distinctions. But the most important part of the work is still to come. No statistician (or student) worth his salt is content with a mere list of figures. He now begins to ask questions, the most important of which is 'What do the figures tell me?' We now begin, to analyse the figures, and statistical techniques are largely methods of extracting the utmost possible information from the data we have available. We could, for example, calculate the average age of students living at home, and compare it with the average age of students living in flats to try to determine whether we are right in assuming that the younger student will tend to live at home and the older students tend to be flatdweller. We can do the same thing for both male and female students to see if they behave differently. Let us say that there are many questions that the statistician can ask even from the simple data we have used so far.

We said earlier that the most obvious way for the union secretary to collect his data was to issue a questionnaire to each and every student. The results of his enquiry would cover every single student in the college — it will refer to what statisticians call the *population* of students. Beware of this term population. In statistics it does not mean the number of people living in a particular area. What it implies is that we have examined or obtained information about every single member of a particular group we are investigating. Thus we can talk of a population of telegraph poles, a population of shaggy-haired dogs, a population of ball-bearings and so on.

But is there any need for us to examine the population of students attending the college? If we wish to save time and money can we not do as so many public opinion polls do and take a sample of students? We could issue the questionnaire to, say 60 or 70 students only, or perhaps to every tenth student, and so reduce our raw data considerably. The *sample results* we obtain can then be applied to the population of students: if 12% of the sample live at home, we will argue that about 12% of all students in the college live at home.

Now, you may well argue that this can lead to wildly inaccurate results; and if you consider some of the results of public opinion polls in recent years it is apparent that things can, and do, go wrong. The sample chosen may be too small; it may not be representative of the population; the error arising as a result may mislead us. At this stage we will merely point out that in taking a sample we are in good company: an extremely high percentage of government statistics such as the statistics of Household Expenditure are based on samples which, on the face of it, appear to be ludicrously small.

If you think back now to the questions we suggested that you might ask about what our tables can tell us, you will realise that most of them involve a more detailed study of one variable only — the age of females living at home; the age of males living in flats. When we do begin to analyse you will appreciate that this is usual. The table presents several variables at once, but we extract just one of them at a time for further examination. In a few cases we will use two variables at once, when we are asking if there is a relationship between them such that one affects the other or that both move in sympathy. But in this foundation course we will never ask you to get involved in the analysis of three or more variables at once — which is indeed a complex matter.

The Grouped Frequency Distribution

Let us examine again the table showing the age of students attending the ABC college.

Age (x)	Number of Students (f)
17	267
18	164
19	96
20	74
21 and over	23
	624

Such a tabel is called a *frequency distribution*. It shows how many students (f) have the stated age (x). For example it tells us that 96 students are 19 years old. We stated that this table was ideal for summarising data dealing with only one variable — in this case age. However, it is not always convenient to arrange single variable data into a table like this. Suppose, for example that a scrap metal dealer buys a job lot of metal pipes. To get some idea of the lengths of pipes he selects 100 and carefully measures them. Now it is highly likely that no two pipes will have *exactly* the same length. It is no use then arranging the data into a distribution like the one we had for the ages of students. It might well be a table consisting of one hundred rows. It would be far more sensible to *group* the lengths together into classes like this:

Lengths of 100 copper pipes

Length (cm)	Frequency
10 but under 20	3
20 30	7
30 40	10
40 50	16
50 60	34
60 70	13
70 80	7
80 90	6
90 100	4

A table such as this is known as a *grouped frequency distribution*.

Certain points about this distribution can be noted.

1. There is no ambiguity about the way the classes have been stated. We are left in no doubt about the class to which a pipe belongs. For example, a pipe with a length of exactly 20 centimetres would go into the second class, i.e. 20 and under 30. However, if we were to state the classes like this

$$10 - 20$$
$$20 - 30$$

into which class would we then put the pipe?

2. Although it is certainly more convenient to put the lengths of pipes into this distribution (much more convenient than a list of 100 lengths), we have lost something. For example, without going back to the original data, we would have no idea of the *actual* lengths of (say) the three pipes in the group 10 and under 20. We have sacrificed detail for the sake of presenting a picture which can be absorbed fairly easily. It might seem, of course, that the use of class intervals will prevent our using the frequency distribution as a basis for further work. Naturally, it does create a problem, and to overcome it we must make an assumption. We will assume that all three pipes in the class 10 and under 20 have lengths precisely at the centre of this class. Now, the smallest length that could be in the class is exactly 10 cms., and the largest length is 19.9999 cms. The centre then (or the *mid-point* as it is called) is

$$\frac{10 + 19.999\dot{9}}{2} = 15$$

3. We stated that the largest pipe in the first group could have a length of 19.9999 cms. and this throws light on the nature of the data we are dealing with. This data is what statisticians call *continuous* data — it can take any value within a particular group. There is, however, no reason why the data should have *particular* values within a group. It may well be that the scrap metal dealer, when taking measurements, has

recorded only certain values. He might, for examples, have recorded only to the nearest half centimetre, 11.0, 11.5, 13.5 and so on. But this way of measuring has been decided by the dealer and not by the *nature* of the data. The pipes may still have any length within the group, no matter how that length has been recorded. Generally speaking, any data that is obtained by measuring rather than by counting is continuous data, and, as we have said, should be listed in classes arranged in the form

x and under y

4. Let us now turn to another example. Suppose we count the number of telephone calls made by 100 firms on a particular day. Clearly this is not continuous data as the number of telephone calls made cannot take any value within a group. Thus, in the group 10 and under 20, we cannot make 10.2 or 14.3 telephone calls. The number of calls made can increase only in uniform steps of 1. Such data is called *discrete* data,[1] and it is usual to arrange it in a frequency distribution like this:

No of Calls   No of Firms
10 – 19   9
20 – 29   14

There is no ambiguity in stating the classes in this way as we cannot have 19.1, 19.2, 19.3 etc. calls. Again we have no precise idea of the exact number of calls made by firms in any particular class, and we assume that all firms have made a number of calls equal to the mid point of the class, i.e.

$$\frac{10 + 19}{2} = 14.5 \text{ calls.}$$

Now you may think that this is nonsense. After all, we have just stated that the number of calls must be a whole number, so how can we assume that 9 firms each made 14.5 calls? Well, don't let this worry you. This is something we assume merely for convenience of calculation.

At this stage it would be useful to give a few words of warning about using continuous and discrete data. We have implied that it is better to use classes of the type

and

10 and under 20
20 and under 30   for continuous data (Type A)

10 – 19
20 – 29   for discrete data (Type B).

Now although it would be wrong to use Type B for continuous data, there is nothing wrong with using Type A for discrete data, though it does create problems. Suppose we had arranged our data on telephone calls like this

10 but under 20   9
20 but under 30   14

---

1. The uniform stepped increase for discrete data need not be 1 — shoe sizes increase by ½, hat sizes by ⅛, and money (in pounds sterling) by £0.005.

The unwary student might be tempted to say that the mid point of the first class is $\frac{10+20}{2} = 15$; but we know it is really 14.5. To avoid problems like like this, never assume that data is continuous merely because it is in a Type A distribution. Read the heading at the top of the table and decide what is the smallest and what is the largest possible value in each class. The mid point is then

$$\frac{\text{smallest} + \text{largest}}{2}$$

Another warning we must give you concerns rounded data (i.e. data given to the nearest whole number, the nearest 10 etc.) Suppose, for example, that the scrap metal dealer had rounded his measurements to the nearest centimetre. Now we know this data is continuous, so we may be tempted to use the Type A distribution.

10 but under 20
20 but under 30

However, suppose we have a pipe with an actual length of 19.6 centimetres. this would be rounded up to 20 and put into the second class *even though, really, it should be in the first class*. Problems such as this can be avoided if classes are carefully stated. It would not occur if, in this case, we had stated the classes as

9.5 but under 19.5
19.5 but under 29.5

The pipe with a length of 19.6 would be rounded up and (correctly) placed in the second class. Now, admittedly, many people use a type B distribution for rounded data, but the implication of doing this is that the class limits are 9.5 to 19.5 — another reason for reading carefully the heading at the top of the table and deciding for yourself whether the data is continuous or discrete.

A final warning concerns the class interval used in a frequency distribution. If it is a Type A distribution say

10 but under 20

then this of course covers a (continuous) range of ten numbers and the class interval is 10.

If it is a Class B distribution, say

10 — 19
20 — 29

you must remember that the groups cover 10 discrete values (10 to 19 inclusive) and the class interval is also 10 in this case.

Finally, you will find, if you look at any published statistical tables, that in many cases no limits are given for the first and last classes. An income distribution showing annual income might begin merely with 'Under £660' and end up with the group 'Over £50,000'. Such open-ended classes create problems and we will give you a few hints on how to handle them later.

One of the most difficult problems you will have in building up a frequency distribution from raw data is to decide on what class intervals to use. Obviously, a great deal will depend on the data you have available, but a few general guidelines may help. Firstly, try not to choose class intervals which will reduce the number of groups below five or six. If you do the data will be so compressed that no pattern emerges. Naturally the rule is not infallible — E.E.C. have published statistics of farm sizes giving only three classes. These three, however, correspond to a generally accepted international definition of small, medium and large farms. Our advice is that you should try not to emulate E.E.C. Equally, at the other extreme, do not have too many classes. About fifteen or sixteen is the maximum. The problem here is not only the difficulty of absorbing lengthy tables, but also the fact that each group will have a very low, or in some cases, even a zero frequency. And this leads to another point. At the upper end of the table, if you stick slavishly to a single class interval you may well find that several consecutive groups have no members while a higher group has a frequency of two or three. In these circumstances you should sacrifice the idea of equal class widths and combine the several classes into a single wider class.

A good general guide is to take the difference between the minimum and maximum value of the variable (which we call the *range*), and divide by ten. This will give you the right class width (or thereabouts) for the majority of classes, provided that you realise that class width of five or ten, or fifty is better than one of four, or seven or sixty-two, and provided that you take care with the extreme values of the variable.

Before we leave this brief description of the frequency distribution it would be an advantage if we show you how to tackle examination questions which ask you to construct a frequency distribution from a mass of figures. For this purpose we will look at a typical examination question.

Example

The following is a record of the percentage marks gained by candidates in an examination:

```
65 57 57 55 20 54 52 49 58 52
86 39 50 48 83 71 66 54 51 27
30 44 34 78 36 63 67 55 40 56
63 75 55 15 96 51 54 52 53 42
50 25 85 27 75 40 37 46 42 86
16 45 12 79 50 46 46 59 57 50
56 74 50 68 52 61 40 38 57 31
35 93 54 26 67 62 51 52 54 61
93 84 28 66 62 57 45 43 47 33
45 25 77 80 91 67 53 55 51 36
```

Tabulate the marks in the form of a frequency distribution, grouping by suitable intervals.

Looking at the figures we find that there are 100 marks given ranging from 12 to 96. We have laid down a principle of aiming at somewhere in the region of 10 classes in our frequency distributions and it certainly seems that the best class width in this example would be 10 marks. If we were to use 5 mark intervals we would end up with some 18 classes which is too many; if we use 15 mark intervals we end up with only 6 classes which is too few.

We now have to find out how many of these marks fall within each class, and we recommend that you should do this in this way. Firstly list every class vertically; now take each candidate's marks in turn, and place a dash or a 1, or some other suitable mark against the class into which it falls. Having done this for every mark we can now take each class in turn and add up how many candidates fall into each class. Your rough working will appear something like this. You will notice that for ease of counting we have divided our dashes into groups of five.

10 – 19	111							3
20 – 29	⊬┼┐	11						7
30 – 39	⊬┼┐	⊬┼┐						10
40 – 49	⊬┼┐	⊬┼┐	⊬┼┐	1				16
50 – 59	⊬┼┐	⊬┼┐	⊬┼┐	⊬┼┐	⊬┼┐	⊬┼┐	1111	34
60 – 69	⊬┼┐	⊬┼┐	111					13
70 – 79	⊬┼┐	11						7
80 – 89	⊬┼┐	1						6
90 – 99	1111							4
								100

The first thing we notice about this data is that it is a list of examination marks. There are extremely few examinations where marks are recorded in anything but whole numbers, and we may safely say that this data is discrete. We will, therefore use what we called earlier a Type B distribution.

Before you do anything else now, check that the total frequency (that is, the number of dashes) in your rough working is the same as the number of items given in the question. Having done this you are now ready to construct your frequency distribution. Remember, though, all the things that are necessary: the heading of the table, the column headings and the source, if it is available. Getting the table correct is only one part of the answer, although it is an important part. Your final frequency distribution will appear something like this:

Percentage Marks gained by Examination Candidates
. . . . Examination . . . . 19 . .

Marks Awarded	Number of Candidates
10 – 19	3
20 – 29	7
30 – 39	10
40 – 49	16
50 – 59	34
60 – 69	13
70 – 79	7
80 – 89	6
90 – 99	4
	100

Source: Examiners' Report 19 .

We will spend some time later on examining frequency distributions such as this to see what further information they can give us. But whatever distribution we study it has one thing in common with all other frequency distributions — it tells us the magnitude of a variable at a given point in time. There are times however when we need to look at the way the magnitude of a variable changes over fairly long periods of time — for example, we might be considering the way in which the volume of British exports to Hong Kong has changed year by year since 1960. Such a table is called a *time series*. It follows from the nature of the time series that it consists of a series of time periods: years, quarters, months, or even days, with the value of the variable given against each time period. Thus, recently, Barclays Bank conducted a survey of the output of motor vehicles quarter by quarter in each of the Common Market countries. The figures for France represent a typical time series with the value of the variable given quarterly.

Output of Motor Vehicles — France
(monthly averages — thousands)

Year 1	1st quarter	150
	2nd quarter	165
	3rd quarter	104
	4th quarter	113
Year 2	1st quarter	173
	2nd quarter	180
	3rd quarter	124
	4th quarter	184

Source: Barclays Bank Briefing No. 12

In much statistical work a time series of this nature may extend over many years, but you will find it surprising what such a series can tell us.

If you have followed carefully the argument of this chapter it should now be apparent that we are concerned with the collection, simplification, presentation and analysis of information which can be expressed quantitatively. If you do your work well you will get an accurate picture of the data you are studying. There will still be some of you who may claim that figures can be made to prove anything, and to be fair, if the data is misused this may be true. If you ignore parts of the data, and conveniently forget to include calculations and information which is inconvenient or does not support your preconceived ideas, most things are possible. But if you follow your analysis through to the bitter end, using all the information which is available, keeping an open mind and interpreting only what your figures throw up, you cannot but be somewhere near the truth. In a world in which so much reliance is placed on the work of statisticians, where government policy often depends on their findings, and where industrial decisions involving millions of pounds are taken on the basis of statistical analysis, it is important to remember that whether you like what your figures show or not, your task is to interpret what the available data tells us as honestly as you are able.

Exercises to Chapter 1

1.1 Describe in detail the steps in a statistical investigation; include in your description a brief summary of the various methods of presenting conclusions. Illustrate your answer with example of statistical investigations that might be carried out in a business environment.

1.2 (a) Prepare a summary of the tasks involved in conducting a statistical investigation.
(b) State three methods of presenting the results of such an investigation.
A.I.A.

1.3 What is the difference between primary and secondary data? Why is it important that statisticians should make a distinction in their use of these categories of data?
A.C.A.

1.4 *Average Weekly Earnings* of Administrative, Technical and Clerical Staff in the Public Sector and in Insurance and Banking*

MALES

October	National and Local Government including Education (Teachers) and National Health Service	Nationalised Industries	Insurance and Banking
	£	£	£
1966	26.69	26.25	26.63
1967	27.88	27.13	27.73
1968	29.65	28.95	29.11
1969	32.03	31.18	30.88
1970	36.00	35.83	34.63

*Including earnings of monthly-paid employees converted to a weekly basis.
Source: Department of Employment.

Write a short report in which you bring out the main features of the data given above. Include appropriate derived statistics.

1.5 In 1951, 207 thousand persons received unemployment benefit, 906 thousand persons sickness benefit, 1437 thousand males retirement pensions, 2709 thousand females retirement pensions, 457 thousand received widows' benefit. 217 thousand persons received other National Insurance benefits. In 1971 the corresponding figures for unemployment benefit was 457 thousand, for sickness benefit was 969 thousand, for male retirement pensions 2611 thousand, for female retirement pension 5196 thousand, for widows' benefit 448 thousand, for other National Insurance benefits 387 thousand. (Source: *Social Trends* 1972.) Tabulate this data, calculate appropriate secondary statistics and include those statistics in your tabulation. Comment briefly on your tabulation.

1.6 An inquiry into the population of a town at 1st April 1974 showed that the total was 297,500 persons of whom 60% were females and 40% males. 50,000 females were aged thirty and under, 60,000 were aged from 31 to 60 and the remainder were over 60 years of age. The corresponding figures for males were 60,000, 39,000 and 20,000. The average family size was 3.5. 75% of the female population lived in the northern area of the town and the remainder in the southern area. Of the males, 20% lived in the southern area. Tabulate the data given showing the analysis of each class of person into areas and age groups. Include actual figures and percentages. Show also the number of families for the town in total.

1.7 The Saturn Finance Company wishes to study for several years (1971, 1972, 1973, 1974) the distribution of its loans according to size of loan (under £100, £100 and under £250, £250 and under £500, £500 and over) and the purpose of the loan (home improvement, car purchase, durable household goods purchase, other).

(a) Prepare a table in which the data can be presented cross-classified by year, size of loan and purpose of loan. Include summary rows and columns for all classifications given.

(b) Insert the following figures in the appropriate cells:

(1) In 1972, 42.7 per cent of all loans made for home improvements were for £100 and under £250.

(2) In 1971, 31.4 per cent of all loans made for purchase of cars were for £250 and under £500.

(3) In 1974, 29.7 per cent of all loans were under £100.

1.8 While the population of the United Kingdom grew from 38.2 millions in 1901 to 55.8 m in 1972, the increase was not uniform between the regions. Seventy years ago, the south-east was the most populous region with 10.5 m. In mid-1972, its population was an estimated 17.3 m. Both the West Midlands and East Midlands have also grown quickly; the former from 3.0 m to 5.1 m and the latter from 2.1 m to 3.4 m. During the same period, East Anglia's population grew from 1.1 m to 1.7 m, whilst the south-west increased from 2.6 m to 3.8 m.

Despite its generally high birth rate, Northern Ireland's population, whilst growing from 1.2 m in 1901 to 1.5 m in mid-1972, fell as a proportion of total U.K. population. A similar trend is revealed in other economically depressed regions, and both Wales and Scotland, because of losses due to migration, have grown much more slowly than average, Wales from 2.0 m to 2.7 m and Scotland from 4.5 m to 5.2 m. (Source: *New Society and Social Trends*, Modified.) *Note:* not all regions have been included.

Arrange the above data in a suitable table, providing additional columns, in blank, appropriately headed, in which could be inserted derived statistics enabling comparisons to be made.

1.9 The total number of employees of Core and Peel Ltd. at 31st December 1971 was 10,590, of which 6721 were men, 3106 women and the rest juniors. During 1971 108 men resigned and 74 men were engaged. The corresponding figures for women were 29 and 87 and for juniors 17 and 23. 1386 men, 976 women and 16 juniors were absent sometime during the year due to illness, 509 men, 876 women due to domestic circumstances and 366 men, 272 women and 3 juniors due to other causes. The average weekly wage rate paid to men was £32.00, to women £20.13 and to juniors £18.25. The company worked 50 weeks in the year. Tabulate these details showing suitable totals and sub-totals. Include also:

(a) the numbers employed at 1st January 1971,

(b) an estimate of the total annual wages paid per grade of labour and for the employees in total.

1.10 (a) The records of the Family Expenditure Survey of 1971 show that in 4642 households interviewed in the survey, the head of the household was in employment. In 1236 households the occupation of the head of the household was 'professional'; 101 earned less than £30 per week; 153 earned £30 but under £40 per week; 194 earned £40 but under £50 per week; the remainder earned £50 or over per week. In 470 households the occupation of the head of the household was 'clerical'. 119 earned less than £30 per week; 129 earned £30 but under £40 per week; 90 earned £40 but under £50 per week; the remainder earned £50 or over per week. In the remaining households the occupation of the head of the household was 'manual'. 930 earned less than £30 per week; 814 earned £30 but under £40 per week; 638 earned £40 but under £50 per week; the remainder earned £50 or over per week. Tabulate this data.

(b) Using your table as an example list the basic rules which should be observed when tabulating statistical data.

1.11 In 1970, 44,000 houses were purchased with a local authority mortgage (total sum advanced = £154m) and 32,000 with the help of insurance companies (£154m advanced) whilst building societies lent £1986 m to 540,000 purchasers. These figures compare with 56,000 (£168 m), 34,000 (£124m) and 504,000 (£1,477m) in 1967, and 19,000, 40,000 and 460,000 houses purchased in 1969 with the help of loans of £69m, £179m, and £1556 m from local authorities, insurance companies and building societies respectively (abridged from *New Society*).

(a) Arrange the above data in concise tabular form.

(b) Prepare a table in blank, with suitable headings in which could be inserted derived statistics from the above figures which would facilitate the making of comparisons.

1.12 (a) List clearly the steps taken in forming a frequency distribution from a set of about 1000 observations giving reasons where necessary.

(b) Construct a frequency distribution using the following 100 observations:

Lives of electric light bulbs, in hours, to the last complete hour.

690	701	722	684	662	699	715	742	726	716
728	705	693	691	688	706	707	691	701	713
740	662	676	738	714	703	695	692	699	685
698	687	703	726	699	692	714	724	664	689
694	705	717	682	717	707	696	697	681	708
712	733	705	673	694	716	745	692	719	701
679	680	654	691	669	685	725	704	724	714
689	702	710	696	697	709	721	677	680	671
685	724	736	696	688	692	728	656	690	695
702	696	708	698	710	682	694	676	700	663

1.13 The following is a record of the heights in centimetres of a sample of 85 servicemen:

169 179 183 186 166 181 177 173 167 193 176 183 162 170 186 174
188 165 168 174 170 176 186 177 185 175 179 166 190 182 182 180
194 177 184 175 168 181 180 172 178 192 175 189 180 175 183 191
172 188 180 176 185 178 179 173 165 170 178 181 181 189 187 191
179 196 179 182 171 169 171 184 198 182 175 190 187 176 164 187
167 185 177 184 178

Tabulate the above data in the form of a frequency distribution, using as intervals 160 cm and under 165 cm, 165 cm and under 170 cm, 170 cm and under 175 cm, and so on.

1.14 The lengths of telephone calls from a certain office were noted and the results are shown below giving the times in seconds.

141	43	203	104	82	63	24	84	41	86	47	43
100	53	139	147	137	186	214	106	150	109	170	172
194	124	175	177	162	129	128	219	40	105	48	65
105	154	154	35	149	54	104	109	119	74	140	104
168	127	191	30	109	88	104	207	38	164	182	120
166	53	145	29	112	143	49	199	130	52	109	77
142	75	146	105	125	112	40	126	67	49	90	140
132	118	134	133	159	123	161	112	157	104	92	112
151	98	156	117	156	190	122	135	116	96	163	116
186	155	106	153	69	105	136	106	131	118	94	121

Arrange these figures in a grouped frequency distribution using the intervals 0–19, 20–39, etc.

1.15 The data below are the times for completion, rounded to the nearest hour, of a sample of fifty houses.

911 902 900 867 897 915 945 940 917 883 874 880 932 919 899 903 872 901 874 925 886 928 917 906 925 913 898 888 912 896 921 908 933 903 920 885 901 892 931 902 893 887 928 907 916 895 907 864 891 890

Classify these data into a frequency distribution.

1.16 The following is a record of marks scored by candidates in an examination

77	59	84	73	51	43	50	81	61	53	69
37	58	63	67	61	90	61	50	60	84	56
77	57	42	43	41	49	37	21	24	35	34
50	11	52	30	16	33	67	87	64	47	59
37	92	88	30	38	22	22	49	46	50	64
23	73	73	48	26	36	51	85	71	57	45

Tabulate the marks in the form of a frequency distribution, grouping by suitable intervals.

1.17 A person's socio-economic status can be classified as either A, B, C1, C2, D or E in descending order. A random sample of 60 individuals taken in 1973 gave the following information on weekly earnings (£) in relation to socio-economic class:

45 (B)	20 (E)	16 (D)	61 (C2)	18 (E)
32 (C2)	22 (C2)	64 (C1)	62 (C1)	33 (D)
49 (D)	28 (C1)	60 (C2)	74 (B)	50 (C2)
49 (C2)	64 (B)	33 (D)	29 (C2)	21 (E)
24 (C1)	48 (C2)	23 (D)	27 (C2)	26 (E)
26 (D)	37 (C2)	85 (A)	18 (C1)	42 (C2)
43 (C1)	37 (C2)	67 (B)	19 (E)	22 (D)
17 (C2)	17 (C2)	19 (D)	23 (D)	50 (C2)
66 (C1)	74 (C1)	17 (E)	37 (C2)	55 (C1)
52 (C2)	37 (C2)	26 (D)	42 (C1)	40 (C2)
79 (B)	23 (E)	24 (D)	31 (D)	44 (C1)
15 (E)	18 (E)	40 (C2)	65 (C1)	17 (E)

Required:

(1) Compile a frequency distribution of earnings with intervals of a suitable width.

(2) By considering earnings to fall into one of three groups of less than £25, £25 – 45, and more than £45, compile a two-way table showing the frequencies in each earnings group/socio-economic class combination.

(3) Describe the main features of the data as observed from the table compiled in answer to (2) above. A.C.A.

Chapter Two

The Presentation of Data I

Numerical data, however well organised, can often fail to fulfil one of its prime functions — to communicate information. This is because many people have a positive hatred of numbers. Be honest — how often have you read a newspaper article and simply skipped over any tables it contains? It is not enough to organise data well: it must also be well presented. One of the main jobs of a statistician is to identify the main features of information given by data, and to present them in such a way that they become intelligible and interesting. In this chapter, we will examine some of the graphs and diagrams that the statistician uses to achieve this end.

Watch any television education programme and almost certainly one of the first things to strike you will be the number of devices used to present information in a vivid and interesting fashion. Graphs, diagrams, blocks of wood, animated cartoons — all play their part in putting the subject over.

Of course, television is the ideal medium for this type of visual presentation, and one of the more important developments of the second half of the twentieth century may well turn out to be the impact of television as a means of imparting information. Yet, in all this, the television producer is doing little more than statisticians have already been doing for a considerable period of time. The scale is larger; the impact is probably many times greater; but the techniques are the same.

We would like, at this point, to be able to say to you that we are going to give you a few simple rules which will enable you to master the technique of presenting diagrammatic information quickly and easily. It is not, however, as simple as that. The diagram you draw depends in part on the information you have, and what you are trying to stress. If you are trying to stress how much the government is spending annually you will produce a very different diagram from one which sets outs to examine what the government spends the money on. It might also depend on the readers you are aiming at. A diagram suitable for the readers of a mass circulation daily newspaper will be very different from one in an accounting journal; a diagram intended to extol the qualities of a particular brand of soap powder will be far removed from one designed to show the changing composition of agricultural output in the Common Market. Effective presentation is a question of flair and experience, and there is only one guideline — does the diagram present clearly and vividly the information it is designed to present?

We cannot pretend that the way in which we illustrate the presentation of particular information is the only way. You may, in fact, think it is not the best way. What we will do is to indicate the weapons you have available. The rest is up to you!

1. *The Bar Chart*

One of the most common of all techniques for presenting data is the use of the bar chart, in which the length of the bar is proportional to the size of the items we are considering. Suppose we are considering the population of some E.E.C. countries; we could be presented with a table like this:

	Population (million)		Population (million)
France	50	Denmark	4.9
Belgium	10	Britain	56
Germany	61	Holland	12.9
Italy	54		

Now, shut your eyes, and without reference to the table state which country has the highest population, which is the lowest, and where Britain comes in the league table. If you have played fair we are prepared to bet that a high proportion of you cannot answer all three questions correctly. Why? Simply because experience shows that people do not absorb lists of figures. A very different result would be obtained if we presented the same information in the form of a bar chart.

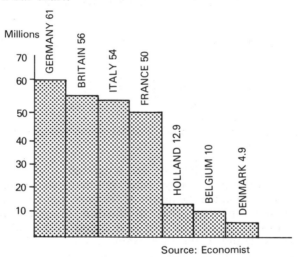

Population of E.E.C. countries (millions)

Diagram 2.01.

The immediate impression we get is of four giant members accompanied by a number of very small members; and it hits the eye that Britain is a giant, second only to Germany.

You will notice that the bar diagram has a heading, as should all diagrams. It is no use presenting information unless the reader knows what the information is. Note, too, that the source of the information is given as we are using secondary data. This enables the reader, if he is interested, to go back to the original figures and delve more deeply.

Here we were looking at one variable factor at a particular point in time, but we may equally use the bar diagram to show how the value of something, say the output of motor vehicles in France, has varied over time.

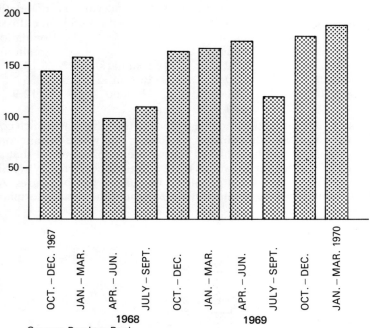

Source: Barclays Bank
Production of motor vehicles — France (monthly average — thousands)
Diagram 2.02

Here you will notice that the bars are separated — this is a matter of personal preference. It is probably neater, but many statisticians say that it is more difficult to compare the height of the bars. Looking at the diagram it is immediately apparent that production in three quarters was exceptionally low; in April–June of year 2 and in the July–September quarter in years 2 and 3. Further examination may show that in the July–September quarter production is low every year, possibly due to its being the holiday season, but the April–June quarter of year 1 is clearly exceptional and one is constrained to make a more detailed examination of what happened at that time.

The bar chart is a most versatile instrument and capable of adaptation to almost any data. Three further modifications will, we think, convince you of this versatility.

Often the data we are examining includes negative figures. Profits in one year, for example, might be converted into losses the next; the balance of payments may be in surplus or deficit. Such negative figures can be represented on the bar chart quite simply by extending the bar below the zero base line as in Diagram 2.03.

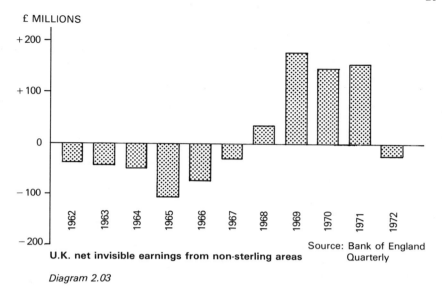

U.K. net invisible earnings from non-sterling areas Source: Bank of England Quarterly

Diagram 2.03

Sometimes, of course, we may be less interested in an absolute total than we are in the way that total is made up. We may wish to find out how the final cost of production is made up, where sales revenue goes to, what the government does with the money it collects. Again the bar chart proves itself equal to the task, since we can always subdivide the bar as shown in Diagram 2.04.

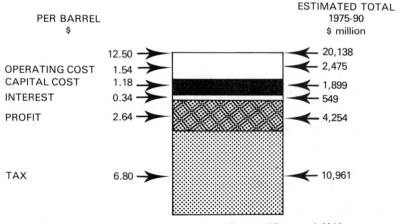

Source: Kitcat and Aitkin Report on North Sea Oil

Costs of North Sea oil

Diagram 2.04

Such a bar chart is often called a compound or component bar chart, since it illustrates the components that go to make a total. Sometimes we may be more interested in expressing our information in the form of percentages. It may be important to our argument that tax on North Sea Oil is 54.4% of total selling price rather than that it is 6.8 dollars. There is nothing to stop us from constructing a component bar chart the length of which represents 100% subdivided into sections to show the percentage that each component item forms of the total. In the case of tax, then, the section of the bar representing tax paid would be just over one half the length of the bar. Such a diagram, for obvious reasons, is called a percentage component bar chart.

As a final example of the use of the bar chart we will try to show how it can be used to derive information which is not immediately obvious. With Britain's entry into E.E.C. a burning question is the efficiency of European agriculture. But how can efficiency be measured? It is fair to say that if 30% of the total labour force is engaged in agriculture but the agrarian output is 10% of the Gross National Product, that country's agriculture is inefficient compared with a country for which the figures are 10% and 12% respectively, and this will be the criterion we will use.

We think you will agree that, using our criterion of efficiency, of the seven countries shown, only three — Holland, Belgium and Britain — can be said to be efficient producers in the agricultural sector, and this may well be a factor in the problems facing the common agricultural policy of the E.E.C.

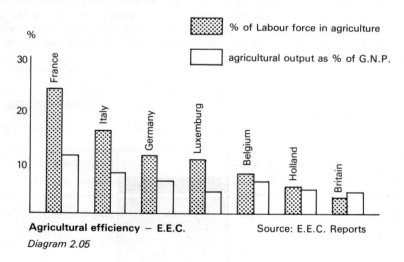

Agricultural efficiency – E.E.C. Source: E.E.C. Reports
Diagram 2.05

2. *The Pie (or Circular) Diagram*

The circular, or pie diagram is a device beloved of those who present statistical data for the general public. It is a rare report, and a still rarer copy of the *Economist* magazine which does not contain several such diagrams. In fact, we firmly believe that the extent to which this technique is used grossly exaggerates its utility.

This diagram has only one real use – to show the relative size of the component parts of a total. A complete circle represents the total and this circle is divided into segments the size of which represents the relative importance of each constituent of the total. Thus, if we were trying to show the nature of road accidents in a particular area, we might find that of 300 accidents occurring last year, 57 involved motor-cycles. We now have to mark off a segment of the circle corresponding in size to the proportion of accidents involving motorcycles, that is, 57/300 or 19/100. Since there are 360 degrees in a circle the appropriate segment must subtend an angle of 19/100 × 360 degrees at the centre, i.e. an angle of 68.4 degrees. Our complete pie diagram will appear something like this:

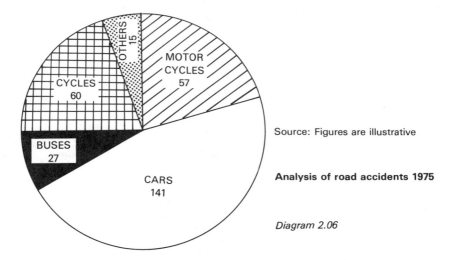

Source: Figures are illustrative

Analysis of road accidents 1975

Diagram 2.06

Even with this single simple example you should find it easy to spot the weaknesses of the pie diagram:

(a) It involves cumbersome calculations.
(b) It is more difficult to compare segments of a circle in a pie diagram than to compare heights in a bar diagram.
(c) It gives no information as to absolute magnitude unless figures are inserted in each segment, whereas the bar diagram is scaled against a single axis.

Our own advice is not to use this method of presentation unless it is forced upon you.

Plotting the Frequency Distribution; the Histogram

You will remember that the frequency distribution examines the frequency of occurrence of different values of a variable at a given point of time, and that sometimes the values of the variable are combined together into classes of a predetermined size. It is this type of distribution that the histogram represents. The histogram is so similar to the bar chart that students often assume that they are one and the same thing. There is, however, a major difference between them. In the bar chart we are interested in only one factor, say output, the magnitude of which can be represented by the height of a bar. It does not matter how wide the bars are (as long as they are all the same width); we look only at the height. In a histogram, however, we are interested in two factors — the width of the bar (which represents the value of the variable) and the height of the bar (which represents the frequency). In the histogram, then, it is the *area* of the bar that is important.

We will now illustrate the histogram with a simple example. Suppose we examine a batch of cars at the end of the production line, and count the number of faults per car. The following results were obtained.

Number of faults	Number of cars
1	18
2	25
3	19
4	8
5	3
6 or more	0
	73

The histogram looks like this:

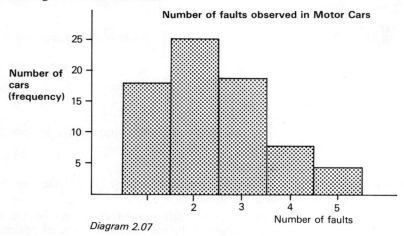

Diagram 2.07

Notice that the rectangles are drawn to 'enclose' the value of the variable (the rectangle representing 1 fault runs from 0.5 to 1.5, the rectangle representing 2 faults runs from 1.5 to 2.5 etc). This prevents any ambiguity. Notice also that the area of the rectangles (1×18 square units, 2×25 square units etc) is proportional to the frequencies.

Let us now examine how we would draw the histogram of a grouped frequency distribution. In Chapter 1 we examined a scrap metal dealer who bought a large job lot of metal pipes. To obtain some idea of the lengths of pipes he selected a sample of 100, measured them, and recorded the results in a frequency distribution.

Lengths of 100 metal pipes

Length (cm)	Frequency
10, but under 20	3
20 ,, ,, 30	7
30 ,, ,, 40	10
40 ,, ,, 50	16
50 ,, ,, 60	34
60 ,, ,, 70	13
70 ,, ,, 80	7
80 ,, ,, 90	6
90 ,, ,, 100	4

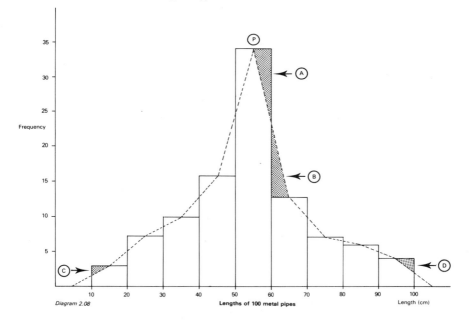

Diagram 2.08 — Lengths of 100 metal pipes

The histogram of this data is shown in diagram 2.08. Here, the rectangles are drawn to enclose the class limits, so the width of each rectangle is the same as the class interval. Notice again that the areas of the rectangles (10×3 sq. units, 10×7 sq. units etc) is in the same proportion as the frequencies. The dotted line superimposed on the histogram is called the *frequency polygon,* which converts the histogram into a simple graph. It is drawn in such a fashion that the area under the frequency polygon is the same as the area under the histogram. Take, for example, the part of the frequency polygon marked PQ. The area under the histogram (but not under the polygon) which is marked A is the same as the area under the frequency polygon (but not under the histogram) which is marked B. We can obtain a frequency polygon for a distribution by joining the mid-points at the top of each rectangle *as long as the class width is constant.* To compensate for the loss of areas C and D, we must start the frequency polygon at 5 (the mid-point of the group 0 and under 10) and end it at 105 (the mid-point of the group 100 and under 110).

Drawing the histogram and frequency polygon for unequal class intervals is rather more awkward. Suppose, for example, our scrap metal dealer had obtained the following distribution

Length (cm)	Frequency
10, but under 20	4
20 ,, ,, 30	6
30 ,, ,, 40	8
40 ,, ,, 50	11
50 ,, ,, 60	10
60 ,, ,, 70	9
70 ,, ,, 80	8
80 ,, ,, 100	7
100 ,, ,, 130	6
130 ,, ,, 170	4

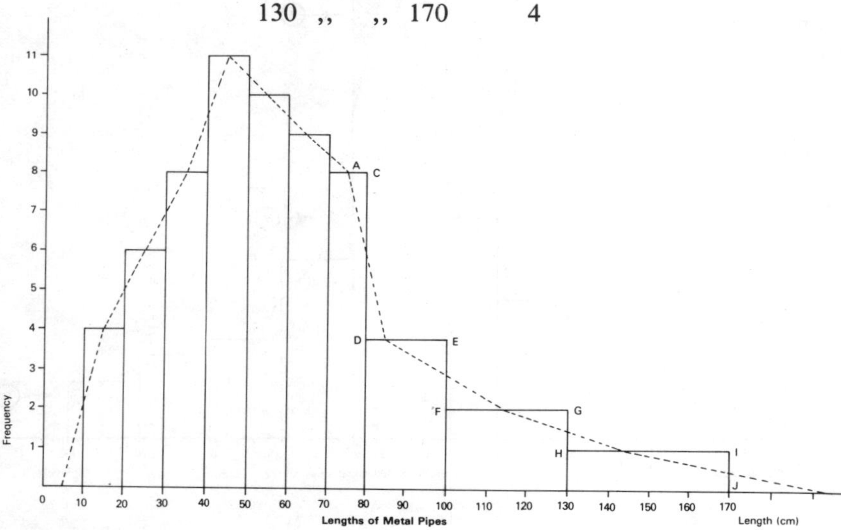

Diagram 2.09

The histogram of the distribution is shown in diagram 2.09. The first seven classes present us with no problems as the class width is a constant 10. The rectangles of the first seven classes have areas of 40, 60, 80, 110, 100, 90, and 80 square units respectively. Now let us turn to the eighth class. This class is for those pipes that have lengths greater than 80 cm but less than 100 cm — a class width of 20. To be in line with other classes, the rectangle representing this class should have an area of 70 square units. As the rectangle must have a width of 20, its height must be $70 \div 20 = 3\frac{1}{2}$. In other words, if we double the class width, we must divide the frequency by 2 to give us a rectangle with an area representative of the frequency. Can you see what we have implied by drawing the rectangle in this way? We know that there are 7 pipes with lengths between 80 and 100 cm, and we have assumed that $3\frac{1}{2}$ pipes are in the group 80 but under 90, and $3\frac{1}{2}$ are in the group 90 but under 100. However, we do not draw a line down the middle of this rectangle — this would be quite wrong as it would imply that there *exactly* 3½ pipes in the group 80 but under 90 and *exactly* 3½ pipes in the group 90 but under 100. We can deal with the remaining classes in a similar fashion. The ninth class (which has a width of of 30) contains 6 pipes. We must assume that 2 pipes are in the group 100 but under 110, 2 in the group 110 and under 120 and 2 in the group 120 and under 130, so we draw a rectangle with a height of 2. Finally, can you see that the correct height for the last rectangle is 1?

Let us now examine the effect of unequal class intervals on the frequency polygon. For the first 6 classes there is no problem — we proceed as before joining the mid-points at the top of the rectangles. Problems arise when we reach point A. If we join point A to the mid-point at the top of the rectangle representing the group 80 but under 100, then the area under the polygon would be greater than the area under the histogram. We can overcome this by drawing the line to it passes through the mid-point of CD. The line should pass through the mid-point of EF, GH and IJ. This assures that the histogram and frequency polygon enclose equal areas.

Before we leave the histogram, we must have a few words to say about grouped, discrete frequency distributions. Often you will see the classes arranged like this;

Number of telephone calls made by 100 firms on a particular day

No. of calls	No. of firms
10 – 19	9
20 – 29	14
30 – 39	18
40 – 49	27
50 – 59	12
60 – 69	10
70 – 79	6
80 – 89	4
	100

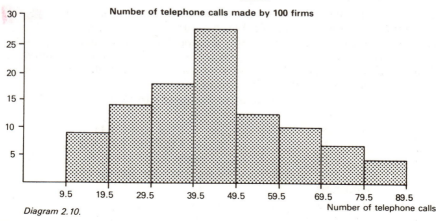

Diagram 2.10.

Diagram 2.10 shows the histogram of this distribution. Notice that the rectangles are drawn from the points 9.5, 19.5, 29.5 etc. The reason for this is that the class width is 10 (10 to 19 inclusive is 10 numbers). However, if we draw the rectangles from 10 to 19, from 20 to 29 etc, then the class width would only be nine.

Finally, it is often useful to present the frequency distribution in a different way altogether. Instead of considering the frequency of each class, we may consider the frequency with which the variable falls *below* a particular value. Consider the following distribution.

Distribution of Earnings of Adults in Full-time Employment

Earnings (£)	No. of workers (million)
10 and under 15	0.1
15 ,, ,, 20	1.1
20 ,, ,, 25	2.1
25 ,, ,, 30	2.3
30 ,, ,, 35	1.9
35 ,, ,, 40	1.4
40 ,, ,, 45	0.8
45 ,, ,, 50	0.5
50 ,, ,, 60	0.4
60 ,, ,, 70	0.3
70 ,, ,, 80	0.1
	11.0

Source: UK Annual Abstract of Statistics

The Chancellor who wishes to help the more poorly paid in society might well propose to introduce a tax exemption bill applying to all those earning less than £35 a week. Naturally, he will wish to know firstly how many people this bill will affect, and, secondly, how many more wage earners than before will now be free of the burden of income tax.

It is, of course, simple to add up the frequencies of the first five classes of our frequency distribution, but there is more than this. For many purposes it is easier to read off directly the frequency we require, and there are measures we will introduce you to in a later chapter which are far, far easier to obtain graphically than to calculate.

To obtain information of this nature the statistician has devised the Ogive or Cumulative Frequency graph. This diagram plots, on the horizontal axis, certain values of the variable, usually the upper value of each group; on the vertical axis it shows the frequency of the items with a value less than this. Thus, in constructing the ogive, we first construct a cumulative frequency table from our frequency distribution in this way:

Wage (£)	Cumulative Frequency (millions)
Under 15	0.1
20	1.2 (i.e. 0.1 + 1.1)
25	3.3 (i.e. 0.1 + 1.1 + 2.1)
30	5.6
	and so on until we reach the final group
Under 80	11.0

It is this table that we plot as the ogive, in Diagram 2.11.

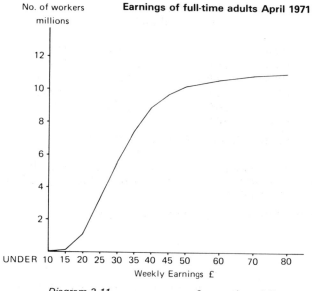

Diagram 2.11 Source: Annual Abstract

The first thing you will notice is that we have joined the points on the ogive by straight lines rather than the smooth curve that you are probably more used to. The reason for this is that we do not know how the items in any group are scattered between the upper and lower limits; so we make the only reasonable assumption that we can — that the items in any group are distributed evenly across the group. It may be a false assumption. We may find, in the group £25 and under £30, that all the 2.3 million members get a wage of £29.50; but it is not likely. So long as there is a reasonable number of members within the group, we are fairly safe in assuming an even distribution. We know that it will not materially affect our results.

4. *Plotting the Time Series*

You will remember from the last chapter that much data is given in the form of a time series, in which we take a variable and show how its magnitude has varied over a period of time. We have obtained from the United Kingdom Annual Abstract of Statistics the following table showing how consumer expenditure at constant prices varied during the period of time when prices in the United Kingdom were rising at an alarming rate, 1964 to 1974.

Consumer Expenditure at Constant Prices
(£ million)

1964	28330	1970	31472
1965	28760	1971	32397
1966	29301	1972	34318
1967	29869	1973	35962
1968	30598	1974	35741
1969	30715		

Source: Annual Abstract of Statistics

Perhaps you will agree that these figures, involving thousands of millions of pounds, mean very little; if we are honest probably very few of you even read the figures in detail.

But suppose we now draw a graph of these figures!

Diagram 2.12

We are sure that even a quick glance at this graph will leave you with a permanent impression of the way in which consumer expenditure has been rising — slowly and steadily from 1964 to 1970, then much more rapidly, reaching a peak in 1973. Since we have plotted expenditure at constant prices, this represents a rising consumption of goods, i.e. a rising standard of living (it may of course be at the expense of past saving).

You will have noticed the break in the graph on the vertical axis and the sudden jump from 0 to 25. This is a device used by statisticians when every figure is high and we do not want to crush our graph into a small space at the top of the graph paper. Remember too that we could have left you with a very different impression of the behaviour of consumer expenditure by adjusting the scales on the axis. Try it for yourself by doubling the scale on the vertical axis and halving the horizontal scale; or halving the vertical scale and doubling the horizontal. We have altered the scales in diagrams 2.13 and 2.14. After looking at three graphs of consumer expenditure can you really draw any conclusions as to the rate at which consumer expenditure has been rising? We know it has gone up — and that is about all.

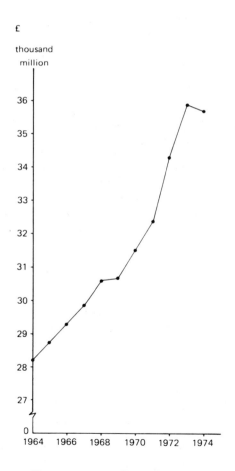

Consumer expenditure change of Scale

Diagram 2.13

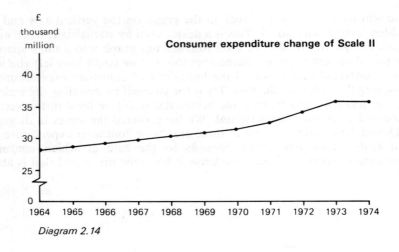

Diagram 2.14

Now, this is all very well, but our friends, the economists, will immediately want to know what we have been spending more on. Are we buying more food and clothes, buying more houses, or wasting our resources in riotous living? For their sakes, and for yours, we add below the way in which consumption of four sub-categories of our expenditure have been behaving in the same period. Whenever we wish to compare the way in which several variables have been behaving over a period of time, it is perfectly permissable to draw as many as four or five time series on the same axes, as long as we clearly distinguish the different graphs and the diagram is not too difficult to interpret.

	Expenditure on			
	Food	Drink	Clothing	Housing
		(£ thousand million)		
1964	6080	1866	2366	3481
1965	6081	1849	2426	3597
1966	6170	1922	2425	3695
1967	6228	2001	2450	3836
1968	6260	2108	2568	3973
1969	6264	2149	2596	4106
1970	6365	2296	2693	4181
1971	6362	2463	2726	4287
1972	6320	2641	2910	4377
1973	6388	2988	3060	4483
1974	6418	3081	3039	4460

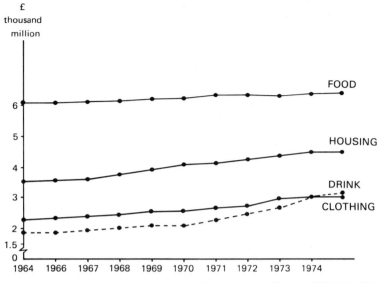

Source: Annual Abstract 1975 Table 338

Consumer expenditure at constant prices 1964-74

Diagram 2.15

You can see from the graphs we have drawn how the pattern of expenditure is changing. Although total consumption was rising fairly rapidly over this period, expenditure on food seems to have risen very slowly indeed — the slope of the graph is almost non-existent. Far different is the case of drink. Particularly from 1969 expenditure on drink rose at an alarming rate. Remember that this expenditure is listed at constant (1970) prices, so the rise of the graph represents increasing real consumption. Economists and sociologists would find this an interesting comment on human behaviour, and, in looking at this diagram, use your knowledge of other subjects to interpret what has been happening.

Do you remember when we discussed the bar chart we showed how it could be used to illustrate the constituents of a given total? Now, there are times when we want to illustrate how these constituents have varied over time. It may be, for example, that over the course of years road accidents involving bicycles have been forming a smaller and smaller part of total road accidents, while accidents involving motor-cars have been gradually forming an increasing proportion of total accidents. There is nothing to stop us from drawing three or four different time series, one for each type of vehicle showing the number of accidents involving that type of vehicle. We could even draw a number of graphs showing the percentage of accidents involving each type of vehicle, and we have no doubt that the diagrams would bring out the changing pattern. For this type of analysis it is far better to use a special type of diagram, the *Strata Graph,* or, if you are dealing in percentages, the *Percentage Strata Graph*. We must, however, keep the number of constituents reasonably few since otherwise the diagram may become difficult to interpret. The great advantage of this type of presentation is, as you will see, that the lines showing the magnitude of each constituent can never cross. Diagram 2.16 is a typical strata graph, showing the constitution of road accidents over a period of time in a large city. In order to stress the points that we would like to bring out, the figures are purely imaginary, so do not think that your own town is abnormal if the pattern of accidents is different from the one we have represented.

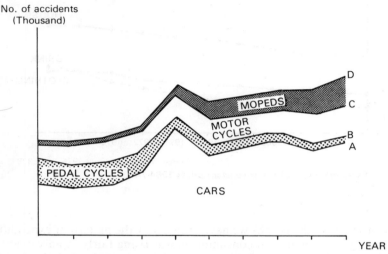

Number of road accidents 19... to 19... analysed by type of vehicle
Diagram 2.16

The principle of the strata graph is that our totals are successively cumulated. Thus the height of line A represents the number of accidents involving motor-cars each year. When we consider those accidents involving bicycles we add together for each year those accidents involving motor-cars and those involving bicycles to produce line B, the height of which for any year tells us the number of accidents involving (motor-cars + bicycles). In the same way line C is the graph of the number of accidents involving (motor cars + bicycles + motor-cycles). Finally when we add to these figures those accidents involving mopeds, we arrive at line D, which is the total number of accidents for all vehicles. You will have realised already that the distance between each line represents those accidents involving a particular vehicle. Thus the gap between A and B shows us accidents involving bicycles and the gap between B and C those involving motor-cycles. It is, of course, desirable to distinguish each type of vehicle by distinctive shading to give the diagram clarity, but we think you will agree that such a diagram is relatively easy to interpret. Even a cursory glance tells us that the total number of road accidents has been on the increase, but within this total three distinct trends are apparent. Bicycle accidents, which were quite numerous in the early years, have dwindled away and are now very few: conversely, moped accidents, which were very few in the early years, probably because there were few moped owners, are now becoming quite an important constituent of total accidents: car accidents have increased, and although it is difficult to judge, probably remain much the same proportion of an increased total. Motor-cycle accidents, it is apparent, have remained fairly constant over the years.

If we wish to draw a percentage strata graph, it is just as easy, although the initial calculations, converting our figures into percentages, are very cumbersome. Can you see that the line D in such a graph would be a horizontal straight line at 100%?

Exercises to Chapter 2

2.1 (a) Statistical information is sometimes shown in the form of ideographs (pictograms). For example, pictures of sacks of different sizes or alternatively pictures of different numbers of sacks of the same size may be used to show the production of flour in different countries. Comment on the advantages and disadvantages of these pictorial forms.

(b) Describe the purpose and construction of a pie chart, illustrating your answer with a simple example of your own invention.

2.3 (a) What are the advantages of using charts and graphs in statistical investigations?

(b) Describe clearly the methods to be employed in constructing:

(i) a pie chart,

(ii) a histogram.

(c) State the kind of diagram you consider most suitable to illustrate:

(i) daily hours of sunshine for a period of one month.

(ii) the number of workers, male and female, employed in a factory at each of three dates.

Briefly give your reasons for your choice in each case.

2.3 Discuss the suitability of a compound bar chart and a pie chart, for the presentation of data to:

(a) top management,

(b) lower management,

(c) the public.

Give examples of data which could be satisfactorily presented by each of these charts. (Do not draw the charts.)

2.4 A daily count of the number of rejects from the assembly line of a local manufacturer has yielded the following data:

138	164	150	132	144	125	149	157
146	158	140	147	136	148	152	144
168	126	138	176	163	119	154	165
146	173	142	147	135	153	140	135
161	145	135	142	150	156	145	128

(a) Using the data, construct a frequency distribution table and from that sketch the corresponding frequency curve.

(b) Comment on the shape of the frequency curve you have obtained and compare it with the sketched shapes of two others with which you are familiar.

2.5 The set of figures below shows the ages at which 50 employees were appointed to a certain grade.

```
28  27  30  27  28  28  26  27  28  28
26  28  29  31  27  28  27  29  27  29
28  27  31  27  27  29  30  27  28  28
28  29  28  29  27  30  27  28  27  29
28  27  28  29  29  28  28  28  28  27
```

(a) Write the data in the form of a frequency table and draw the frequency curve.

(b) Draw also the cumulative relative frequency curve.

2.6 Estimates of Gross Domestic Product (GDP) in current prices from 1951 to 1972 in £m are given below.

	£m		£m
1951	12,639	1962	25,279
2	13,790	3	26,878
3	14,877	4	29,187
4	15,726	5	31,156
5	16,867	6	33,057
6	18,264	7	34,835
7	19,369	8	37,263
8	20,196	9	39,667
9	21,248	1970	43,303
1960	22,633	1	48,675
1	24,213	2	53,940

Source: National Income and Expenditure, 1973

Construct a frequency distribution and a histogram of these figures and comment.

2.7 The following is a record of marks scored by candidates in an examination:

```
77  59  84  73  51  43  50  81  61  53  69
37  58  63  67  61  90  61  50  60  84  56
77  57  42  43  41  49  37  21  24  35  34
50  11  52  30  16  33  67  87  64  47  59
37  92  88  30  38  22  22  49  46  50  64
23  73  73  48  26  36  51  85  71  57  45
```

(a) Tabulate the marks in the form of a frequency distribution, grouping by suitable intervals.

(b) Construct a histogram from your frequency distribution.

(c) Explain the essential differences between a histogram and a bar chart.

2.8   *Stocks of Coal — Great Britain, July 1972 to June 1973*
       (thousand tons)

	Total	Opencast Sites and Central Stocking Grounds	Collieries
1972			
July	8,839	3,419	5,420
August	9,282	3,530	5,752
September	9,764	3,528	6,236
October	10,030	3,473	6,557
November	10,471	3,458	7,013
December	10,934	3,376	7,558
1973			
January	11,130	3,201	7,929
February	11,455	3,224	8,231
March	11,972	3,244	8,728
April	12,470	3,268	9,202
May	12,925	3,332	9,593
June	13,292	3,417	9,875

Source: Department of Trade and Industry.

(a) Write a short report stressing the main features of the data given above and include any derived statistics which may be appropriate.

(b) Prepare a suitable diagram or graph which will illustrate your report.

2.9

Household Expenditure in 1971

	£	Per Cent of Total
Housing	3.98	12.8
Fuel, light and power	1.85	6.0
Food	8.02	25.9
Alcoholic drink	1.46	4.7
Tobacco	1.30	4.2
Clothing and footwear	2.81	9.0
Durable household goods	2.01	6.5
Other goods	2.32	7.5
Transport and vehicles	4.26	13.7
Services	2.90	9.4
Miscellaneous	0.09	0.3
Total weekly household expenditure	30.99	100.0

Draw carefully and neatly a chart, graph or diagram to represent in visual form the household expenditure in units of actual money spent. Draw carefully and neatly a second chart, graph or diagram to represent in visual form the household expenditure in percentage terms. State the reasons for your choice.

2.10 *Unemployed in Great Britain Receiving Unemployment Benefit*
(thousands)

	Total	Unemployment Benefit Only	Unemployment Benefit and Supplementary Allowance
1970			
February	332	260	72
May	303	238	65
August	286	226	60
November	305	245	60
1971			
February	401	312	89
May	406	310	96
August	427	321	106
November	494	379	115
1972			
February	514	391	123
May	451	339	112
August	385	291	94
November	344	261	83

Source: Department of Employment.

(a) Prepare a suitable graph or chart to represent the data given above.

(b) Write a short report on the main features revealed by the table and your graph or chart using derived figures where appropriate.

2.11 *National Insurance — Great Britain: New Claims –*
Weekly Averages
(thousands)

Year	Sickness and Invalidity Benefits	Injury Benefit	Unemployment Benefit
1966	206.1	18.5	50.6
1967	193.2	18.8	63.0
1968	204.0	18.0	58.8
1969	219.4	17.9	59.6
1970	204.5	15.8	60.7
1971	169.3	14.0	68.7

Source: Department of Health and Social Security.

Prepare a suitable graph or chart to show the data given above relating to National Insurance claims in Great Britain. Write a short report on the main points revealed by your graph, using derived figures where appropriate. Explain carefully what is meant by 'Weekly Averages' and why this term is used rather than that relating to any other time period.

2.12 *Subjects studied at Britannia College of Commerce 1972*

Subject	Number of Students
Professional:	
Management	240
Banking	120
Accountancy	980
Languages:	
Spanish	20
French	220
German	100
General:	
G.C.E.	350
O.N.C.	225
H.N.C.	115

(a) Depict the data given above in the form of:
    (i) a simple bar chart,
    (ii) a component bar chart (actuals),
    (iii) a percentage component bar chart,
    (iv) a pie chart.

(b) Comment on the effectiveness of using the pie chart and the component bar chart as a means of illustrating data classification.

Chapter Three

The Presentation of Data II

The diagrams that we have examined so far are in fairly common usage. In fact we would be prepared to bet that, with the exception of strata graphs, you will have already seen all of the diagrams in the previous chapter. Basically, all of these diagrams are attempting to show the relative quantities in the data in an interesting and easily grasped manner. Yet some diagrams show much more than just the relative sizes, and we will now turn our attention to such diagrams. The chapter is concluded with a discussion on how diagrams can be made to misrepresent data.

1. *Logarithmic Graphs*

As we have seen, one of the problems with ordinary graphs is that they tend to give a false impression of the way in which figures are changing. Using a vertical scale of 1 inch = 100 units, a change from 100 to 200 units is represented by the same upward movement as a change from 1000 to 1100 units, even though it is 100% change as compared with a 10% change. The slope of the graph is the same — yet the rate of change is very different. Now, in most cases this does not matter. We may be interested only in the swings of the absolute figures; or the range of the figures may be such that the difference in the rates of change is negligible. If this is the case, we can get all we need from the graphs we have drawn. However, if we are considering inflation, for example, it is the rate at which prices are rising which is important, rather than the actual price increases.

Fortunately, if we plot the logarithms of our figures rather than the figures themselves, we can produce a graph the slope of which represents the real rate of change. Why should this be so?

Consider a case in which prices are rising by 20% a year. We can easily construct a table to show what happens to prices.

Period	Price	Log of Price	Difference of Logs
1	100	2.0000	
2	120	2.0792	0.0792
3	144	2.1584	0.0792
4	172.8	2.2375	0.0792
5	207.36	2.3167	0.0792

Don't you find it rather frightening that an inflation rate less than that of Britain in the mid 1970's will more than double prices in five years?

As you can see, the log of each price rises by 0.0792 irrespective of the magnitude of the price change, and hence we obtain a straight-line graph if we plot the logarithms. This of course tells us that the rate of change of prices is constant. It does not tell us what the rate of change is, but merely enables us to compare rates of change over time.

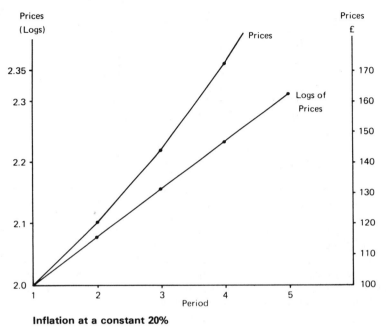

Inflation at a constant 20%

Diagram 3.01.

Strictly such a graph is known as a *semi-logarithmic* graph, as we plot logarithms only on the vertical axis. On the horizontal axis we do not plot the logarithms of the time periods.

As you can see from the diagram, if we plot the prices themselves the slope of the curve gets steeper and the fact that the rate of change is the same is completely hidden.

Often a logarithmic graph can be used also when the figures with which we are concerned range so widely that it is inconvenient to use a normal graph. Such a situation would occur if we tried to plot German price movements during the great inflation of the 1920's when prices were increasing by hundreds of times each month.

Now, although few students find difficulty in mastering the principles involved in the construction of semi-logarithmic graphs, this is such a common question in examinations that we would strongly recommend you to work through the following example with us, making sure that you understand each step.

Example

Sales of Two Companies
Sales (£000)

Year	Company A	Company B
1961	2240	980
1962	2460	1082
1963	2680	1205
1964	2915	1289
1965	3136	1382
1966	3362	1476
1967	3590	1580
1968	3821	1687
1969	4049	1787
1970	4280	1891

(a) Plot the time series on (i) an arithmetic scale graph,
 (ii) a semi-logarithmic graph.

(b) Interpret the results.

Well, the first part of this question will cause you no difficulty at all. As you well know, an arithmetic scale graph is a normal graph, in which we let a given distance (usually a centimetre) represent a given change in the magnitude of the variable on both the horizontal and the vertical axes. We then plot the figures given in the question according to the scale we have used. Thus, in diagram 3.02 on the horizontal axis we could let two centimetres represent one year, while on the vertical axis it represents £500,000 sales. We have plotted the graphs showing the variation in the sales of the two companies in the normal way.

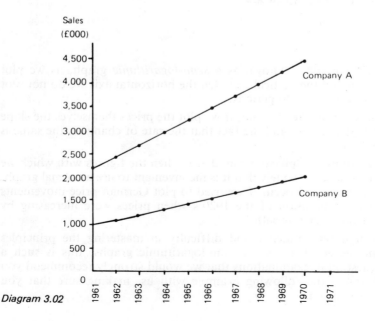

Diagram 3.02

When we now come to consider the semi-logarithmic graphs, we have to be careful. In some examinations you will be issued only with the normal graph paper we have used so far; in others with graph paper already designed for semi-logarithmic use. Let us suppose firstly that you have been issued only with the normal arithmetic scale graph paper. In this case, as you know, we will plot the years along the horizontal axis as normally, but along the vertical axis you will have to plot the logarithms of the sales figure for each year. Our first step then must be to obtain these logarithms. We strongly advise you to do this as part of the answer and not on a piece of scrap paper. It is easy to make a mistake, and if you do examiners are far more inclined to be generous if they can trace easily where the mistake has arisen. So we proceed as follows:

	Company A		Company B	
Year	Sales	Log	Sales	Log
1961	2240	3.3502	980	2.9912
1962	2460	3.3909	1082	3.0342
1963	2680	3.4281	1205	3.0810
1964	2915	3.4646	1289	3.1103
1965	3136	3.4964	1382	3.1405
1966	3362	3.5266	1476	3.1691
1967	3590	3.5551	1580	3.1987
1968	3821	3.5822	1687	3.2271
1969	4049	3.6073	1787	3.2521
1970	4280	3.6314	1891	3.2767

Although this seems to be a cumbersome process, almost any of the small pocket calculators you are normally allowed to use will give you the logarithms you require in a matter of seconds. All you have to do now is to scale the vertical axis to accommodate the logarithms you have obtained, and draw the graphs. We have done this in diagram 3.03, but you will notice that we have also inserted on the vertical axis on the right some of the absolute values of the sales. While modern management is well aware of the use of semi-logarithmic graphs they also wish to be able to see at a glance the value of the sales turnover without having to refer to a book of log tables.

Even a quick glance at diagrams 3.02 and 3.03 brings out the advantages of the semi-logarithmic graph. The graphs of the sales figures on arithmetic scale imply that the two companies increased their sales at a steady rate throughout the whole period. But as you are aware, such graphs show absolute changes, not *rates* of change, and equal absolute increases imply a declining *rate* of increase. This is well brought out in the semi-logarithmic graphs, where a close look at company A shows a marked tendency of the rate of change to fall.

More important, the two graphs give completely different impressions of the relative performance of the two companies. The arithmetic scale graph implies that company A is expanding more rapidly than company B because the slope of the graph of the former's sales is steeper. But again the use of absolute figures is deceptive. The semi-logarithmic graphs show that company B's performance is at least as good as that of company A in that it is expanding at the same rate, while on a very close examination we find that the rate of growth over the whole period is marginally better than that of company A, and from 1964 onwards growth certainly seems to be steadier and more sustained than that of company A. Thus, we may come to the conclusion that, although both companies have grown substantially in the period, company B, although smaller, appears to have done marginally the better.

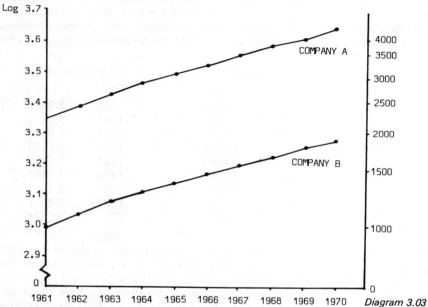

Diagram 3.03

Now this is all very well, but suppose you are issued with semi-logarithmic graph paper already ruled. The great advantage of this paper is that you can insert the *absolute* figures of sales on the graph without first having to look up the logarithms. Probably, however, few of you have had the opportunity yet of seeing semi-logarithmic graph paper, so let us explain the principles on which it is ruled. One axis is ruled normally, in centimetres and millimetres (after all, it is *semi* logarithmic). But the other axis is quite differently ruled. It is drawn on the basis that equal distances represent equal percentage changes rather than equal absolute changes. Thus, if one centimetre represents a change from 10 to 100 (an increase of 10 times), the next centimetre will represent a change from 100 to 1000 (also an increase of 10 times).

There is more to it than this, however. No-one would expect you to calculate the percentage change for each figure that you have to plot on the graph. We have merely stated the obvious − that one axis of the semi-logarithmic graph paper is scaled logarithmically. Let us take it further. If we take the first centimetre on the vertical scale to represent an increase of 1 unit (from 1 to 2), the next centimetre would represent an increase from 2 to

4. We could have obtained exactly the same effect by plotting logarithms. The log of 1 is 0.0, so you will realise that the logarithmic graph can range upwards only from an absolute value of 1 (not 0 as can the arithmetic scale graph). Now the logarithm of 2 is 0.30103, and the logarithm of 4 is 0.60206. Similarly the logarithm of 8 is 0.90309. Can you see that, in allowing each centimetre to represent the same proportional increase in the absolute figures, we are scaling according to the logarithms of the numbers? In this case one centimetre represents an increase in the logarithm of .30103; and when we add .30103 to a logarithm, we are in fact multiplying the previous number by two. So each successive centimetre represents an increasing change in the absolute figures – 2, 4, 8, 16 and so on.

But the logarithmic scale is not drawn in centimetres as is the arithmetic axis. Suppose once again that we take the first vertical division on the graph paper to be one centimetre long (whether it is or not depends on the graph paper with which you are issued). We can let this centimetre represent any absolute magnitude that we wish. Let us again assume that it represents an increase from 1 to 2. The next main division on our graph paper will also represent an absolute increase of one unit, from 2 to 3, but this division will not be a centimetre deep. We have already said that the second centimetre represents an increase in absolute values from 2 to 4, so you would naturally expect the second division representing an increase from 2 to 3 to be less than a centimetre deep. It will in fact be about .585 centimetres deep only. Thus each successive unit increase on the logarithmic axis will entail a smaller and smaller vertical rise, and the vertical scaling will look something like Diagram 3.04.

In practice, of course, the graph paper you will be given will have each main division subdivided as normal graph paper is into ten subsections to enable you to plot the intermediate figures. If you have not seen such graph paper before, we strongly recommend that you immediately study carefully diagram 3.05 where we have drawn the two graphs we are concerned with on semi-logarithmic paper, and, most important, obtain a stock of such paper of your own and practice drawing semi-logarithmic graphs using the exercises at the end of this chapter.

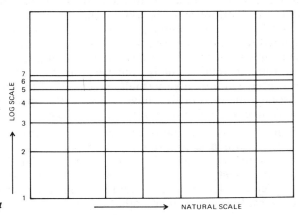

Diagram 3.04

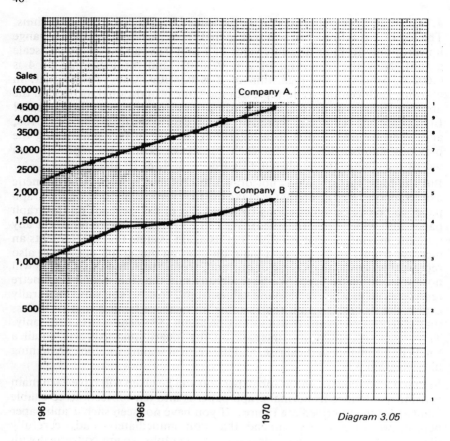

Diagram 3.05

2. *The Lorenz Curve*

If you look at statistics of income one of the first things that strikes you is the inequality in the distribution of incomes in most countries. Not only is the range of incomes wide, from under £1000 a year to £20,000 a year and more in the United Kingdom, but we also find that a very small percentage of the income recipients at the top of the scale receive a disproportionately large share of total income. Equally, the very large percentage of low income earners receive in total a very small percentage of the total income. You must, at some time in your life, have met such statements as 'the top 5% of income recipients receive over 70% of total income'.

Now any economist will tell you that one purpose of our taxation system, or any taxation system which is progressive, is to reduce the inequality of incomes, and naturally we would all like to know how far the system is succeeding in this objective. Statisticians have derived a diagram, the Lorenz curve, which enables us to show graphically the extent of inequality, not only of incomes, but also of many other things.

In this diagram, considering income distribution, we measure on the horizontal axis the percentage of population, and on the vertical axis the percentage of income, see diagram 3.06.

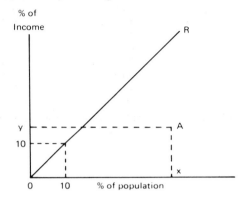

Diagram 3.06

Now, obviously, if incomes are distributed equally, the bottom 10% of income earners will receive 10% of total income, the bottom 20% of income earners will receive 20% of total income and so on. The graph representing such a distribution of income will be the straight line OR passing through the origin and at 45 degrees (provided, that is, that the horizontal and vertical scales are the same), and any divergence from this line will indicate some degree of inequality. The point A for example would be interpreted as 'the bottom x% of income earners receive y% of total income', and, since x is greater than y, would be derived from a situation such as 'the bottom 65% of income earners receive only 32% of total incomes'.

Let us illustrate the use of Lorenz curves by applying them to the following income statistics of the United Kingdom.

Income Class	No. of Incomes		Pre-tax Income		Post-tax Income	
£	(000)	%	£m	%	£m	%
50 – 249	5070	18.6	991	4.9	990	5.5
250 – 499	6570	24.2	2590	12.8	2486	13.8
500 – 749	6155	22.6	4143	20.5	3844	21.4
750 – 999	4830	17.8	4580	22.6	4168	23.2
1000 – 1999	4145	15.2	5849	28.9	5113	28.5
2000 – 3999	353	1.3	1305	6.4	940	5.2
4000 – 5999	59	0.2	469	2.3	281	1.6
6000 and over	18	0.1	330	1.6	129	0.8
	27200		20257		17951	

Source: National Income and Expenditure

The first step in the construction of our diagram is to calculate the percentages appropriate to each group and each column. Thus the 5,070,000 individuals in the income class £50 – £249 comprise 18.6% of all income recipients and they received 4.9% of all pre-tax income. We have inserted the relevant percentages in the body of the table, although in an examination you would have to calculate each of them from the original figures given to you.

The next step is to cumulate the percentages you have calculated in this way:

Income Earners (%)		Pre-tax Income (%)	Post-tax Income (%)
18.6	receive	4.9	5.5
42.8		17.7	19.3
65.4		38.2	40.7
83.2		60.8	63.9
98.4		89.7	92.4
99.7		96.1	97.6
99.9		98.4	99.2
100.0		100.0	100.0

It is now easy to draw the Lorenz curves relating to pre-tax and post-tax income:

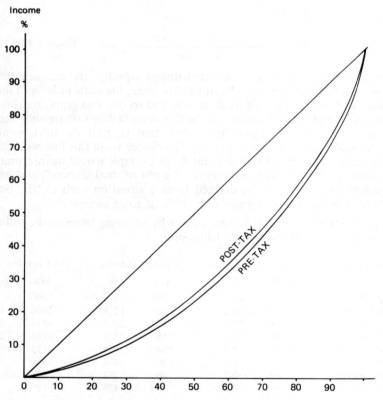

Diagram 3.07   Distribution of pre-tax and post-tax incomes U.K.

Now any divergence from our straight line of equal distribution of income indicates that there is inequality in the distribution of incomes, and the further the Lorenz curve is from this line of reference, the greater is the degree of inequality. It is worth noting that if the Lorenz curve is below the straight line the inequality is in favour of the upper income groups in that a high percentage of low income earners receive a small percentage of income.

If, on the other hand, the Lorenz curve is above the straight line, it implies that a given percentage of the bottom income earners receive a higher percentage of total income; the inequality, that is, works in favour of the poor.

Looking at our diagram, it is apparent that at this time there was a great deal of inequality in the income structure of the United Kingdom. Although the taxation system did reduce it, the reduction seems to have been minimal. We have not, however, asked what was done with the tax revenue. Much of it was returned to the lower income groups in the form of social benefits — family allowances, supplementary benefits, subsidised housing and so on. It may well be that when we allow for this transfer of income, the effects of taxation would appear very different.

3. *The Z Chart*

A diagram which is often used in industry and commerce, although it seems to be less popular among statisticians, is the Z Chart, so called because the completed diagram takes the form of the letter Z. This is merely a device to enable management to show concisely three different aspects of a time series plotted on the one graph. On the bottom bar of the Z we plot the time series of monthly (or weekly) sales, or output, or whatever variable we are considering. On the diagonal bar of the Z we plot the cumulative total to date, that is, the total sales or output we have achieved since the beginning of the year. Finally as the top bar of the Z we plot the total sales achieved in the last year: the first or January figure is the total sales achieved during the period 1st February last to 31st January this year; the February figure is the total from last March until the end of February this year, and so on.

Output – ABC Limited

	Last Year	Current Year	Cumulative Total	Moving Annual Total
January	9	11	11	146
February	8	14	25	152
March	9	12	37	155
April	13	15	52	157
May	14	16	68	159
June	18	19	87	160
July	16	18	105	162
August	15	14	119	161
September	12	14	133	163
October	10	13	146	166
November	9	11	157	168
December	11	15	172	172

We will illustrate by plotting the figures of output for the firm ABC Ltd. The figures are simplified to enable you to follow the calculations more easily. Obviously, although we are plotting figures for the current year only, if we are to obtain a running total of sales over the last twelve months we will need the figures for two years.

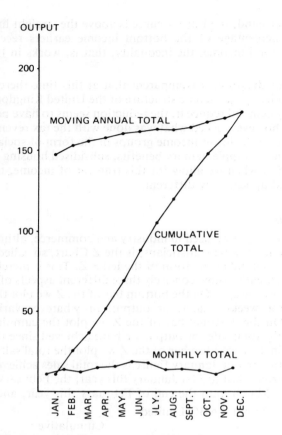

Diagram 3.08

A glance at diagram 3.08 shows what a vast amount of information is given by this simple diagram. The cumulative total tells us performance to date, and it is a simple matter to superimpose on the diagram a line showing planned or expected performance. Thus we can see at a glance whether our plans are being realised or whether we are falling behind. The moving annual total enables us to compare performance this year with that at a comparable time last year. If this chart is rising it means that figures in that month are higher than they were in the same month last year. Thus we have the means of making a direct comparison with last year. Finally the monthly total enables us to keep a direct check on what is happening now; and if we are falling behind our plan we can usually spot a month with low figures which has caused this and so discover why. Is it any wonder, with this wealth of information to be had, that the Z chart is so popular in industry?

4. *The Scatter Diagram*

The final diagram we will introduce you to is not a graph at all. It looks at first glance rather like a series of dots placed haphazardly on a sheet of graph paper. But it is anything but haphazard as we will see. The basic aim underlying the scatter diagram is to try to ascertain if there is a relationship between two factors, such that when one is high the other is high, when one is low the other is low. Or perhaps the relationship is inverse — when one variable is low, the other is high and vice versa. Suppose we were examining the relationship between the level of employment and the level of industrial investment, see Diagrams 3.09a, b and c.

Firstly we will have figures over a considerable period of time giving us the level of industrial investment and the percentage employment rate associated with that level of investment. Let us just take four of those pairs of figures. The first pair tells us that the level of investment is high and associated with it was a high level of employment. This position is indicated by point A in diagram 3.09a. Another pair of figures tells us that when investment was slightly lower employment was considerably lower — point C. Still a third pair of figures tells us that at a time when investment was low employment was also low. This is indicated by point B. If we examine the last figures we find that although investment was the same as before, the level of employment was, in fact, very much higher — point D. If we plot sufficient pairs of figures we may well get a series of dots such as those on diagram 3.09b which show a pattern. Generally the higher the level of investment the higher the level of employment, and this we have indicated by inserting freehand a dotted line rising upwards from left to right. The same sort of pattern is seen when we examine savings and the rate of interest in Diagram 3.09c. Of course, as you can see from the scatter of the points the relationship is not perfect. We cannot forecast the exact level of employment from the level of investment. It would be fine if we could do this, but at the moment all we are interested in is the general tendency. Before you leave the scatter diagram, experiment for yourself. Draw a scatter diagram showing the relationship between investment and *un*employment. You should get an inverse relationship — the points *falling* from left to right. And finally, try to draw a scatter diagram in which there is no clear relationship shown between two variables.

Some Pitfalls to be Avoided

We must not leave the subject of graphs and diagrams without giving you some advice on what not to do. Unfortunately you will find examples every day of diagrams which illustrate what we are about to say should not happen.

A device beloved of advertisers today is to represent their information in the form of little pictures or ideograms. The sales of a particular brand of beer may be represented by the size of a foaming tankard, the amount of washing-up liquid you get for a penny by the height of liquid in a test-tube. Now, many advertisers using this technique are merely doing their job to the best of their ability; but some advertisements we have seen seem designed to mislead the reader.

Look at diagram 3.10 which represents the sales of 'Whizz' by varying the size of the packet.

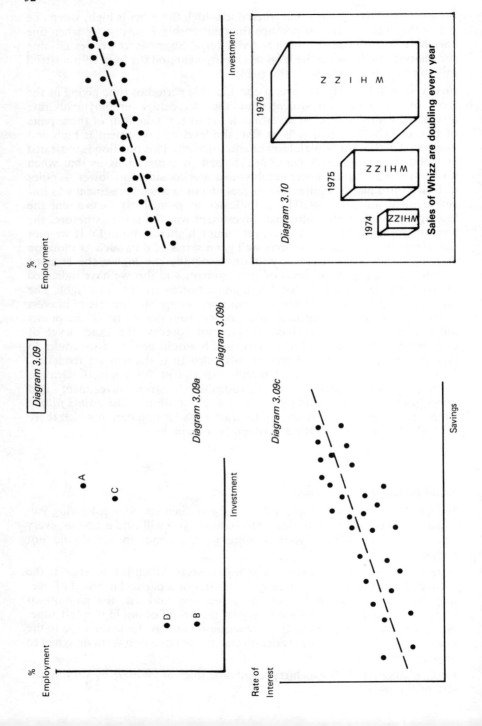

There is nothing wrong with this technique — it can put over the idea of increasing sales forcibly and effectively. But look at the size of the packets. Doubling the dimensions of a packet does not indicate, of course, that sales have doubled. The volume of the second packet is in fact eight times that of the first, and the volume of the last packet is, believe it or not, sixty-four times that of the first. Thus, in spite of the writing saying that sales are doubling every year, the reader is left with a completely false impression of the rate at which they are rising.

Even more unpardonable, in our opinion, is the situation shown in diagram 3.11.

You know that before you draw any graph, you should clearly mark the scales on the axis. Before you indignantly retort that this is an obvious point and will never be forgotten, you must realise that many people do quite deliberately omit the scales, intending to mislead. You must have seen this situation in television commercials. A mysterious line runs across the screen — showing absolutely nothing, but still leaving the impression that sales are skyrocketing and that we are missing the chance of a lifetime by not buying the product. Couple this with the statement that this is the housewives' choice and we are caught. No-one likes to be out of step with one's friends and neighbours.

Equally bad, if not worse, is the invention of units that do not exist. The scales should tell us something; they should use units that are real and that can be understood. But look at Diagram 3.12.

Imagine this graph on your television screen, and the smooth voice of the announcer 'proving' by pointing to the graph that daily brushing with Gritto is bound to make our teeth whiter. But what is a unit of brightness? We are afraid that we do not know, and we very much suspect that the advertising agencies do not know either.

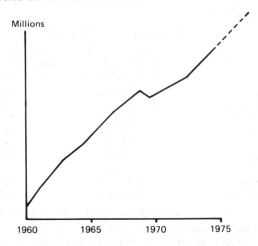

Sales of Bang show it to be the housewives' choice

Exhibit 3.11

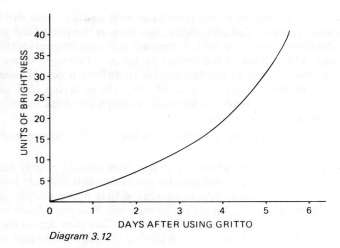
Diagram 3.12

Students often ask us which is the best diagram to use, for a particular purpose. There is no real answer to this and we suggest that you do not waste time looking for one. Any diagram should present the salient features of the data simply and vividly. Equally, if the reader has to spend a great deal of time 'sorting out' what a diagram means, it is a bad diagram.

But do not let it stop here. Certain features of the data should be obvious from the diagram. Now ask questions. Why has the variable behaved in this manner? Why has A behaved differently from B? Is there a relationship between C and D, and, if so, what is this relationship? Only by asking questions such as these will you get full value from your diagrams.

Exercises to Chapter 3

3.1

Transport: Great Britain 1973

	Number of New Registrations of Road Vehicles	Index of Vehicle Distances Travelled (av. 1963 = 100)
January	206,295	153
February	198,531	142
March	223,353	177
April	192,364	180
May	194,446	184
June	186,845	197
July	172,407	214
August	254,438	212
September	159,388	197

Source: Department of the Environment.

(a) The table of transport in Great Britain given above contains two sets of data expressed in different terms. By means of a ratio-scale graph show how these two sets of data may be compared. Use natural scale paper.

(b) Comment on the situation revealed by your graph.

3.2 (i) Describe four diagrammatic methods of presenting numerical data.

(ii) Explain how semi-logarithmic paper differs from arithmetic (or difference) graph paper and give an example of a situation in which semi-logarithmic graph paper would normally be used.

(iii) The following hypothetical data shows the state of weather in London on ten consecutive days.

Day	1	2	3	4	5	6	7	8	9	10
Noon temp. (°C)	19	18	17	20	21	19	18	20	22	19
	C	W	W	S	S	C	W	S	S	C

(C – cloudy, W – wet and S – sunny)

By using appropriate diagrammatic methods, illustrate:

(a) the proportion of days which were cloudy, wet and sunny;

(b) how the noon temperature varied from day to day.

3.3 (a) What advantages has semi-logarithmic graph over a natural scale graph?

(b) Plot the following two series on the same diagram using semi-logarithmic graph paper to show their relative movements and comment on the results.

How could a similar comparison be achieved without using semi-logarithmic paper?

Growth of a Company

	1970	1971	1972	1973	1974
Turnover (£)	70,000	400,000	1,200,000	1,800,000	4,100,000
Cost of materials (£)	30,000	160,000	500,000	750,000	1,400,000

3.4 (a) For what reasons do we use semi-logarithmic or ratio-scale graph paper rather than the more usual arithmetic scale paper?

(b) Plot the data given below, on semi-logarithmic paper.

(c) Comment briefly on what your graph shows.

Consumer's expenditure in the United Kingdom in £m at current prices from 1960 to 1970

Item/Year	1960	1964	1968	1970
Total consumer expenditure	16,900	21,500	27,200	31,300
Food	4,200	4,900	5,700	6,400
Housing	1,660	2,340	3,290	3,900
Running costs of vehicles	450	780	1,400	1,700

Source: Annual Abstract of Statistics.

3.5 The following table shows the sales of natural gas in the United States, and the natural gas sales of the Metropolitan Gas Corporation, for the period from 1956 to 1968:

Year	United States Natural Gas Sales Cubic Feet (billions)	Metropolitan Gas Corporation Natural Gas Sales Cubic Feet (billions)
1956	7,500	850
1957	8,013	1,030
1958	8,502	1,090
1959	8,750	1,150
1960	9,501	1,250
1961	10,095	1,275
1962	10,700	1,300
1963	11,030	1,350
1964	12,050	1,350
1965	12,800	1,280
1966	13,250	1,302
1967	13,900	1,390
1968	15,500	1,500

(a) Graph this data on a semi-logarithmic scale.

(b) State what the graph indicates about the comparative natural gas sales of the Metropolitan Gas Corporation and the industry as a whole, and explain the advantage of using a semi-logarithmic scale.

A.C.A.

3.6 Using a slide rule or a table of logarithms you are required to construct semi-log graphs and plot on them the following data:

In Great Britain

Year	Total expenditure on highways (£ million)	Total number of cars licensed (thousands)	Total number of goods vehicles licensed (thousands)	Total casualties in road accidents (thousands)
1959	228.0	4,972	1,378	333
1960	238.0	5,532	1,448	348
1961	270.7	5,983	1,503	350
1962	301.2	6,560	1,522	342
1963	342.4	7,380	1,582	356
1964	405.8	8,252	1,633	385
1965	421.2	8,922	1,661	397
1966	457.4	9,522	1,639	392
1967	528.2	10,312	1,692	370
1968	580.9	10,825	1,640	349

Source: Annual Abstract of Statistics.

Interpret the graphs you have drawn and give the advantages, if any, of this type of presentation.

I.C.M.A.

3.7 (a) Plot the following data using arithmetic scale graph paper.

Period	1	2	3	4	5	6	7	8
Data	200	320	640	1180	2080	4050	6480	9030

(b) Describe two methods of drafting a semi-logarithmic curve when no semi-logarithmic graph paper is available. Using one of those methods plot the data and state the advantages and disadvantages of using the semi-log scale over the more usual arithmetic scale.

I.C.M.A.

3.8 *Stoppages of Work due to Industrial Disputes*

	Number of Stoppages beginning in 1973	Aggregate Number of Working Days Lost in these Stoppages (to nearest thousand)
Under 250 days	1200	125
250 and under 500	453	161
500 and under 1000	402	282
1000 and under 5000	592	1250
5000 and under 25,000	180	1844
25,000 and under 50,000	24	859
50,000 days and over	22	2625

Illustrate the above data by means of a Lorenz curve. Why is this form of graph the most suitable for displaying the above information?

3.9 (a) Construct a Lorenz curve in respect of the following data concerning the net output of manufacturing industry X:

Manufacturing Industry X

Average Number of Employees	Number of Firms	Net Output (£ million)
25 and under 100	205	16
100 and under 300	200	60
300 and under 500	35	18
500 and under 750	30	26
750 and under 1,000	20	26
1,000 and under 1,500	10	54

(b) Explain the purpose of presenting information in the form of the Lorenz curve, and comment on your answer to (a) above.

A.C.A.

3.10 A sales department of a firm might plot a Z chart to check or monitor the annual performance of its sales staff. Describe such a chart with an example and explain how it is used.

A.C.A.

3.11 The production figures for a company were as follows:

	Jan.	Feb.	Mar.	Apr.	May	June
1973	60,746	57,071	63,621	66,014	71,736	80,213
1974	68,123	58,983	60,693	61,247	67,778	76,567

	July	Aug.	Sept.	Oct.	Nov.	Dec.
1973	91,780	92,314	76,770	67,123	71,512	87,490
1974	92,124	99,632	104,210	99,634	88,241	107,870

(a) Round the above production figures to the nearest thousand.
(b) Use the rounded figures to produce a Z chart for 1974.
(c) Comment on your chart.

3.12   *Imported Softwood Deliveries (monthly averages)*
        (thousand cubic metres)

Month	1972	1973
January	532	882
February	667	527
March	731	828
April	629	654
May	791	940
June	746	838
July	828	870
August	553	851
September	800	833
October	919	1004
November	900	806
December	576	535

Source: Department of Industry

From the information given, construct a Z chart for 1973. Explain when you might find such charts used and comment briefly on the advantages of representing data in this way.

3.13   *Tonnage and Service Speed of Passenger Liners of*
        the X-Y Company

Gross Tonnage	Service Speed (knots)
26,000	25.0
21,700	24.0
26,000	25.0
21,630	24.0
20,300	19.4
15,020	21.6
15,010	21.6
20,100	20.0
26,300	24.6
21,660	20.5
10,470	19.4
20,450	22.2
11,730	20.1
12,150	19.9
16,990	19.6
19,930	19.4
14,440	19.0
9,420	19.3

(a) What is the purpose of analysing data by means of a scatter diagram?
(b) Draw a scatter diagram to illustrate the above figures.
(c) Comment on what the diagram reveals.

Chapter Four

Approximation and Error

Does it surprise you that in a course on Statistics we should devote a whole chapter to the subject of error? Surely the aim of a course such as this should be to teach you about accuracy, not about errors. You may believe, in fact, that we should devote our time to teaching you how to avoid error.

This apparently strange chapter becomes less strange if we consider what we mean by error. To the majority of people trained in answering mathematical problems, the very idea of error implies that we have made a mistake. Certainly, if we say that $4 + 3 - 2 = 6$, we have made an error in our calculations, and our answer is wrong. It would be futile to pretend that this sort of thing does not happen in statistical work. No-one is perfect, and even writers of textbooks sometimes make mistakes that would cause a first-year student to feel ashamed.

Even if our calculations are accurate, our answers may still be wrong because of deficiencies in our raw material. We doubt very much if even the Census of Population taken, in the United Kingdom, every ten years, can claim to be completely accurate. The investigator might not ask the right questions; the respondent may tick the wrong answer; the statistician might analyse the data wrongly, or he might misread his figures, writing 796 instead of 769 when constructing his frequency distributions. There are so many possible mistakes that can be made that one sometimes wonders whether any statistician can ever guarantee one hundred per cent accuracy.

When, however, we talk of statistical error, it is not in this sense that we are using the word. Often it is inconvenient, or impossible to give figures which are correct to the last unit. When we say that 1.7 million children die every year in a particular area, we obviously do not mean that exactly 1,700,000 children die every year. What we are saying is that approximately 1.7 million children die, and a note at the head of our table will tell us the extent of our approximation. It is very likely that a figure such as this would be accurate to the nearest 0.1 of a million. The annual number of deaths, that is, could be anything between 1,650,000 and 1,750,000. In quoting 1.7 million, we could be as far as 50,000 different from the absolutely accurate figure. It is this difference between the approximate figure and the true figure that the statistician calls 'error'. It is, then, a very different concept from that of a mistake. When we make a mistake we do not know the amount by which our results are wrong; when we speak of error we are qualifying the degree of accuracy of our result. Then, too, a mistake is usually involuntary and something to be avoided; approximations are made to improve the presentation of our figures and in this sense the creation of statistical error is deliberate. So useful is the idea of approximation that you will find very few statistical tabulations which give figures with absolute accuracy, and this is understandable. Which of the following statements do you think that you could remember more easily? The output of steel was 110 million tons' or 'The output of steel was 109,764,397 tons.' The widespread use of approximations, however, means that any conclusions drawn from our figures are themselves subject to error, and we must look very carefully at the effects of approximating our figures.

Approximation

Suppose we come across a table telling us the number of cattle in our country at a particular date. We might find the heading of the column in which we are interested reads, 'No. of Cattle (million)'. What does this mean precisely? The fact that we have 'million' in brackets tells us that the person who has constructed the table has approximated his figures to the nearest million. Thus, if he tells us that the number of cattle in the country was 25 millions, we know that the true figure must lie between 24,500,000 and 25,500,000. if it were any less than 24,500,000 the approximated figure would be 24 millions, if it were any more than 25,500,000 the approximated figure would be 26 millions. A problem of course arises if the true figure were exactly 24.5 or exactly 25.5 millions. Do we raise or lower the figure in order to approximate? A good general rule is to look at the number of millions, and if it is an odd number raise it, if it is an even number lower it.

If we were approximating to the nearest thousand we would naturally look to see whether we had an even or odd number of thousands, and so on.

In tabular work it is often enough to put the degree of approximation in brackets like this: (millions). For other purposes it may be much more convenient to give the approximated figure and append to it the maximum amount of the possible error. Thus we could express exactly the same information as before in the form:

$$\text{No. of cattle} = 25 \text{ million} \pm 500{,}000$$

Occasionally too it is convenient to express the error in percentage form, and we could say:

$$\text{No. of cattle} = 25 \text{ millions} \pm 2\%$$

Each of these three statements says exactly the same thing. It makes little difference which way you put it. What is important is that whenever you approximate you should indicate clearly the basis of your approximation.

If you are approximating a number of figures and then totally them, a peculiar result may arise. It is possible that when you add the approximated figure you do not get the same result as when you approximate the exact total.

For example:

	Actual Output (tons)	Output (million tons)
Firm A	102,428,621	102
Firm B	204,364,147	204
Firm C	673,424,961	673
	980,217,729	980 NOT 979

You can see that, because of the approximations, the addition of the approximated totals is only 979. But the total output to the nearest million tons is 980 million. This is the figure that goes in the total of the approximated figures and you must *never* adjust it merely to make your additions seem to be right.

Biassed and Unbiassed Errors

When we are dealing with a number of figures that we are totalling, the amount of error in our total depends to a very large extent on the way in which we are approximating. If we are approximating to the *nearest* hundred (or thousand, or million), some of our approximated figures will be too high, others will be too low. But the excesses and the deficits will tend to cancel each other out and the aggregate error will probably be very small. In fact, the greater the number of items we consider, the smaller is the total error likely to be relative to the actual total. Such an error is said to be unbiassed, because it is not consistently in the same direction. Suppose we are approximating the following figures.

Actual Figures	To nearest thousand	Error
81,171	81	−171
316	0	−316
3,521	4	+479
2,638	3	+362
34,451	34	−451
5,467	5	−567
95,711	96	+289
90,501	91	+499
842	1	+158
Total 314,618	315	+382

As you can see, the error in the approximated total is only +382 in a total of 315,000, an error of only 0.12%; it is equally apparent that the reason for such a small error is that the positive and negative errors offset each other.

But suppose we were to approximate to the nearest thousand above, or to the nearest thousand below. Here the picture is very different. Far from offsetting each other, the errors are all in the same direction and the aggregate error is cumulative. We can see what a tremendous difference it makes to the error in our total if we use the same figures as above, but approximate them differently.

	Approximation to the nearest thousand	
	Above	Below
	82	81
	1	0
	4	3
	3	2
	35	34
	6	5
	96	95
	91	90
	1	0
	319	310
Error	+4382	−4618
Percentage error	+1.39%	−1.46%

This type of error is know as a biassed error — for obvious reasons, and, as you can see, the aggregate error is dramatically larger than the aggregate unbiassed error. Why, then, should we ever concern ourselves with biassed errors? Often, when we are collecting statistics, an element of error may creep into every figure we collect. Many people argue, for example, that if we are collecting ages every woman will tend to understate her age. Whether this is correct or not, it is certainly true that if we are recording temperature or atmospheric pressure on an instrument which is recording wrongly, we will consistently overstate or understate the true reading. The error that results in any of these cases will be a biassed error, and we must learn to be aware of it.

You should also beware of data that has built-in biassed error — this is not as rare as you might think. There are many examples of data expressed correct to the last complete unit — age is expressed correct to the last completed year, the electricity and telephone companies bill us to the last complete unit consumed.

Absolute and Relative Error

You will have noticed that in the last section we said that the unbiassed error in the figures we were discussing was +382 or +0.12%. When we state it as +382 we are measuring the arithmetic difference between the actual total and the approximated total, and this we call the *absolute* error. Unfortunately, the size of the absolute error does not give us any indication of the importance of the error involved. An error of 5 in a total of 50 is of much greater importance relatively than an error of 500 in a total of 50,000. To enable us to compare the extent of the error when widely differing totals are involved, the *relative* error is used. This is merely the absolute error expressed as a percentage of the approximated total.

Calculations Involving Error

In both theory and practice we often have to make calculations in which error is present. We might wish to know the number of children likely to be born next year in our local town in order to plan maternity services. But the population statistics may well be available only to the nearest thousand, and the birth rate only accurate to one place of decimals. Similarly if we are planning the building of schools two or three years ahead, we need to know the number of children likely to be born in the relevant years, and also the number of these who will die before reaching school age. We will also want to know the number of children who will move out of the area with their parents, and the number of children who will move into the area. All these figures are usually available, but every one of them is approximate, and the potential amount of error involved in any calculation can be very large.

Fortunately, provided we know the amount of error involved in each of our figures, it is simple to calculate the probable degree of error arising in the final result. Let us look first at the simple processes of addition and subtraction. Suppose we are adding together 50,000 correct to the nearest 1000 and 3,600 correct to the nearest 100

$$\text{that is, } (50,000 \pm 500) + (3,600 \pm 50)$$

This maximum result we could possibly obtain would be when we added together the two largest figures, i.e. $50,500 + 3,650 = 54,150$. The smallest results we could get is when we add together the two smallest numbers, i.e. $49,500 + 3,550 = 53,050$. Thus, our result would lie between 53,050 and 54,150 which we may state as $53,600 \pm 550$.

Let us now subtract 727 ± 15 from 926 ± 20. Here the maximum result is obtained when we subtract the smallest total from the largest: $946 - 712 = 234$. The minimum result is obtained when we subtract the largest figure that 727 could reach from the smallest figure that 926 could reach: $906 - 742 = 164$. Our result lies between 234 and 164, and again we can express this as 199 ± 35.

We will now put these two results together and look at them carefully.

$$(50,000 \pm 500) + (3,600 \pm 50) = 53,600 \pm 550$$
$$(926 \pm 20) - (727 \pm 15) = 199 \pm 35$$

It is apparent that when we are adding or subtracting we perform the normal arithmetic process ignoring the fact that the figures are approximated, and that the maximum possible error to which this total is subject is plus or minus the sum of the individual *absolute* errors. Be careful when performing this calculation. You will often find that one or more errors are given as relative errors, but the rule applies only to the sum of the absolute errors.

This result could be of importance to any manufacturer, especially during a time when his costs are likely to vary. Suppose that a garment manufacturer wishes to know what price he should quote for cotton shirts to be delivered in six months' time. He knows that his average fixed cost will be 35 pence per shirt but that his supplies of cotton may cost up to 20% more or 20% less than the current cost of 85 pence per shirt. Similarly, his wage cost will be £1.20 ± 5 pence per shirt depending on changes in productivity and wage rates. His total cost per shirt would be:

[35 + (85 ± 20%) + (120 ± 5)] pence or [35 + (85 ± 17) + (120 ± 5)] pence and this equals £2.40 ± .22.

He could not then be sure of making a profit unless he sold his shirts at a price greater than £2.62 each. Of course, if all went extremely well he might be able to break even if he sold at a price as low as £2.18, but he cannot be sure of this. You will notice that in this calculation, in the case in which a percentage error was given we immediately changed it into the absolute error.

When we come to multiplication and division we find that the assessment of error in the answer is not quite as simple as this. We find that the error that arises in one direction is different, and often very different, from that arising in the other. Suppose we are multiplying 40 ± 10% by 20 ± 5%. The result is 800 plus and minus an error. The largest figure we could get is 44 × 21 = 924, an error of +124 or +15.5%. To our surprise, however, when we calculate the smallest figure it is 36 × 19 = 684, an error of only −116 or −14.5%.

When multiplying two approximate values, it is customary to state the *largest* error that could occur, so the result above would be given as 800 ± 15.5%. The same thing is true when we are dividing. If we are dividing 1000 ± 50 by 100 ± 10 the quotient is 10. The largest error that could occur is when the result of our division is a maximum, and this occurs when we divide the largest dividend by the smallest possible divisor, i.e. 1050 ÷ 90 = 11.667. Thus the largest error could be +1.667 or +16.67%, and our result is stated as

$$(1000 \pm 50) \div (100 \pm 10) = 10 \pm 16.67\%$$

You should not try to remember the formulae for these operations devised by some earlier writers. Remember the basic principle − the maximum error is always possible and it is this, therefore, that should be quoted. Remember too that no amount of juggling with figures can increase accuracy when you are dealing with approximated figures. The accuracy of your result can never be better than, and can seldom be even as good as, the least accurate of the figures used in your calculation.

Let us conclude this chapter by looking at an exercise set by the Institute of Statisticians.

A motorist whose car has a broken mileage recorder and who measures distance travelled by map reading, wishes to assess his petrol consumption in miles per gallon, on a continental trip. His petrol gauge indicates his fuel consumption in litres with an error of ±10%. He also uses approximate conversion figures of 1 gallon = 4.5 litres and 1 mile = 1.6 kilometres instead of the more accurate 1 gallon = 4.55 litres and 1 mile = 1.61 kilometres.

Estimate his total fuel consumption in miles per gallon if he claims to be getting 31 miles per gallon after a trip of exactly 615 miles.

Notice that here only the amount of petrol consumed is subject to error in the sense in which we have been discussing it. The remaining errors are errors in converting litres into gallons and kilometres into miles. So firstly let us eliminate the errors made when converting metric into Imperial measurement.

When our motorist says that he has travelled 615 miles, he is in fact saying that he has travelled $615 \times 1.6 = 984$ kilometres. But, of course, this is not really 615 miles since he used the wrong conversion factor. The distance travelled in miles is really $984 \div 1.61 = 611.18$.

Now a motorist who says he is getting a petrol consumption of 31 miles per gallon on a trip of 615 miles, is saying that he has used $615 \div 31 = 19.84$ gallons of petrol, or, as his gauge reads in litres, 19.84×4.5 litres of petrol, or 89.28 litres. We know, however, that this reading is subject to a 10% error, so the true petrol consumption lies somewhere between $89.28 + 8.928$ litres, and $89.28 - 8.928$, between 98.208 and 80.352 litres. In gallons then his petrol consumption for the journey of 611.18 miles lay somewhere between $98.208 \div 4.55$ and $80.352 \div 4.55$ gallons, between 21.584 and 17.66 gallons.

We now have the motorists's true mileage and the true limits of his petrol consumption. Thus his fuel consumption lies between $611.18 \div 21.584$ and $611.18 \div 17.66$ miles per gallon, or between *28.32 and 34.61 miles per gallon*.

This seems to be quite a wide range and is possibly not a very satisfactory answer to our problem. What you must remember is that when we say between 28.32 and 34.61 we are quoting extreme limits. It does not seem half as bad if we put the solution as 31.465 ± 3.145 miles per gallon. Here the implication given is that the motorist will get about 31½ miles to the gallon subject to a small error of about 3 miles per gallon either way. The concentration is switched from extreme figures and quite a wide range to a central figure with only about 10% error.

Enough has been said in this chapter to warn you against reading too much into published statistics. In all but a few cases, figures are quoted approximately and hence are subject to error. You can imagine the size of the error in your conclusions if you begin to calculate using figures approximated to the nearest hundred million. Fortunately, as you will see, most calculations of error in statistics do not involve magnitudes as great as this, and the calculation of the probable error in our results can indeed be extremely useful.

Exercises to Chapter Four

4.1 Write an essay on the importance of accuracy and approximation when using statistics.

4.2 (a) Explain what is meant by the term 'error' in statistics and the circumstances in which it may arise.
 (b) 2.91; 3.473; 8.2; 4.005.
   (i) Add the above rounded figures and state the answer as accurately as possible, showing the limits of error.
   (ii) Calculate the percentage relative error in (i).
 (c) (i) The death rate of the population of a certain town was stated to be 15.5 per thousand when the population was known to be 22,000 expressed to the nearest thousand. Calculate, as accurately as possible, the number of deaths during the year.
   (ii) Calculate the percentage relative error in (i).

4.3 An inventor plans to set up a business to manufacture and market a product which incorporates his recently patented device. He considers the product will sell for £10 and that in the first year 120,000 plus or minus 20% will be sold. The cost of the materials in the product is estimated at £4, a figure which is considered to be correct to within 10%, similarly, the labour cost is estimated at £1.50 but it is thought this could increase by 20%. The first year's fixed costs are expected to be £100,000 plus or minus 5%. You are required to calculate for the first year,
 (a) the estimated profit,
 (b) the minimum expected profit, and
 (c) the maximum expected profit. A.I.A.

4.4 Explain what is meant by the statistical terms:
 (i) unbiassed error,
 (ii) biassed error,
 (iii) absolute error,
 (iv) relative error.

A builder has a plot of land on which he can build 350 houses. His costs are estimated as follows:

	£
Land	2,000,000
Materials	750,000 to the nearest £10,000
Labour	1,000,000 ±2%
Overheads	600,000 ±5%

He intends to sell the houses at £14,000 ±£250 each. Assuming he sells all the houses, estimate his profit from the contract (express the answer in the form of £$x \pm$error).

4.5 Statistical data is frequently rounded in some way before calculations begin or schedules compiled.
   (a) What advantages are obtained from rounding?
   (b) Explain what is meant by the terms:
       (i) unbiassed error,
       (ii) biassed error,
       (iii) absolute error,
       (iv) relative error,
   and illustrate with the calculation of each by using the following actual figures:

56,321	998
1,465	15,431
43,501	9,702
2,786	11,100
12,157	7,009

I.C.M.A.

4.6 (a) Discuss the various types of error in statistics.
   (b)   *Return of twelve American banks at 31st December 1964*
          Total advances $2.830 million
          Ratio of advances to deposits 25.8%
          Ratio of cash to deposits   12.4%
   Given that the above figures are correct to three significant figures, estimate (a) the amount of deposits and (b) the amount of cash. State the limits of error in each case.

4.7 Calculate the answers required in each of the following cases. Give the limits of error. The degree of approximation is given.
   (a) (i) $(283 \pm 3\%) + (146 \pm 1\%)$,
       (ii) $(490 \pm 3\%) - (230 \pm 1\%)$,
   (b) Total bank advances £984m.
       Ratio of advances to deposits 15.5%.
       Both figures are correct to 3 significant figures.
       Estimate the amount of deposits.
   (c) Acreage sown 3,982,000 to the nearest 1000 acres.
       Yield per acre 21 cwt correct to the nearest cwt.
       Estimate the total yield in tons.

4.8 (a) If $A = 5600$ to the nearest hundred and $B = 120$ to the nearest ten calculate
       (i) $A + B$
       (ii) $A - B$, giving the absolute and relative errors in both cases.
   (b) Using logarithms or a slide rule calculate the value of each of the following correct to 3 significant figures.
       (i) $$\frac{11 \times 136 - 3 \times 5}{\sqrt{[(11 \times 167 - 3^2)(11 \times 148 - 5^2)]}}$$
       (ii) $\sqrt[4]{(268 \times 343 \times 496 \times 534)}$

4.9   *Imports of Motor-cars and Motor-cycles into a Country in 1966*

	Unit	No. of Units Imported	Total Value (£m)
Motor-cars	thousands	10.3	8.6
Motor-cycles	thousands	42.6	2.8

Each entry is correct to the number of figures given in the table.
(a) Calculate the average import price per vehicle of each type and state the limits of error.
(b) What is meant by the term 'error' in Statistics?

4.10 (a) What are the advantages of using approximate amounts?
(b) The total output of 86 businesses amounted to 3,829,000 tons. Estimate the possible error in this figure if, in arriving at the total,
  (i) the output of each business had been rounded off to the nearest thousand tons,
  (ii) the output of each business had been rounded off to the nearest thousand tons above the actual figure.
(c) (i) Factory A produced 28 million tons in 1973. The number of workers was 206,000. Calculate the average output per worker and indicate the possible absolute error of your answer if it is suspected that both of the values have been rounded off.
  (ii) Factory B produces 1250 (to the nearest ten) tins of pink powder per batch and 2700 (to the nearest ten) tins of white powder per batch.
  Each tin contains between 80 and 100 ounces of powder. Calculate the total pounds' weight of powder in a consignment of two batches of pink powder and one batch of white powder sent to a customer. Show the relative error of your total.

4.11   *Value of United Kingdom exports of chemicals during 1971*

	£ thousands
Chemical elements and compounds	19,295
Dyeing, tanning and colouring materials	7,385
Medicinal and pharmaceutical products	14,036
Essential oils and perfume materials	6,021
Explosive and pyrotechnic products	1,001
Plastic materials and artificial resins	13,326
All other	12,601
Total	73,665

Source: Department of Trade and Industry.
(a) Explain the terms (i) *absolute error* and (ii) *relative error*.
(b) Approximate the above data to
  (i) the nearest million £
  (ii) the nearest hundred thousand £.
(c) Estimate the absolute and percentage relative errors of the totals obtained in (b).
(d) Using (c) state the answers in (b) as accurately as possible showing the limits of error.

4.12 A firm's estimated sales for 1967 are 83,000 units at a price of £5 each. These figures are liable to errors of 1% and 5% respectively. The estimated costs are:
  Wages £72,500 ± 2%,
  Materials £125,000 ± 4%,
  Other expenses £41,500 ± 3%.
  (a) Calculate the approximate net profit for the year giving the limits of possible error.
  (b) Calculate the relative errors in the estimate of net profit.

4.13

Personal Incomes U.K. 1966

	£000m	%
Consumer expenditure		
Food	5.3	17
Other	18.8	60
Income taxes	3.7	11
National Insurance contributions	1.8	6
Savings	1.9	6
Total Personal Income	31.5	100

Source: National Income and Expenditure 1967.

The data given in the table are rounded to the nearest £100m, i.e. to 1%. Calculate, giving the error in each case:
(a) Consumer expenditure.
(b) Income after tax.
(c) The aggregate amount saved given total personal incomes and % savings.
(d) % savings given aggregated savings and total personal incomes.
C.I.P.F.A.

4.14 (a) What do you understand by the terms
  (i) biassed rounding off,
  (ii) unbiassed rounding off,
  (iii) compensatory error,
  (iv) cumulative error?
  (b) The following figures, taken from the annual abstract of statistics, show the value of U.K. exports in 1970 (in £ millions) to E.F.T.A. and E.E.C. countries

E.F.T.A. countries	£ millions
Finland	128.9
Sweden	364.1
Norway	173.8
Denmark	221.2
Switzerland	209.3
Portugal	88.6
Austria	90.7
E.E.C. countries	
Western Germany	502.9
Netherlands	377.8
Belgium and Luxembourg	294.3
France	339.2
Italy	239.7

(i) Obtain totals for both E.E.C. and E.F.T.A. countries.
(ii) Round off each figure in an unbiassed sense to the nearest £10 millions and again obtain the totals.
(iii) Calculate the relative error in total resulting from your rounding for E.F.T.A. countries.
(iv) If you wish to take the difference between exports to E.F.T.A. and E.E.C. countries, how would you round your totals if you wished your rounding errors at that stage to be compensatory?

4.15 A company is marketing a new product to sell at £5. The sales for the first year are expected to be $50,000 \pm 10\%$. The costs are calculated to be:
   Fixed overheads £40,000 $\pm 5\%$.
   Variable costs £3 per unit sold.
Calculate:
(a) The net profit which can be obtained, giving the limits.
(b) The rate of net profit to sales and the limits.
(c) What would be the net profit is there were no variation?

4.16 (a) Define the terms
   (i) absolute error,
   (ii) relative error,
   (iii) compensating error.
(b) A firm works a nominal 40-hour week but with overtime and short time its actual working week varies by as much as ½ hour from the nominal figure. The firm produces (50 ± 1) articles per hour. If the production costs and selling prices are £2.00 and £3.00 respectively per unit rounded off to the nearest 10 pence, estimate
   (i) the weekly profit,
   (ii) the percentage profit, based on cost price, per unit sold (to the nearest 0.5%).

4.17 (a) Use logarithms or a slide rule to calculate the value of each of the following, correct to 3 significant figures.
   (i) $\sqrt{468} + (24.3)^2$
   (ii) $1 - \dfrac{6 \times 728}{16(16^2 - 1)}$
   (iii) $37.5 + \dfrac{117 - 89}{83} \times 4$
(b) If $x = 90.8956$, write the value of x correct to
   (i) 4 significant figures,
   (ii) 3 decimal places,
   (iii) the nearest integer,
   (iv) the nearest tenth.

Chapter Five

The Averaging of Data

So far, we have concentrated our attention on organising and presenting data. What we shall now do is attempt to summarise the data at our disposal into a single statistic. This will certainly ease the task of making comparisons, though we must always remember that summarising data is bound to mean that something will be lost.

If you think carefully about it, you will realise that statisticians spend a great deal of their time making comparisons, and the conclusions they reach are often of fundamental importance to every one of us. Comparing income today with income ten years ago is an indicator of how our living standards have changed. Comparing incomes between regions helps the government in its regional policies. Comparing how our prices are changing over time in relation to price changes abroad indicates how competitive our industries are, and comparing our balance of payments over time indicates our ability to pay our way in the world.

Many of the comparisons made present no problems, as we are comparing a single figure value. For example, we can state that I.C.I. earned a certain profit in 1979, and compare this with the profit earned in 1980. Any problems involved in this comparison will be concerned with the calculation of the profit and the rate of inflation, and no-one will dispute that it is legitimate to compare profit in 1979 with profit in 1980 as long as inflation is taken into account. In some cases, however, the things we are trying to compare vary in themselves. Suppose, for example, we attempt to compare incomes on Merseyside with incomes in London: not only is there a variation between the regions, but also a variation within each region. In a previous chapter you saw that we could put raw data into a frequency distribution, and present it in the form of a histogram etc. No doubt we could draw histograms, showing the distribution of incomes in London and Merseyside, and this would enable us to state that incomes earned in London exceeded incomes earned on Merseyside. But is this good enough?

You can probably identify two problems in the statement above. Firstly, do we mean that *all* incomes in London exceed *all* incomes on Merseyside? Obviously not! Many people on Merseyside will earn much more than, say, a porter on the Underground. Secondly, given that incomes earned in London exceed incomes earned on Merseyside, we would wish to know by how much. In other words, we wish to *quantify* the differences in income, and we certainly will not be able to do this by just looking at a frequency distribution or a histogram. What we need is some figure that is *representative* of income in London. We can then obtain a representative figure for Merseyside incomes, compare them, and draw some conclusions as to the size of the difference in incomes.

How are we going to obtain this representative figure? Probably you have guessed already, especially in the current economic climate with our preoccupation with income levels. The TV newscaster does not state that incomes have risen by 10% over the last twelve months: he states that *average* incomes have risen. We are given information on *average* hourly rates of pay, *average* rainfall levels, batting and bowling *averages,* the *average* amount we spend on drink — and so on. The advertising men are very fond of telling us what the *average man* buys, where he goes for his holidays, and what he does with his leisure.

Three conclusions can be drawn from this last paragraph. Firstly, an average is obviously meant to tell us something about the matter under consideration, and unless it is representative of the data, it obviously cannot do this efficiently. Secondly, the word 'average' is one that we meet daily in our conversation, and is a word that is used in a very loose manner. How many times have you heard people use such remarks as 'I *think* that on average I use *about* five gallons of petrol per week'? Here the idea of an average and an *estimate* are shading into each other. We must avoid this at all costs. An average is capable of being calculated from data, and so it is precise. Thirdly, averages are used to describe a wide variety of data, and we must be really sure that we know what an average is. In fact there are many types of average, and we must be sure to select the right one for the right job. If we don't do this, then there is a great danger that the average we quote will not be representative of the data.

The Arithmetic Mean

Most people will tell you to calculate an average something like this: total all the numbers in the group and divide by how many numbers there are in the group. So the average of 5, 7, 9, and 10 is

$$\frac{5+7+9+10}{4} = 7.75$$

In fact, it is easier to demonstrate how to calculate an average than it is to explain how to calculate it. Now mathematicians have developed two useful symbols to overcome this problem. Suppose we put this group of figures we wish to average in a column, and give this column a heading x. The group we considered above would look like this:

$$\begin{array}{c} x \\ 5 \\ 7 \\ 9 \\ 10 \end{array}$$

If we wish to total these figures, the mathematician would state $\Sigma\ x$ — meaning take the sum of the column headed x. (Σ is a Greek capital letter pronounced *sigma*.) Also, we state that there are n figures in the column (in this case $n = 4$). So the expression

$$\frac{\Sigma x}{n}$$

tells us precisely how to calculate the average. We stated earlier that there are many forms of average, and the average we have just calculated is called the *arithmetic mean*. Statisticians use the symbol $\bar{x}$ (pronounce it '*x*-bar') to stand for the arithmetic mean, so we can write

$$\bar{x} = \frac{\Sigma x}{n}$$

The arithmetic mean is certainly the most widely used average, both by statisticians and laymen. It will be useful, then, to examine in what sense it is representative. Returning to our example, we have

$$x = 5, 7, 9, 10 \quad \bar{x} = 7.75$$

Notice that the arithmetic mean represents not one single item in the group, so it cannot be representative in the sense that it is typical. It follows, then, that 'representative' means something other than typical. If we subtract the mean from each of the items in the group, we have

$$-2.75, -0.75, +1.25, +2.25$$

We call each of these differences a *deviation* from the arithmetic mean. Notice that the sum of these deviations is zero. Using the sigma notation we have

$$\Sigma(x - \bar{x}) = 0$$

So the arithmetic mean tells us the point about which the values in the group cluster ('mean' in fact means centre, and statisticians call averages measures of central tendency.) This is what we imply when we state that the mean is representative. So we now have a definition of the arithmetic mean − a measure chosen such that the sum of the deviations from it is zero.

For the moment, we shall postpone judging whether this meaning of representative is valid, and concentrate on this important definition of the mean. In fact, the definition enables us in many cases to simplify our calculations of the arithmetic mean. Suppose we guess a value for the arithmetic mean (call this guess x_0). Now if our guess is correct, then the sum of the deviations from x_0 would be zero. If it isn't, then our guess was wrong, and we can adjust our guess to give the true value of the mean. Suppose, for example, we wish to find the arithmetic mean of the group

$$100.1, 100.2, 100.4, 100.8$$

If we guess the mean to be 100, then the deviations are

$$+0.1, +0.2, +0.4, +0.8, \text{sum} = 1.5$$

Clearly, our guess was too low, and we must adjust our guess by

$$\frac{+1.5}{4} = +0.375$$

so the true value of the mean is

$$100 + 0.375 = 100.375$$

(You should check this value by calculating the mean directly.) We can again use the sigma sign to show precisely how to use this method to calculate the arithmetic mean.

$$\bar{x} = x_0 + \frac{\Sigma(x - x_0)}{n}$$

In other words, the arithmetic mean is the assumed mean plus a correction factor.

Now let us examine another factor of the arithmetic mean which will simplify calculations. We can multiply or divide the group of numbers we wish to average, and find the average of this adjusted group. We can then adjust the average we have calculated to the true value. Suppose, for example, we want to find the arithmetic mean of the group

0.0002, 0.0005, 0.0012, 0.0015

Multiplying this group by 10,000 we have

2, 5, 12, 15

which has a mean of 8.5. To obtain a true value for the mean, we now divide by 10,000 giving

$$\bar{x} = 0.00085$$

Of course, in many cases you would not bother to use either of the simplifications mentioned — especially if you have access to a calculating machine. Later, though, we shall meet cases where they speed up our calculations considerably, and also lessen the risk of arithmetic error.

The Arithmetic Mean of a Frequency Distribution

Earlier we recommended that you put raw data into frequency distributions wherever possible, so we must now examine how to find their arithmetic mean. If we consult the General Household Survey, we would learn that 1000 couples married eight years ago would be expected to have the following number of children now.

Number of children	0	1	2	3	4
Number of families	364	362	226	44	4

We require to know the mean number of children per family. First, we will write the data into two columns, one headed x (the number of children — this is our variable) and the other headed f (frequency). If we multiply the columns together (fx) this will give the total number of children for each family size. So, for example, we see that there are a total of 132 children from 3 children families.

x	f	fx
0	364	0
1	362	362
2	226	452
3	44	132
4	4	16
	1000	962

Adding up the *fx* column, there are a total of 962 children in the 1000 families, which gives an average of $\frac{962}{1000} = 0.962$ children per family. Now we know that Σf means total the column headed f (this gives the total number of families) and Σfx means total the column headed fx (which gives the total number of children). So we now know how to find the arithmetic mean of a discrete frequency distribution.

$$\bar{x} = \frac{\Sigma fx}{\Sigma f}$$

Now this is all very well, but much data is in the form of grouped, continuous frequency distributions. Suppose we wished to find the average age of the male labour force. We could obtain the data we require from the Annual Abstract of Statistics.

Age	Number in employment (millions)	Age	Number in employment (millions)
15 – 19	1.1	45 – 49	1.5
20 – 24	1.7	50 – 54	1.3
25 – 29	1.5	55 – 59	1.4
30 – 34	1.3	60 – 64	1.1
35 – 39	1.3	65 – 69	0.4
40 – 44	1.4		14.0

Now we have a problem here: look at the age group 20 to 24 years. Within this group there are 1.7 million men, but we have no idea what their *actual* ages are. So we are going to have to make some assumption about their ages. Probably the most sensible assumption to make is that the 1.7 million men in this group have an average age of 22.5 years — the mid-point of the group.[1] If we do this for all groups, then we can use the formula we obtained earlier to calculate the arithmetic mean.

Mid-point x	Frequency f	fx
17.5	1.1	19.25
22.5	1.7	38.25
27.5	1.5	41.25
32.5	1.3	42.25
37.5	1.3	48.75
42.5	1.4	59.50
47.5	1.5	71.25
52.5	1.3	68.25
57.5	1.4	80.50
62.5	1.1	68.75
67.5	0.4	27.00
	14.0	565.00

1. Although the data appears to be discrete, remember that a person aged 24·999 would report his age as 24 years, so it is better to consider the data as continuous with class limits 20-25, 25-30 etc.

$$\bar{x} = \frac{\Sigma fx}{\Sigma f} = \frac{565}{14} = 40.36$$

So the average age of working males is 40.36 years.

Now let us see if we can simplify the arithmetic involved in these calculations. Well, the first thing we can do is to multiply all the frequencies by 10: this will remove the decimal fractions and not make any difference to our answer. We can take an assumed mean (in this case we will take 42.5) and calculate the deviations of the items x from 42.5. Notice that all the deviations are divisible by 5, and if we do this we obtain a column that we head d.

x	$(x-42.5)$	d	f	fd
17.5	−25	−5	11	−55
22.5	−20	−4	17	−68
27.5	−15	−3	15	−45
32.5	−10	−2	13	−26
37.5	−5	−1	13	−13
42.5	0	0	14	0
47.5	5	1	15	15
52.5	10	2	13	26
57.5	15	3	14	42
62.5	20	4	11	44
67.5	25	5	4	20
			140	−60

So the average deviation from the mean is $-\frac{60}{140} = -.4286$. But as we divided the deviation by 5, we must multiply the average deviation by 5 to give -2.143. In other words, our value for the assumed mean is $\frac{15}{7}$ greater than the true value of the mean. So the true value of the mean is

$$42.5 - 2.143 = 40.36 \text{ years}$$

which agrees with the value we obtained earlier.

Now if we say that d is the deviation from the assumed mean divided by a constant c then we can write a formula for our simplified method like this:

$$\bar{x} = x_0 + \frac{c\Sigma fd}{\Sigma f}$$

Notice that the constant c is normally equal to the class width, though you can choose any value for c which is convenient.

It will be interesting to examine the accuracy of the two means we have calculated. Considering first the mean number of children per family, the distribution gave us the exact number of children in each of the 1000 families. So our calculation of the mean number of children gave us the exact value for the mean: the same result would have been obtained if we had used the frequency distribution or the raw data. However, this is not so with the mean age of the male working population as we did not know the exact age of any of the individuals. In fact, for the members in each class we found it necessary to make an assumption about their ages, so we cannot guarantee the accuracy of the mean. You should realise that although it is much more convenient to consider grouped frequency distributions than raw data, the price we pay for this convenience is the loss of accuracy. In most cases, however, the loss in accuracy is not too serious.

So far we have considered frequency distributions with constant class widths. Now let us look at a case where the class width is uneven.

Distribution of Earnings of Weekly Paid Adults aged 21 and over

	Frequency (millions)	Mid-point (x)
£12, but under £16	0.2	14[1]
£16 –	0.4	17
£18 –	0.6	19
£20 –	0.7	21
£22 –	0.9	23
£24 –	1.0	25
£26 –	0.9	27
£28 –	0.9	29
£30 –	1.9	32.5
£35 –	1.4	37.5
£40 –	0.8	42.5
£45 –	0.5	47.5
£50 –	0.4	55
£60 –	0.3	65
£70, but under £100	0.1	85

Source: Annual Abstract of Statistics.

It will certainly be worthwhile multiplying the frequencies by 10, and probably taking an assumed mean (in this case we have chosen £30). However, there is no suitable constant for simplifying the deviations. So we have:

x	$(x-30)$	f	$f(x-30)$
14	−16	2	−32
17	−13	4	−52
19	−11	6	−66
21	−9	7	−63
23	−7	9	−63
25	−5	10	−50
27	−3	9	−27
29	−1	9	−9
32.5	2.5	19	47.5
37.5	7.5	14	105
42.5	12.5	8	100
47.5	17.5	5	87.5
55	25	4	100
65	35	3	105
85	55	1	55
		110	238

$$\bar{x} = 30 + \tfrac{238}{110} = £32.16 \text{ per week}$$

1. Strictly speaking, the data is discrete and the class limits for the first class are £12 – £15.995, giving a mid point of $\dfrac{12 + 15 \cdot 995}{2}$ = £13.9975, but this is splitting hairs.

So sometimes it is not practical to use the simplified method, and you should always weigh up carefully whether it is worth the labour involved. Better still, buy an electronic calculator (they are very reasonably priced these days) and it will always be worthwhile using the first method.

Limitations on the Use of the Arithmetic Mean

Earlier, we postponed judgement on just how representative was the arithmetic mean. We simply stated that 'representative' certainly doesn't mean typical. The time has now come to put the arithmetic mean through a series of tests, and see how it performs.

If you examine sources of published statistics, you will be surprised how often frequency distributions have 'open-ended' classes. We reproduce below a frequency distribution showing the estimated wealth of individuals in Great Britain.

Wealth	% of Total
Not over £5000	78.38
£5000 – £15,000	16.17
£15,000 – £25,000	2.52
£25,000 – £50,000	1.88
£50,000 – £100,000	0.70
£100,000 – £200,000	0.23
£200,000 –	0.11

Source: Annual Abstract of Statistics.

When we calculated the mean from a frequency distribution, we had to make assumptions about the values of the items in each class. We took the mid-point as the representative value of each class. But what assumptions can we make about the average wealth of individuals in the final class? It would be a brave man indeed who would estimate the upper limit for this group. Likewise, what would be a reasonable lower limit to put on the first class? Unless we have adequate information about the first and last groups, we cannot calculate the mean for this type of distribution. To obtain a mean, we must refer to the raw data (and this just is not available for government published statistics). Clearly, we need a measure that does not depend upon the adequate knowledge of extreme values of the distribution. This problem of open-ended statistics is one you will meet continually in analysis of published statistics — in fact we had quite a job finding distributions that were not open-ended! Before we leave this point, however, it should be stated that if you *must* estimate the missing limit for open-ended distributions, make sure that the frequencies in the open-ended classes are small in relation to other frequencies. Any inaccuracies due to estimates shouldn't, then, be too serious.

The second point about the arithmetic mean is that it often produces results that are not suitable from a communication viewpoint. Earlier, we found that the average number of children per family was 0.962. Most people would consider this to be a ridiculous statement. In situations like this, people expect an integer (i.e. a whole number) to be representative of the number of children per family. If we are going to ensure that all averages are integers, then we must change our idea of an average.

The third point is perhaps the most important of all: the mean is highly sensitive to extreme values. Consider the case of Fred, who at an interview for a job is told that the average income of salesmen in the company is £8000 per year. He accepts the job as he considers the firm to be very progressive with excellent prospects for himself. Although his starting salary is only £2000 a year, his salary will obviously climb very quickly. You could imagine how cheated Fred felt when he found that the sales force consisted of just five men: the sales director on £30,000 a year and four salesmen on £2500 a year! The extreme value (in this case the sales director's salary) has caused the arithmetic mean to be most unrepresentative. To examine this point further, let us return again to the distribution of weekly incomes of males over the age of 21. The cumulative frequency distribution would look like this:

	Cumulative Frequency (millions)
Under £16	0.2
Under £18	0.6
Under £20	1.2
Under £22	1.9
Under £24	2.8
Under £26	3.8
Under £28	4.7
Under £30	5.6
Under £35	7.5
Under £40	8.9
Under £45	9.7
Under £50	10.2
Under £60	10.6
Under £70	10.9
Under £100	11.0

Consulting the ogive for this distribution (Diagram 5.01), we see that 6.45 million people earn less than the average wage of £32.16, which represents nearly 60% of all workers in this category. On this basis, many would claim that the mean is not representative of this data. Had we taken *all* incomes rather than weekly incomes only, then the average would have been much higher (and hence even more unrepresentative). This is the reason that so many people scoff at the average income figures that are quoted from time to time.

Before we search for an alternative to the arithmetic mean, we must say something in its defence. The main redeeming feature of the arithmetic mean is that its calculation involves the use of *all* the data. We will see later that this is not a characteristic of alternatives to the arithmetic mean. In other words, the weaknesses of the arithmetic mean are also its strengths. Because all the data is used, and because the arithmetic mean is capable of precise calculation, most statisticians still prefer it to other measures. In fact the arithmetic mean is capable of, and is the basis for, advanced analysis. Any alternative to the arithmetic mean cannot be used for advanced analysis, so their uses are descriptive rather than analytical.

The Median

Suppose we arranged a group of numbers in ascending order, then the median would be the value of the item in the middle. So the median of the group 3, 4, 7, 9, 11 is 7. Notice that the median is quite unaffected by extremities at either end of the group. Take the case of the sales force considered earlier: the incomes are

$$£2500, £2500, £2500, £2500, £30,000$$

and the median income is £2500. The actual size of the director's income makes no difference whatsoever to the median: if the director's salary was doubled or halved then the median would still remain £2500. So the median is quite insensitive to extreme values in the group or distribution, and certainly overcomes the major objections to the arithmetic mean. If you arrange a group of n items in ascending order, then the value of the $\frac{(n+1)}{2}$ th item is the value of the median. If we had an even number of items, then there will be *two* middle items. Conventionally, we take the median to be the arithmetic mean of the two middle items. So the median of 4, 9, 13, 17, 30, 32 is $\frac{13+17}{2}$ = 15. Now if n is large, the difference between $\frac{(n+1)}{2}$ and $\frac{n}{2}$ is negligible — so take $\frac{n}{2}$ in such cases. Finally, then, we can see that half the items in a distribution will have a greater value than the median, and half the items a smaller value.

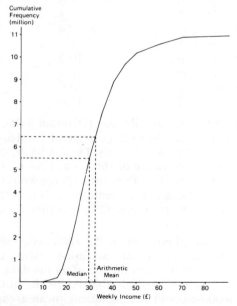

Cumulative frequency distribution of weekly incomes of males over 21 years old
Diagram 5.01.

Look again at the cumulative frequency distribution of weekly incomes. There are 11 million workers, so if we wish to find the median income, this will involve finding the income of the 5.5 millionth worker. Reading off the income of this worker from the ogive, we find that the median income is £29.8, compared with a mean income of £32.16. You would probably find that most people would accept the median as being more representative than the arithmetic mean.

It is often quite difficult to read off the value of the median from the ogive, and to overcome this problem it is useful to magnify that part of the ogive containing the median. From the cumulative frequency distribution, we can see that 5.6 million people earn less than £30 per week, and 4.7 million earn less than £28 per week. So the median lies somewhere between £28 and £30 per week. We can plot these points on a graph and join them with a straight line. Reading off from Diagram 5.02, we see that the median income is £29.78.

If you look carefully at Diagram 5.02, you will realise that it is possible to calculate rather than read off the value of the median. Do you remember the similar triangle theorems? If you do, then you will realise that

$$\frac{AB}{AC} = \frac{ED}{AE}$$

$$\text{so} \quad ED = \frac{AB \times AE}{AC}$$

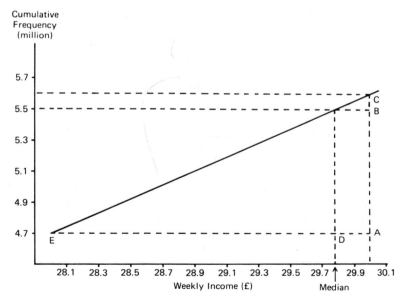

Cumulative frequency distribution of weekly incomes of males over 21 years old 1971

Diagram 5.02

Now we can see that ED is the amount by which the median exceeds 28, so the median is

$$28 + \frac{AB \times AE}{AC}$$
$$= 28 + \frac{0.8 \times 2}{0.9}$$
$$= 29.78$$

It might be useful to see how we could obtain these figures directly from the cumulative frequency distribution. The relevant parts of this distribution are:

	Cumulative Frequency
Under £28	4.7
Under £30	5.6

so we see that the median is in the group £28—£30, and the median is:

$$28 + \frac{0.8 \times 2}{0.9}$$

Clearly, 28 represents the lower class boundary (LCB) of the median group, and AE represents the class interval of this group. The quantity AC is the frequency of the median group $(5.6-4.7)$ and the quantity AB is the median item minus cumulative frequency up to the median group $(5.5-4.7)$. So we can calculate the median like this:

$$\text{LCB} + \frac{\text{class interval} \times ([\frac{n+1}{2}] - \text{cum. frequency to median group})}{\text{frequency of median group}}$$

Now let us see if we can calculate the median directly, i.e. without reference to graphs.

Size of Companies Acquired.

Cost	Frequency	Cumulative Frequency
Not more than £100 thousand	279	279
£100 thou. but under £200 thou.	157	436
£200 thou. but under £500 thou.	166	602
£500 thou. but under £1 mill.	117	719
£1 mill. but under £2 mill.	58	777
£2 mill. but under £5 mill.	66	843
£5 mill. but under £10 mill.	30	873
£10 mill. but under £20 mill.	12	885
£20 mill. but under £50 mill.	5	890
Over £50 million	1	891

Source: *Board of Trade Journal*

We wish to find the median cost of companies acquired, i.e. the cost of the $\frac{891+1}{2} = 446$th company. The median company is in the group £200 thou. — £500 thou., which contains 166 companies. Hence the median is

$$£200,000 + \frac{300,000 \times (446-436)}{166} = £218,072$$

Notice how difficult it would be to calculate the arithmetic mean of this distribution. It would be extremely difficult to put a lower limit on the first group, or an upper limit on the last group. The median, then, has decided advantages over the arithmetic mean; it can cope with open-ended distributions and is unaffected by extremities at either end of the distribution. Its disadvantage is that it ignores the bulk of the data presented to us, and this disadvantage really is critical! We would like to emphasise again that you should always attempt to use the mean rather than the median, especially if there is not much difference between them. In fact, the difference between them depends on the skewness of the distribution (this is illustrated in Diagram 5.03). The median splits the area under the frequency curve into two halves. If the distribution is symmetrical, then the mean and median will coincide. With a negatively skewed distribution the median exceeds the mean, and with a positively skewed distribution the mean exceeds the median. The more pronounced is the skew, the greater will be the difference between the mean and median. Let us now attempt to summarise this into a simple rule: if you require a representative measure and the distribution is markedly skew, use the median — otherwise use the mean.

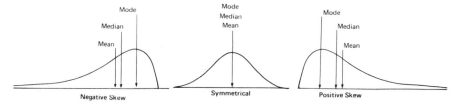

Diagram 5.03

One final problem concerned with the median is that we cannot pool two medians to find an overall median. If we know that a factory employs 100 women at a median wage of £32 per week and 200 men at a median wage of £43 per week, we cannot calculate the median wage for all workers without consulting the raw data. But this is not the case with the arithmetic mean. Suppose the data referred to mean rather than median wages, then the total wages earned by women would be £32 × 100 = £3200, and the total wages earned by men would be £43 × 200 = £8600. So the labour force of 300 earns £11,800 per week, which gives an average wage of $\frac{11,800}{300}$ = £39.33 per week. This pooling of arithmetic means is extremely useful.

The Mode

The mode is the value or attribute that occurs most often, so is an extremely simple concept. Look again at the incomes of the sales force described earlier and you will realise that the modal income is £2500. Straight away, though, we can see a snag with the mode: if all the numbers in the group are different, then we cannot have a modal value.

With a frequency distribution, the mode is the value with the greatest frequency. Let us examine again the distribution of the number of children of 1000 young married couples.

Number of children	0	1	2	3	4
Number of families	364	362	226	44	4

The mean number of children per family we calculated to be 0.962, and the median number of children is one (i.e. the number of children in the 500th family). But the modal number of children is zero, because more families are childless than have any other number of children.

Finding the mode of a grouped frequency distribution will not be quite so easy. Consider again the distribution of weekly paid adults: the class width is not constant so the modal class cannot be obtained by inspection. Probably the best way to deal with this is to draw the histogram of the distribution. If you examine the distribution carefully, you will see that incomes in excess of £35 are falling off rapidly, so we will concentrate our attention on incomes less than this figure. Consulting diagram 5.04 we can see that the modal class is £24 – £26 per week, and as the adjacent classes have the same frequency, we might be justified in saying the modal income was £25 per week.

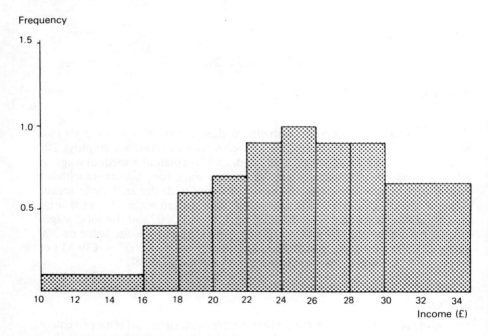

Weekly incomes of males over 21 years old

Diagram 5.04

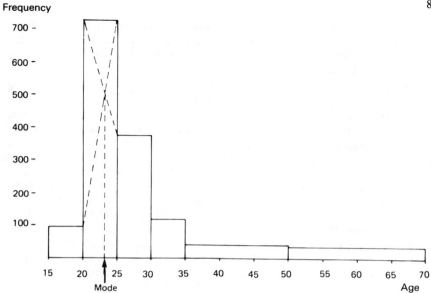

Age distribution of drug addicts known to the Home Office, 1972 (Source: Home Office)
Diagram 5.05

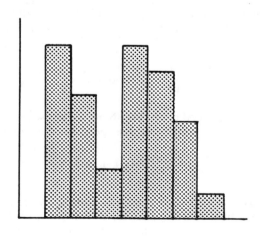

A bimodal distribution
Diagram 5.06

Suppose the adjacent classes do not have the same frequencies; then we would estimate the mode by splitting the modal class in the ratio of the frequencies of the adjacent classes. This is often done geometrically as illustrated in diagram 5.05.

The mode has the same weaknesses as the median: it ignores the bulk of the data and is not capable of being pooled. We have also seen that in certain groups of numbers, a mode might not be present. Also, in some distributions it is possible to have more than one mode — a distribution like the one in diagram 5.06 is called bimodal. To quote two modes is just clouding the issue. The strength of the mode is that it is extremely easy both as a concept and as a measure of calculation. It is particularly useful to describe attributes — when we state that the average family cleans their teeth with 'Gritto', we mean more families use Gritto than any other toothpaste. You can readily understand why the mode is so popular with market researchers!

Before we conclude measures of central tendency, there is one final point we should consider. Why did we call the mean the *arithmetic* mean? This surely implies that there are other means we could consider. This is indeed the case, and if we calculate a mean, we must be absolutely certain that we are using the right one! Let us, therefore, consider some alternative means (called minor means) and discuss their appropriate usage.

The Harmonic Mean

Suppose a motorist told you that he travelled from Liverpool to Manchester at an average speed of 40 kph, and returned at an average speed of 50 kph. If then he claimed that his overall average speed was 45 kph, you would probably agree with him, wouldn't you? However, you would be quite wrong. Let us examine the problem in more detail. The distance from Liverpool to Manchester is 55 km, so the time taken at 40 kph is $\frac{55}{40} = \frac{11}{8}$ hours, and the time taken at 50 kph is $\frac{55}{50} = \frac{11}{10}$ hours. The time taken for both journeys is

$$\frac{11}{8} + \frac{11}{10} = \frac{99}{40} \text{ hours}$$

and the total distance travelled is 110 km. Now we know that average speed is distance divided by time, so the average speed for the round journey is

$$110 \times \frac{40}{99} = 44\frac{4}{9} \text{ kph}$$

So you see that averaging the two speeds simply by adding them and dividing by two gives us the wrong answer. Why is the arithmetic mean not appropriate here? Well the reason is that the different speeds are maintained for the same *distance* and not the same *time*. If they were maintained for the same time then the arithmetic mean would have been appropriate. What we should have used here is the *harmonic mean,* which is used for averaging rates and prices. We calculate the harmonic mean as follows

$$H.M. = \frac{n}{\sum \frac{1}{x}}$$

or using our example

Average speed
$$= \frac{2}{\frac{1}{40} + \frac{1}{50}}$$
$$= \frac{2 \times 200}{9}$$
$$= 44\tfrac{4}{9} \text{ kph}$$

Now let us examine an example using prices. Suppose we made a mixture using four ingredients costing £1, £2, £3 and £4 per kilogram respectively, and spent an equal amount on each ingredient, then the price per kilogram of the mix is

$$\frac{4}{1 + \tfrac{1}{2} + \tfrac{1}{3} + \tfrac{1}{4}} = £1.92 \text{ per kilo}$$

However, had we mixed equal quantities *worth* of ingredients, then the price would have been £2.50 per kilo. You should check these conclusions by direct calculation.

The Geometric Mean

Let us suppose we invest £100 at 10% per annum compound interest. At the end of the first year we would have £100 + £10 = £110 invested. At the end of the second year we would have £110 + 10% of £110 = £121 invested. Now we can continue in this fashion and calculate the sum available at the end of each year for 10 years. The graph of the sum available against time is shown in Diagram 5.07.

The sum available at the end of the 10th year is £259.37. Now suppose we knew the sum available after 10 years, could we use the arithmetic mean to estimate the sum available after 5 years? The arithmetic mean would be

$$\frac{100 + 259.37}{2} = £179.69$$

which, as you can see from the graph, is a rotten estimate. Once again, we have used the wrong mean. The arithmetic mean would have given a good estimate had the sum invested earned a constant *amount*. In other words, it would have worked if the sum invested had earned simple interest – the graph would have been a straight line. In this case, however, the sum invested increases by a constant *proportion*, and a much better estimate is achieved if we use the *geometric mean*.

$$G.M. = \sqrt[n]{x_1 \times x_2 \times \ldots \times x_n}$$

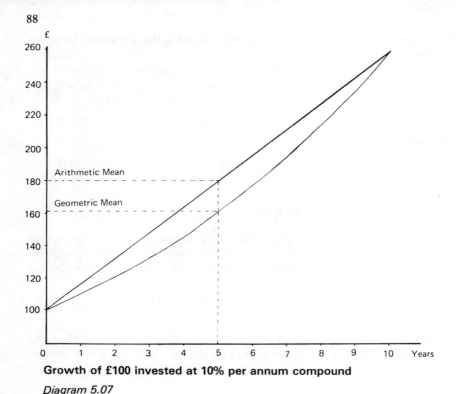

Growth of £100 invested at 10% per annum compound

Diagram 5.07

which simply states that we multiply the n numbers together, and then take the nth root. So our estimate of the sum available after 5 years is

$$\sqrt{100 \times 259.37}$$
$$= £161.05$$

which agrees exactly with the value obtained from the graph.

This method of estimating a 'mid-point' value by using an average is called *interpolation*. To attempt to clarify which mean should be used, consider this problem: suppose you started work earning £1500 per year, and you were told what your salary would be in 10 years' time. You want to estimate your salary after 5 years. If your salary rises by constant annual increments, then the arithmetic mean would be appropriate, but if it rises by a constant percentage of the previous year's salary, you would have to use the geometric mean. Again, a census is taken every 10 years, and it would be useful to use these figures to estimate the population on the 5-year intervals. As populations do not grow by constant amounts, the geometric mean would give a better estimate. But even using a geometric mean might not give a good estimate: can you think why this might be so?

Now a word or two in conclusion. We have seen that there are many forms of 'average', and it is vitally important that you choose the right one for the job. We must state, though, that the arithmetic mean is by far the most important and is most commonly used in preference to the other alternatives. As you work through this book, you will meet the arithmetic mean over and over again (from now on we shall refer to it simply as the mean) — but always bear in mind that it does have limitations!

Exercises to Chapter 5

5.1 Explain why it is frequently necessary to summarise masses of data by using representative or typical values. Describe two such measures and their methods of calculation.   A.C.A.

5.2 If we examine the batting average table for Coalshire we find that Fred Sloggin has an average of 51 runs, and D.E.M. Bones has an average of 39 runs. Would you conclude that Sloggin was the more effective batsman?

Sloggin's average of 51 runs has been obtained in 39 innings. The next innings will be his last of the season. How many runs must he score in the last innings if his final average is to be at least 55 runs?

5.3 Given that $\Sigma (x - \bar{x}) = 0$, show that $\bar{x} = \dfrac{\Sigma x}{n}$

5.4 No. of children per 1000 families in which couples were married

No. of children	0	1	2	3	4	5	6+
Frequency (%)	5.3	5.2	55.4	11.4	18.9	1.5	2.3

Source: General Household Survey.

Calculate the median.

Calculate the mean number of children per family for the above data, given that the average size of a family with 6 or more children is 6.5.

5.5 A company employing 60 people found that the number of sick days taken by its employees last year were as follows:

10	5	12	0	2	35	11	12	4	9
12	17	3	7	8	8	8	10	11	29
44	4	9	3	6	6	7	13	18	4
15	25	5	2	7	20	9	16	10	9
5	2	31	6	0	7	10	9	22	1
3	1	23	9	12	18	6	9	31	0

Group the above figures into intervals of five days. Calculate the mean. Draw the cumulative frequency curve of the distribution and comment on the distribution..

5.6 *Orders Received*

Value of Order (£00)	Number of Orders Received
0 and under 5	20
5 and under 10	51
10 and under 15	139
15 and under 20	116
20 and under 25	31
25 and under 30	14
30 and over	5
	376

(a) From the above table of orders received calculate the mean value of orders received.

(b) Comment briefly on your results for (a)

(c) Suppose the value of the median to be less than the mean. What would this indicate?

5.7 The rateable values of 120 houses were found and the results are shown below.

Rateable Value (£)	Number of Houses
70 – 79.99	3
80 – 89.99	15
90 – 99.99	30
100 – 109.99	36
110 – 119.99	18
120 – 129.99	12
130 – 139.99	6

Calculate the mean of the distribution.

5.8 The table below gives the age distribution of the management of a large company.

Age	Frequency
Under 20	2
20 – 29	12
30 – 39	31
40 – 49	39
50 – 59	26
Over 60	10

Calculate the mean of the distribution. List the assumptions you made in carrying out the calculation and explain why you think you were justified in making them.

5.9 *Age Distribution of the Members of a Golf Club*

Age (in completed years)	No. of Members
10 – 19	185
20 – 29	263
30 – 39	325
40 – 49	442
50 – 59	368
60 – 69	134
70 and over	83

(a) Calculate the arithmetic mean age for the data. (Use mid-value of Group 40—49 as a working origin.)

(b) Define and distinguish between discrete and continuous variables.

5.10 The frequency table gives the age distribution of the estimated male population of Northern Ireland on 30th June 1971.
Calculate the arithmetic mean for this distribution.

Age	Number (000's)
0 – 9	164
10 – 19	141
20 – 29	108
30 – 39	83
40 – 49	83
50 – 59	77
60 – 69	59
70 and over	41

5.11 Distinguish between discrete and continuous data. The data below shows the number of local telephone calls made by 75 subscribers during a certain interval of time.

No. of Calls Made	No. of Subscribers
1—10	9
11—15	12
16—20	24
21—25	16
26—40	14
	75

(a) Calculate the average number of calls made per subscriber.

(b) Calculate the maximum error in your average due to grouping in the data given.

5.12 *Mileages Recorded by 60 Commercial Travellers in the Course of One Week*

515	611	530	557	586	528
533	516	519	560	572	509
520	543	556	532	512	605
559	549	539	609	589	537
524	521	513	541	581	618
544	545	535	568	583	521
555	552	579	581	558	539
562	578	563	598	594	560
595	507	562	532	590	578
526	533	574	531	584	543

(a) From this date tabulate directly a grouped frequency distribution using equal class intervals and starting with 500-519.

(b) What is the direction of skew of this distribution?

(c) Calculate the arithmetic mean of the grouped frequency distribution.

(d) Explain why the arithmetic mean of the ungrouped data would be different from the mean obtained in (d).

5.13 Values of orders taken by representatives employed by a wholesale firm during 1967. The values are rounded off to the nearest £1.

Value of Order (£)	No. of Orders
5 – 9	85
10 – 19	120
20 – 29	225
30 – 49	135
50 – 99	105
100 and over	20

(a) Illustrate the data by means of a histogram.
(b) Calculate the median and mark it on your histogram.

5.14 The following table shows the age distribution of employees in two factories A and B. Estimate the median age in each factory using an appropriate graph and check the results by calculation.

Age of Employees	Number of Employees A	B
15 – 19	79	5
20 – 24	98	23
25 – 29	128	58
30 – 34	83	104
35 – 39	39	141
40 – 44	19	98
45 – 49	11	43
50 – 54	7	19
55 – 59	3	6

5.15 Explain the difference between a continuous variable and a discrete variable.

The table below gives the earnings (to the nearest £) of 150 employees in a large factory

Income	Frequency	Income	Frequency
£5000 – £5499	1	£2500 – £2999	8
£4500 – £4999	0	£2000 – £2499	34
£4000 – £4499	1	£1500 – £1999	38
£3500 – £3999	2	£1000 – £1499	42
£3000 – £3499	4	£500 – £999	20

(a) What are the class boundaries of the class with frequency 8?
(b) Draw a cumulative frequency curve of the distribution.
(c) What percentage of the group earn less than £2200?
(d) Calculate the median income as accurately as you can.
(e) Under what conditions is the median a better measure of central tendency than the arithmetic mean?

5.16 The income distribution of a sample of about 7000 households is given as follows:

Household Weekly Income in the U.K. in 1969

Income (£ per week)	% of Households
under 8	7.1
8 and under 10	4.4
10 and under 15	9.0
15 and under 20	8.2
20 and under 25	10.8
25 and under 30	12.0
30 and under 35	11.9
35 and under 40	9.7
40 and under 50	12.7
50 and under 60	6.7
60 and over	7.5
Total	100.0

Source: Family Expenditure Survey, Report for 1969.

(a) Obtain the cumulative frequencies and plot the cumulative frequency curve.

(b) A firm proposes to market a household product in two qualities — standard and de luxe — which it expects to sell in the income brackets £24-£36 and £32-£48 per week respectively.

If each 1% of the sample can be taken to represent 180,000 households in the population at large, estimate the market size for each quality separately and for the two grades together.

5.17 *Distribution of Personal Incomes before Tax in the U.K. 1969-70*

Income Range (£)	Number of Incomes (00,000's)
below 400	7
400 and under 600	26
600 and under 800	27
800 and under 1000	26
1000 and under 1250	32
1250 and under 1500	30
1500 and under 2000	42
2000 and under 3000	19
3000 and over	8
Total	217

Source: Social Trends, 1971 modified.

(a) Obtain the cumulative frequencies.

(b) By calculation or graphically, obtain the median.

(c) Would you expect the arithmetic mean of the distribution to be greater than, equal to, or less than the median?

Explain you answer. (Do not carry out any further calculations.)

5.18 (a) Define the median, mean, and mode listing the advantages and disadvantages attributable to each.

(b) (i) Calculate the median, mean and mode for the following:

Wage Groups (hourly rate in pence)	Number of Employees
50 and under 60	5
60 and under 70	25
70 and under 80	134
80 and under 90	85
90 and under 100	69
100 and under 110	43
110 and under 120	34

(ii) Illustrate a use for each our the three statistics calculated.

I.C.M.A.

5.19 The marks scored by students in an examination are shown below. (Marking is in whole marks only.)

Marks Scored	10-19	20-29	30-39	40-49	50-59	60-69	70-79	80-89	90-99
No. of Students	4	12	23	37	43	32	19	8	2

(a) Draw a histogram to illustrate the above distribution.

(b) What are the class boundaries of the modal class?

(c) Estimate by calculation the median mark.

(d) If $16\frac{2}{3}\%$ of the pupils are to be given a pass with distinction, what mark will be necessary to achieve this?

(e) What proportion would fail if the pass mark were 48?

5.20 The following figures give the age at 1st June 1979 of a sample of 100 students taking O.N.C. in Business Studies examinations in 1979. The ages are rounded to the next lowest complete month.

Years	Months	Years	Months	Years	Months	Years	Months	Years	Months
17	7	18	4	17	0	16	5	18	8
18	2	18	3	17	11	18	8	18	7
17	9	17	3	17	10	18	9	17	3
16	5	17	1	18	9	17	2	17	10
19	2	18	9	19	5	16	10	18	1
17	9	17	8	17	3	17	3	18	4
17	3	18	11	17	4	16	9	17	11
16	11	19	0	18	0	16	8	18	0
18	5	19	8	17	9	17	1	18	3
18	4	16	4	17	11	18	4	18	1
17	11	16	10	16	10	19	0	16	10
17	10	18	1	17	2	18	11	16	11
18	3	17	8	17	2	17	11	17	2
16	9	17	7	17	1	18	0	17	3
17	2	16	11	17	7	17	10	17	8
18	0	17	3	16	10	17	3	17	11
17	0	18	2	18	3	16	9	18	2
17	8	18	11	17	2	16	6	17	10
17	9	17	5	17	3	17	5	17	10
18	1	18	5	17	9	17	8	18	3

(a) From these figures compile a frequency distribution table. You are advised to work in half-yearly class intervals, paying particular attention to the way you select these intervals and their class boundaries.
(b) Draw a histogram to represent the frequency distribution.
(c) From the histogram find a value for the modal age.

5.21 A random sample of the accounts of 50 construction companies gave the following frequency distribution of profit.

Profit (£ million)	Number of Companies
−10 and under −5	2
−5 and under 0	0
0 and under 5	2
5 and under 10	4
10 and under 15	8
15 and under 20	11
20 and under 25	13
25 and under 30	6
30 and under 35	4

Required: Construct an ogive of the cumulative frequency distribution and use it to estimate:
(1) the median profit,
(2) the profit exceeded by 75% of companies,
(3) the number of companies out of 500 similar ones whose profit for that year was between 8 and 18 million pounds.

A.C.A.

5.22 (a) Required: Place the arithmetic mean, median and mode in order of merit as averages for the following frequency distributions, and briefly explain your rankings in each case.
(1) Incomes, taken from a wages survey.
(2) Ladies' shoe sizes, based on sales data.
(3) Percentage of defective products, based on batches examined.

(b) The following table shows the number of hours of sunshine recorded during July at Bournpool for the year 1965-73:

Hours of Sunshine	Number of Days
0 and under 1	1
1 and under 2	2
2 and under 3	4
3 and under 4	11
4 and under 5	24
5 and under 6	35
6 and under 7	43
7 and under 8	49
8 and under 9	54
9 and under 10	31
10 and under 11	15
11 and under 12	10
	279

Required: Calculate:
(1) the mean number of hours of sunshine.
(2) the median number of hours of sunshine.

A.C.A.

5.23 The table below gives the distribution of marks of a group of candidates taking a statistics examination.

Examination Results

Mark	Number of Candidates
30 and under 40	2
40 and under 50	5
50 and under 60	7
60 and under 70	13
70 and under 80	15
80 and under 90	5
90 and under 100	3

You are required to:
(a) calculate the mean and the median marks, and
(b) explain what the values calculated in your answer to (a) indicate about the distribution of examination results, particularly with reference to the difference between the mean and median values. A.C.A.

5.24 In a survey of the value of orders received, a manufacturing company obtained the following grouped frequency table:

Value of Orders (£)	Number of Orders
100 and up to but less than 200	169
200 and up to but less than 300	176
300 and up to but less than 400	75
400 and up to but less than 500	32
500 and up to but less than 600	8

Draw the histogram for this distribution and from this, or by other means, find the value of the modal order, to the nearer £1. Calculate the mean value of orders, and thus estimate the total value of all the orders. I.C.S.A.

5.25 By taking random samples over a period information has been collected giving the following age distribution of young persons attending a sport and recreation centre:

Age in Years	Number of Young Persons
12 and under 13	3
13 and under 14	11
14 and under 15	29
15 and under 16	99
16 and under 17	203
17 and under 18	258
18 and under 19	247

(a) Calculate to the nearest month the arithmetic mean age and the median age of young persons attending the centre.
(b) Find the modal age to the nearest month of young persons attending the centre.

Discuss the relative merits and possible uses of these three measures of central position.

5.26   *Personal Incomes before tax 1967-8 U.K.*

Income Range	Number (thousands)	Income for Class (£ millions)
less than £350	903	283
£350 – £450	1346	539
£450 – £600	2388	1251
£600 – £800	3058	2140
£800 – £1000	3134	2821
£1000 – £1200	2953	3239
£1200 – £1400	2510	3255
£1400 – £1750	2874	4471
£1750 – £3000	2007	4247
over £3000	558	3026
Total	21731	25272

Source: Abstract of Regional Statistics 1970.

Find the mean, median and modal incomes. Which, in your opinion, is the best indicator of the average income?

5.27 (a) Discuss the importance of averages in statistical analysis, and described the characteristics and methods of calculation of two *commonly* used averages.

(b) Using the geometric mean, calculate the average annual ratio of change in family income from the data in the following table

Year	Family Income as a Ratio of Preceding Year's Income
1967	1.05
1968	1.17
1969	1.33
1970	1.25

(c) Explain the use of the geometric mean as an average and what your answer indicates about the change in family income over the period.   A.C.A.

5.28 Given that the population of the United Kingdom in 1962 was 53,266,000 and in 1972 was 55,933,000; estimate the population in 1967. Comment on your estimate given that the population in 1967 was 54,746,000.

5.29 Suppose we decided to make a general-purpose fertiliser by mixing together in equal values 5 ingredients costing £1, £5, £6, £9, and £11 per kilo. Find the cost per kilo of the mixture.

5.30 Prove that the sum of squares of the deviations of the items from the arithmetic mean is a minimum (you will need a knowledge of differential calculus to be able to show this).

Chapter Six

The Time Series — I

The components of a Time Series

If Plumpton Rovers were drawn against Arsenal in the third round of the F.A. cup, there is no doubt who would be expected to win. Past records show that non-league clubs seldom win against First Division teams. If, however, the non-league club had won such a tie on eighteen out of the last nineteen meetings, then in spite of their relative strength we would expect Arsenal to lose. We tend to base our forecast of the match result on what has happened in the past. This is as true of the accountant as it is of the football fan. In trying to budget for the cost of raw materials or for labour cost in the immediate future, he will look carefully in the first instance at what has happened in the recent past — expecting that the past behaviour of costs will be carried forward into the future.

The statistical series which tells us how data has been behaving in the past is the *Time Series*. It gives us the value of the variable we are considering at various points in time — each year for the last fifteen years; each quarter for the last five years. Yet, when you look at a typical time series, the data fluctuates so much that it seems unlikely that it can help us a great deal. Let us take a series, then, and begin by assessing the factors which cause this fluctuation in the value of the data.

Imports of Raw Material into Ruritania
(19-0 = 100)

Year	Quarter			
	1	2	3	4
19-2	114	142	155	136
19-3	116	150	153	140
19-4	128	158	169	159
19-5	137	180	192	172
19-6	145	194		

The figures are given in Index Number form, and they are plotted as a graph in Diagram 6.01.

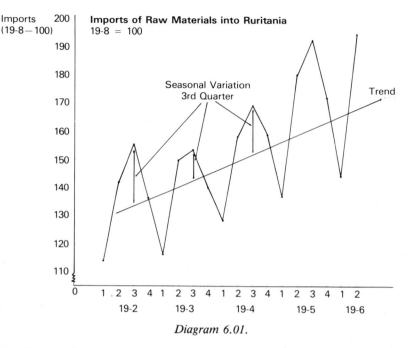

Diagram 6.01.

Look carefully at this diagram. Although the figures are fluctuating quarter by quarter two things are immediately apparent. There is a general upward *trend* in the figures as a whole. It is not an exceptional rise, but it is quite marked. Apart from the third quarter of 19-3, the figures in any quarter are higher than those in the same quarter of the previous year.

Secondly, although the figures fluctuate, there is a pattern. The index of imported raw materials is always highest in the third quarter of the year, and always lowest in the first quarter. There is, then, a very marked *seasonal fluctuation* in the figures.

Although it is not apparent in the graph, there is likely to be a third influence on the data. We would not expect import figures to remain unaffected by day to day happenings such as disputes in the docks, exchange rate fluctuations or fuel shortages. It is probably something of this nature that has reduced the figure for the third quarter of 19-3. Thus, our data is likely to be affected by what we can call *residual* or *random influences* which cannot be foreseen or pinpointed without a great deal of outside knowledge — but which may be of great importance.

The series we are looking at is, of course, a short one. We could link the trend perhaps to a general expansion of the economy during the upward phase of the trade cycle. If we were to extend the series and consider a far longer period, the upward expansion might reach a peak and the trend would begin to turn downwards as the economy moves towards depression. Thus, the series we examine might be affected to a great extent by *cyclical influences* resulting in a trend such as that in Diagram 6.02.

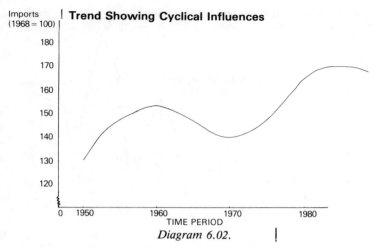
Diagram 6.02.

Thus, data such as this which fluctuates quite markedly over time, may be responding to any, or all, of four sets of forces:

(a) The TREND or general way in which the figures are moving. It is important to distinguish between a trend resulting from cyclical influences on the economy and one resulting from, for example, a change in tastes or consumer buying habits. It is likely that there would have been a downward trend in the sales of black and white television sets, even though the economy as a whole was expanding and the sales of all television sets (including colour) were rising.

(b) SEASONAL VARIATIONS. It is common knowledge that the value of many variables depends in part on the time of year we are considering. Every housewife knows that the price of flowers rises as we approach Mothers' Day; or that the price of tomatoes is higher in winter than it is in summer. You can multiply these examples indefinitely from your own experience, and anyone concerned with planning must take account of them.

(c) RESIDUAL INFLUENCES. These are random external events which affect our variables. Sometimes the effect is negligible; at other times it is great; some occurences will increase our figures; others will reduce them. We cannot see what is going to happen in the future, and so we cannot forecast such events. It is, however, reasonable to assume that in the long run they will tend to cancel each other out, and that in our analysis we may initially ignore their impact.

(d) CYCLICAL INFLUENCES. As the economy expands during a period of boom we would expect to find that such data as sales, output or consumer expenditure also show a rising trend; and during a period of slump we would expect them to show a downward trend. Thus a wavelike motion may be observed in the pattern of our data. If we were to take a sufficiently long period, covering several cycles, we might even find a trend superimposed on the cyclical pattern — each cycle being generally higher, or lower, than the last.

Now, it is one thing to identify the factors which may cause the fluctuations in our data: it is quite another matter to disentangle one from the other and to measure its influence. We will try, in the next section, to isolate and measure the trend.

The Calculation of Trend

In Diagram 6.01 we have drawn a straight line through the middle of the series and said that the trend is rising "something like this". For serious statistical work, however, it is no use proceeding by guesswork. If we are going to use the trend figures we must calculate them, not guess them.

Now there are several methods of calculating trend, but the method we will explain first is a general all-purpose method — the method of *moving averages*. We will illustrate this method by calculating the trend of unemployment in the United Kingdom during a sixteen year period.

Unemployment in the United Kingdom

Year	Unemployment %	5 year Total	5 year Moving Average = Trend
1	1.3		
2	1.1		
3	1.2	7.1	1.42
4	1.4	8.0	1.60
5	2.1	8.5	1.70
6	2.2	8.8	1.76
7	1.6	9.4	1.88
8	1.5	9.8	1.96
9	2.0	9.2	1.84
10	2.5	9.0	1.80
11	1.6	9.0	1.80
12	1.4	9.4	1.88
13	1.5	9.3	1.86
14	2.4	10.1	2.02
15	2.4		
16	2.4		

Source: Employment & Productivity Gazette

A moving average is a simple arithmetic mean. We select a group of numbers at the start of the series, in this case the first five, and average them to obtain our first trend figure,

$$(1.3 + 1.1 + 1.2 + 1.4 + 2.1) \div 5 = 7.1 \div 5 = 1.42$$

This trend figure is placed opposite the centre of the group of five numbers, that is, opposite year 3.

To obtain our second trend figure we drop the first number of this initial group (1.3) and include the next number in the series (2.2). So our second group of five numbers is

$$1.1 + 1.2 + 1.4 + 2.1 + 2.2 = 8.0$$

and our second trend figure is

$$8.0 \div 5 = 1.60$$

which is placed opposite year 4. We now drop the figure for year 2 (1.1) and include that for year 7 (1.6), giving us a third trend figure of 1.70. We carry on in this way until all our data is exhausted.

You will notice that, using this method, there is no trend figure for the first two years nor for the last two. Do not forget this if you are asked to plot the original data and the trend on the same graph. Look carefully at Diagram 6.03 where the data have been plotted.

Unemployment in the U.K.

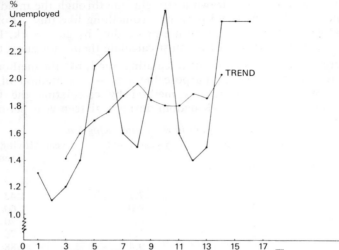

Source: Employment and Productivity Gazette

Diagram 6.03

The trend is probably not as smooth as you thought it would be. True, the large scale fluctuations have been eliminated, and up to year 8 it is fairly smooth; but from this point on there is no clear indication of which way the trend is moving. This was probably a result of changes in government policy as the United Kingdom struggled to counter inflation, prevent large scale unemployment and promote economic expansion.

You are probably wondering why we chose a five year moving average rather than a three year or a seven year one. There is no infallible rule you can follow. The correct time period to use is the one that gives the smoothest trend. A good working rule, however, is to look at the peaks and the troughs in the series. Troughs occur in years 2, 8 and 12; peaks occur in years 6, 10 and 14-16. It looks as if unemployment was subject to a five year cycle, and, in fact, economists confirm that this was so. Hence we chose a five year moving average.

The Trend of a Quarterly Series

A great deal of the data statisticians deal with is given quarterly or monthly, rather than annually. We saw in the previous section that one of the characteristics of such a series is that peaks usually occur in the same quarter of every year, as do the troughs; so the correct moving average to use is the four quarterly moving average. This does not demand any new concepts, but it does raise a practical problem. When we calculate the four quarterly moving average and it is placed opposite the mid point of the group of numbers to which it refers, it falls *between* the second and third figures. Thus we cannot relate our figures of trend to any particular quarter. To see how we overcome this let us calculate the trend of the quarterly series of Imports of Raw Material into Ruritania.

Year	Quarter	Imports	Sum in 4's	Sum of two 4's	Trend
19-2	1	114			
	2	142			
	3	155	547	1096	137.0
	4	136	549	1106	138.25
19-3	1	116	557	1112	139.0
	2	150	555	1114	139.25
	3	153	559	1130	141.25
	4	140	571	1150	143.75
19-4	1	128	579	1174	146.75
	2	158	595	1209	151.125
	3	169	614	1237	154.625
	4	159	623	1268	158.5
19-5	1	137	645	1313	164.125
	2	180	668	1349	168.625
	3	192	681	1370	171.25
	4	172	689	1392	174.0
19-6	1	145	703		
	2	194			

Notice that when we add up in groups of four, the total is placed midway between the relevant quarters. In itself there is little wrong with this, but it prevents any comparison between the level of imports and the trend — a comparison we will have to make if we are to continue the analysis. To eliminate the problem, we use a technique known as *centreing*. At present we have a trend figure of 547 ÷ 4 = 136.75 placed opposite quarter 2½, and one of 137.25 placed opposite quarter 3½. If we take the average of these two it would be placed opposite quarter 3, and we have centred the trend so as to relate it to a specific quarter, i.e.

$$\frac{136.75 + 137.25}{2} = 137.0$$

The easiest way to do this in practice is to total successive pairs of four quarterly totals (eg. 547 + 549; 549 + 557) and divide the resultant totals by eight to give the trend. This is done in the last two columns of the table.

Exercises to Chapter 6

6.1 Describe the components into which statistical data relating to business and economic events may be analysed by the use of time series analysis.

A.C.A.

6.2 Specify and describe the various movements which may be identified when analysing economic and business statistics over a long time span.

A.C.A.

6.3 In the analysis of a time series:
    (a) explain what is meant by a trend line and describe very briefly the method of calculating it.
    (b) explain what is meant by residual variations and why they are important;
    (c) sketch a graph of a time series which shows a trend and also cyclical and seasonal movements.
A.C.A.

6.4 The table shows the number of parcels carried by Rapid Carriage Ltd. to destinations within the United Kingdom:

Number of Parcels Carried (000's)

Year	Quarters			
	1	2	3	4
19-5				800
19-6	690	1580	2250	940
19-7	830	1900	2670	1130
19-8	1040	2160	3070	1360
19-9	1190	2580		

Plot the data and obtain the moving average trend. Plot the trend line on the same graph. What conclusions may be deduced from your graph?

6.5 (a) State four components of a time series.
    (b) The following table shows the quarterly sales of a company in thousands of tons for a period of four years:

Sales	Quarter			
	1	2	3	4
Year 1	70	41	52	83
2	78	44	48	85
3	83	54	51	96
4	85	49	54	89

You are required to use the information given above to:
    (a) plot on a graph the quarterly sales;
    (b) derive and plot the appropriate moving average.

6.6 A traffic census taken at the same time each day during July on a busy road approaching a holiday resort revealed that the number of vehicles passing per hour was as follows (to nearest 10).

	Week				
	1	2	3	4	5
Monday		840	840	830	820
Tuesday		860	860	830	
Wednesday		1190	1200	1220	
Thursday		840	830	840	
Friday		970	1020	1080	
Saturday	1800	1860	1950	2100	
Sunday	1460	1480	1520	1550	

    (a) Plot the data on a graph.
    (b) Calculate a suitable moving average and plot the trend line. Explain your choice of time period.
    (c) If the survey had been carried out in say November, what do you think the figures would have looked like then? (You may illustrate by a rough sketch or by providing possible figures for one or two weeks.)

6.7

Year	Cost of Raw Materials Cost (£ per ton)
1	3.38
2	4.11
3	4.17
4	3.87
5	3.26
6	3.81
7	3.89
8	4.45
9	4.30
10	3.65
11	5.00

(a) Construct a five year moving average, graphing the trend line against the original data. Comment on your results. If the interval for the moving average had not been suggested, how would you have determined it? What is the purpose of a moving average?

(b) With which characteristic movement of a time series would you mainly associate each of the following?
    (i) a minor fire delaying production for three weeks;
    (ii) an increase in unemployment during the winter months;
    (iii) the increasing demand for small cars;
    (iv) a recession.

6.8 Unemployment as a Percentage of Total Employees

	March	June	September	December
19-5	2.3	2.4	2.6	2.6
19-6	2.9	3.2	3.6	3.7
19-7	3.9	3.5	3.6	3.2
19-8	2.8	2.6	2.4	2.1
19-9	2.4	2.5		

(a) Smooth this time series by means of a centred four-quarterly moving average.

(b) Plot the original and the moving average figures on the same graph.

(c) Discuss whether this moving average has any value for forecasting future percentage unemployment.

Chapter Seven

The Time Series — II

The Calculation of Seasonal Variation

We have defined seasonal variation as an upswing and downswing in the value of our data. If we are to measure the magnitude of these fluctuations we must have a point of reference from which to measure. After all, when we say a mountain is 3000 feet high, what we really mean is that it is 3000 feet *above sea level*. It seems logical that we should measure the magnitude of the seasonal swing as the deviation from our calculated trend figure. So seasonal variation is not merely an upswing and downswing, — it is a swing around the trend line. Armed with this definition we can now calculate the quarterly variation in imports from our figures. The variation in any particular quarter will be:

(Original data *MINUS* calculated trend)

Year	Quarter	Imports	Trend	Deviation from Trend
19-2	1	114		
	2	142		
	3	155	137.0	+ 18.0
	4	136	138.25	− 2.25
19-3	1	116	139.0	− 23.0
	2	150	139.25	+ 10.75
	3	153	141.25	+ 11.75
	4	140	143.75	− 3.75
19-4	1	128	146.75	− 18.75
	2	158	151.125	+ 6.875
	3	169	154.625	+ 14.375
	4	159	158.5	+ 0.5
19-5	1	137	164.125	− 27.125
	2	180	168.625	+ 11.375
	3	192	171.25	+ 20.75
	4	172	174.0	− 2.00
19-6	1	145		
	2	194		

The meaning of the deviations we have calculated in the last column is that in that particular quarter seasonal and other influences have caused imports to vary from trend by the calculated amount. Thus, in the 3rd quarter of 19-2 these influences cause imports to rise 18.0 points above trend; in the 3rd quarter of 19-3, 11.75 points above trend; in the 3rd quarter of 19-4, 14.375 points above and so on. Why should the deviations be so different in the same quarter of the year? The answer lies in the fact that seasonal and *other* influences have been at work. The deviations have been caused by both seasonal and residual influences.

While we cannot separate these two influences, it is reasonable to believe that while seasonal influences will always operate in the same direction, residual influences will sometimes raise the figures and at other times will lower them. If, then, we take a sufficiently long series and take the *average* deviation for any particular quarter, the residual influences will tend to offset each other and we will be left with the purely seasonal variation. This is the rationale behind the calculation of seasonal variation, and the reason why it is often called *average seasonal variation*.

We will now pick up the quarterly deviations from trend and tabulate them in order to calculate seasonal variation.

Year	Quarter 1	Quarter 2	Quarter 3	Quarter 4	
19-2			+18.0	−2.25	
19-3	−23.0	+10.75	+11.75	−3.75	
19-4	−18.75	+6.875	+14.375	+0.5	
19-5	−27.125	+11.375	+20.75	−2.0	
Total	−68.875	+29.0	+64.875	−7.5	
Average	−22.958	+9.667	+16.219	−1.875	= +1.053
Adjust	−0.263	−0.263	−0.263	−0.263	
	−23.221	+9.404	+15.956	−2.138	
S.V.	−23	+9	+16	−2	

There are several things to note about this table. Firstly, you will see that there are four deviation figures for the third and fourth quarters, but only three for the first and second. This is common and depends simply on the lengths of the series we are examining, but we must not forget it when we are calculating the average quarterly deviation. Secondly, the total of the average deviations should be zero, (remember the definition of the arithmetic mean). In fact, the total is +1.053, and since we know it should be zero an adjustment is necessary. We do not know where the difference springs from, so we adjust each quarterly average by the same amount, 0.263 (i.e. 1.053 ÷ 4). Since the total deviations are too many we will have to subtract 0.263 from each quarterly average. Thirdly, when we have made all these adjustments, we are left with average deviations correct to three places of decimals. Now this is silly! In the course of our calculations we have made a number of assumptions. Each one is a logical, reasonable assumption, but we must not pretend to a degree of accuracy we cannot guarantee. Since the original data is given in integers only, it is better to give average seasonal variation in integers also. So we will quote the seasonal variation as −23, +9, +16, −2. Notice that the total still comes to zero.

What do these figures mean? Using the figures for the first quarter as an example, it merely says that economic conditions are such that the index of imports will tend to fall below the trend, and that on average it will be 23 points below. On the other hand, in the second quarter of the year, the economic climate is different, imports tend to rise above trend and on average are 9 points above it.

Series with seasonal variation eliminated

Useful as a knowledge of seasonal variation may be to a firm planning for the future, there is little doubt that the constant fluctuations tend to hide the underlying behaviour of the variable. Comparisons and assessments of performance over time are difficult to make. When a particular salesman returns with a full order book, is it because he has made a superhuman effort, or merely because the seasonal swing is in his favour. The need to make assessments of this nature has resulted more and more in data being produced "seasonally adjusted", or "with seasonal variation eliminated".

Adjusting data to eliminate seasonal variation is a relatively simple matter. If seasonal variation is +23, we are, in effect, saying that our data is 23 points above the trend because of seasonal influences, and we must, therefore, reduce our figures by 23 to eliminate these influences. Similarly, if seasonal variation is −23, we must add 23 to our figures. If we remember that (--) = (+), we may formulate the rule that

Series with seasonal variation eliminated = original data − seasonal variation.

Applying this rule to the data we are analysing:

Year	Quarter	Import Index	Seasonal Variation	Series with seasonal variation eliminated
19-2	1	114	− 23	137
	2	142	+ 9	133
	3	155	+ 16	139
	4	136	+ 2	138
19-3	1	116	− 23	139
	2	150	+ 9	141
	3	153	+ 16	137
	4	140	− 2	142
19-4	1	128	− 23	151
	2	158	+ 9	149
	3	169	+ 16	153
	4	159	− 2	161
19-5	1	137	− 23	160
	2	180	+ 9	171
	3	192	+ 16	176
	4	172	− 2	174
19-6	1	145	− 23	168
	2	194	+ 9	185

We now have three different series obtained from the same data, the original figures, the trend and the series with seasonal variation eliminated. When you eliminate the seasonal variation, you might, of course, expect to be left with a series approximating to trend. Do not forget, however, that residual influences will still be found in this series, and the more important the residual influence, the more the series will differ from trend. In Diagram 7.01 we have plotted the three series on the same graph. What conclusions can we draw? Firstly, the trend shows us that there has been a steady and unbroken rise in the volume of imports over the whole period, which shows little signs of ceasing. This may be a result of an expanding industry requiring more and more raw materials, or it may be a result of the domestic consumer buying more and more foreign consumer goods. The graph will not tell us why something happens — but it will lead us to ask the questions.

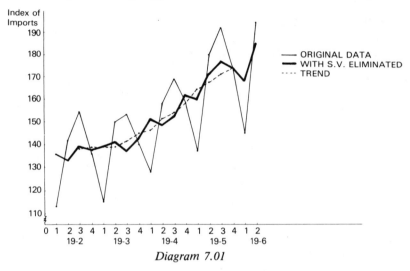

Diagram 7.01

Secondly, the trend and the series with seasonal variation eliminated tend to run closely together. There are differences, but they are small. So we can come to the conclusion that residual influences are fairly unimportant and do not drastically affect the level of imports. This may seem strange, but in fact events such as a dock strike which largely cuts off the flow of goods are very rare.

Nor do such residual influences as there are always work in one direction. Sometimes they raise the level of imports above trend; at other times they lower them; and the pattern is regular. In fact, so regular is the pattern that one can speculate that residuals here have a double effect. At first they slow down the level of imports, but fairly quickly a back-log of orders builds up, and as a result, in the next quarter imports have to rise as the back-log is cleared.

The importance of residuals

You can find out a great deal about residual influences from a graph such as Diagram 7.01, but it does not tell us why residual influences are important.

All planning, whether financial or not, is based on forecasting, and if the forecast cannot be relied on planning is a waste of time. Now, the characteristic of a residual is that it cannot be included in our forecast, so, when its influence is felt, the forecast is upset. In analysing our series of imports we have already discovered that the influence of residuals is fairly small. So, it is probable that though our forecast may be upset, we will not be very far wrong. On the other hand, if the residual influence is great, our forecasts may be badly upset.

Before we come to any decisions, however, we must obtain a quantitative measure of the extent of residual variations. We know that the original data is composed of
 Trend + Seasonal Variations + Residual Variations
and it follows then that,
 Residuals = Original Data − Trend − Seasonal Variations.

Let us, then, bring together the calculations we have made so far:

Year	Quarter	Imports	Trend	Seasonal Variation	Residual Absolute	As % of imports
19-2	1	114		−23		
	2	142		+ 9		
	3	155	137.0	+16	+2	1.3
	4	136	138.25	− 2	−0.25	0.18
19-3	1	116	139.0	−23	0.0	—
	2	150	139.25	+ 9	+1.75	1.17
	3	153	141.25	+16	−4.25	2.78
	4	140	143.75	− 2	−1.75	1.25
19-4	1	128	146.75	−23	+4.25	3.32
	2	158	151.125	+ 9	−2.125	1.34
	3	169	154.625	+16	−1.625	0.96
	4	159	158.5	− 2	+2.5	1.57
19-5	1	137	164.125	−23	−4.125	3.01
	2	180	168.625	+ 9	+2.375	1.32
	3	192	171.25	+16	+4.75	2.47
	4	172	174.0	− 2	0.0	—
19-6	1	145		−23		
	2	194		+ 9		

Our calculations confirm what we suspected from the graph. The residual influence never affects our figures by more than 4.75 points, and, what is more important, on only two occasions is there more than a 3% influence. Thus, if our forecast is otherwise accurate, we would expect it to be correct to within 3% in spite of the residual influences.

Forecasting from the time series

The key words in the last paragraph are "if our forecast is otherwise accurate". Naturally, in looking to the future, absolute accuracy is difficult to attain, but nevertheless, very good forecasts can be made.

The basis of the forecasting we will do lies in our knowledge of the behaviour of trend. The trend will not suddenly change direction. As you can see, in Diagram 7.02 the trend is beginning to rise rather more slowly.

Over the next few quarters the rate of rise may continue to fall and eventually the trend will reach a maximum and perhaps begin to fall. Or it may slowly change direction upwards with an accelerated rate of rise. The point is that it is not going to change *suddenly*. So, we can extend (or project) the trend forward over the next two or three quarters, confident that our projection will be fairly accurate. The further we project the trend, of course, the greater is the possibility of inaccuracy, and great care is needed in making the projection.

In diagram 7.02 the trend is drawn and a suggested projection made, extending over the next two quarters. We make no claim, of course, that this projection is perfect, and you may make a far more accurate one yourself. Our new trend values are:

19-6 Quarter 1   176.5
         2   178.75

To understand just how we can use these figures we will go back to the calculation of trend and reproduce the last part of the calculations.

Year	Quarter	Imports Col. 1	Sum in 4's Col. 2	Sum of two 4's Col. 3	Trend Col. 4
19-5	3	192	681	1370	171.25
	4	172	689	1392	174.0
19-6	1	145	703	($8x = 1412$)	($x = 176.5$)
	2	194	($8x - 703 = 709$)	$8y =$	($y = 178.75$)
	3	($a = 198$)			
	4	$b =$			

We will call the forecast values for 19-6, quarter 3, *a*, and that for quarter 4, *b*. Suppose the trend value we read from the graph is x, – this is placed in column 4. But we obtain x by dividing column 3 by 8, so in column 3 we must place $8x$. In its turn this $8x$ is the sum of 703 plus an unknown number, which will have therefore a value $8x - 703$, which is the next figure in column 2. Column 2 is the sum of four import indices and $8x - 703 = 145 + 172 + 194 + a$. It follows then that

$$a = 8x - 703 - 172 - 145 - 194 = 8x - 1214$$

Since $x = 176.5$

$$a = (8 \times 176.5) - 1214 = 1412 - 1214 = 198$$

This seems to be a reasonable forecast since the third quarter is always slightly higher than the second quarter (which was 194). The values we have calculated have been placed in the appropriate columns in the table. Now, using these figures, and given that the value of the trend y is 178.75, calculate for yourself the value of the import index for the fourth quarter of 19-6.[1] The answer is given in the footnote, and you will probably agree that the two figures we have forecast are entirely consistent with the pattern of the series.

1. Answer: $y = 178.75$ $b = 8y - 709 - 145 - 194 - 198 = 8y - 1246 = 1428 - 1246 = 182$.

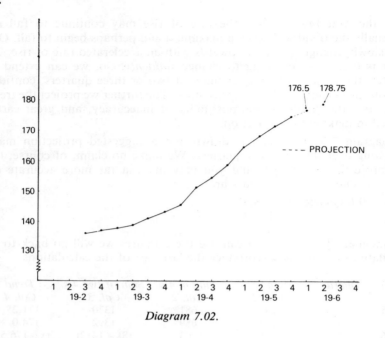

Diagram 7.02.

Forecasting by projecting the trend of an annual series is exactly the same as for a quarterly series. Suppose we are taking a five year moving average, and the last few rows of our calculations are:

Year	Output	5-year total	Trend
19-2	15	70	14.0
19-3	18	74	14.8
19-4	13	76	15.2
19-5	14	?	?
19-6	16		
19-7	?		

The first step in forecasting a value for 19-7 would be to project the trend forward. We could then read a value for 19-5. Suppose this value of the projected trend were 15.6. This figure is the average of a five year total in the previous column, so that total must be $5 \times 15.6 = 78$. This in itself is the sum of $18 + 13 + 14 + 16 +$ (output in 19-7). Thus output in 19-7 would be forecast as

$$78 - 18 - 13 - 14 - 16 = 17.$$

The calculation of seasonal variation in this example is based on the assumption that what we call the Additive Model is the correct one to use.

We assume, that is, that

Data = Trend + Seasonal Variation + Residuals

or, as it is more often put,

$$Y_t = T_t + S_t + R_t$$

Now, in making this assumption we arrive at a figure for seasonal variation which is taken to be applicable to that particular quarter in any and every year. There is little harm in this if the magnitude of the swing around the trend is fairly constant. But suppose that over time the magnitude of the swings is getting larger — we have a problem. We might find that in measuring the deviations from trend for the first quarter they read +4, +8, +14, +23. The average seasonal variation is +12.25. If, however, the deviations continue to be of the order of 23 or more, it is nonsense to use a figure of +12.25 in order to forecast data in the future. You must use the additive method only when the deviations from trend which you obtain are fairly stable.

In their calculations of seasonal variations many firms find that the trend of their data is a linear one. Numbering each quarter in order, 1, 2, 3, 4, 5, 6 and so on, they can then express the equation of the trend line in the form y = a + bx. They may find, for example, that the trend of their output data is given by Trend = 6.0 + 2.5t, where t is the number given to the appropriate quarter. Let us study an example using this principle.

Alpha Products Ltd. have produced figures for their export sales over the last four years as follows:

Sales (£ million)

Quarter			
1	2	3	4
8	17	10	18
19	24	17	29
23	38	31	33
34	45	42	48

The firm knows that the trend line is given by the equation

Trend = 6.0 + 2.5t

Calculate the average seasonal variation and use your figures to forecast sales for each quarter of next year.

If you remember that t is the number given to the particular quarter, the first quarter of the first year being numbered 1, the trend is easy to calculate for that quarter as:

$$6.0 + 2.5 \times 1 = 8.5.$$

In the same way the trend value for the second quarter of that year is

$$6.0 + 2.5 \times 2 = 11.0$$

Our calculations now proceed as follows:

Quarter	Sales	Trend	Deviation from trend
1	8	8.5	−0.5
2	17	11.0	6.0
3	10	13.5	−3.5
4	18	16.0	2.0
5	19	18.5	0.5
6	24	21.0	3.0
7	17	23.5	−6.5
8	29	26.0	3.0
9	23	28.5	−5.5
10	38	31.0	7.0
11	31	33.5	−2.5
12	33	36.0	−3.0
13	34	38.5	−4.5
14	45	41.0	4.0
15	42	43.5	−1.5
16	48	46.0	2.0

Seasonal Variation can now be calculated.

	\multicolumn{4}{c}{Quarter}			
	1	2	3	4	
	−0.5	6.0	−3.5	2.0	
	+0.5	3.0	−6.5	3.0	
	−5.5	7.0	−2.5	−3.0	
	−4.5	4.0	−1.5	2.0	
Total	−10	20	−14	4	
Average	−2.5	5	−3.5	1	= 0

In using these figures to forecast for the next year we remember that the figure for sales is

$$\text{Trend} + \text{Seasonal Variation}.$$

For the first Quarter of next year our forecast would be

$$6.0 + 17 \times 2.5 - 2.5 = 46.0$$

For the second quarter

$$6.0 + 18 \times 2.5 + 5.0 = 56.0$$

For the third quarter

$$6.0 + 19 \times 2.5 - 3.5 = 50.0$$

And for the fourth quarter

$$6.0 + 20 \times 2.5 + 1.0 = 57.0$$

Alpha Products then would predict:

Seasonal Variation		Next Years Sales
Quarter 1	−2.5	46.0
Quarter 2	+5.0	56.0
Quarter 3	−3.5	50.0
Quarter 4	+1.0	57.0

Additive or Multiplicative Model?

In our analysis of time series, we have assumed that

Actual data = Trend + Seasonal Variation (+ Residuals)

This is the so-called *additive model.*: it assumes that there is a constant absolute difference between the actual data and the trend, and we call this difference the seasonal variation. Now most statisticians think this is not a reasonable assumption: it does not seem reasonable to adjust every (say) first quarter by adding or subtracting a constant amount. It would be much better to adjust every corresponding quarter by a constant percentage. This gives the more satisfactory *multiplicative model.*

Actual data = Trend × seasonal variation (+ Residuals).

Instead, then, of taking seasonal variation to be the average of the (actual − trend) values, we will take it to be the average of the (actual/trend) values. Let us rework our example of the import index, this time using the multiplicative model. The trend is calculated in exactly the same manner as for the additive model − by using a four quarter, centred, moving average.

Year	Quarter	Import Index	Trend	Actual/Trend
19-2	1	114		
	2	142		
	3	155	137.000	1.1313
	4	136	138.250	0.9837
19-3	1	116	139.000	0.8345
	2	150	139.250	1.0771
	3	153	141.250	1.0831
	4	140	143.750	0.9739
19-4	1	128	146.750	0.8722
	2	158	151.125	1.0454
	3	169	154.625	1.0929
	4	159	158.500	1.0031
19-5	1	137	164.125	0.8347
	2	180	168.625	1.0674
	3	192	171.250	1.1211
	4	172	174.000	0.9885
19-6	1	145		
	2	194		

	Quarter			
Year	1	2	3	4
19-2			1.1313	0.9837
19-3	0.8345	1.0771	1.0831	0.9739
19-4	0.8722	1.0454	1.0929	1.0031
19-5	0.8347	1.0674	1.1211	0.9885
Total	2.5414	3.1899	4.4284	3.9492
Average	0.8471	1.0633	1.1071	0.9873

We would predict the total of the average ratios to be 4, whereas it is 4.0048. So we must adjust the averages by multiplying each average ratio by $4/4.0048$ to give seasonal variation ratios

	Quarter		
1	2	3	4
0.846	1.062	1.106	0.986

If we wish to find a deseasonalised series, we would divide the actual data by the seasonal variation.

Year	Quarter	Index	Seasonal ratio	Deseasonalised Index
19-2	1	114	0.846	135
	2	142	1.062	134
	3	155	1.106	140
	4	136	0.986	138
19-3	1	116	0.846	137
	2	150	1.062	141
	3	153	1.106	138
	4	140	0.986	142
19-4	1	128	0.846	151
	2	158	1.062	149
	3	169	1.106	153
	4	159	0.986	161
19-5	1	137	0.846	162
	2	180	1.062	169
	3	192	1.106	174
	4	172	0.986	174
19-6	1	145	0.846	171
	2	194	1.062	183

The Adaptive Forecasting

The use of trend to provide a forecast is an example of what a statistician would call *adaptive forecasting*. Think for a moment about what we have done in the previous sections. We have projected the trend forward for one or two time periods, taking into account all the information which we currently have available. We then used our projected trend to forecast actual data values. Time, however, does not stand still, and very shortly we will have the accurate value of the data for the next time period. Now possession of this data gives us two valuable pieces of information. It enables us to check the accuracy of the forecast we made; and it enables us to calculate the value of the next trend figure. If this new calculated value is very close to the value we projected our forecast will probably be fairly accurate, and there will be little reason for us to change our opinion of the way in which trend is going to behave. But suppose our new trend figure differs markedly from the projected figure. We would be well advised to reconsider our opinion of the future behaviour of trend and to make some adjustments to the projection. In this adjustment to the projected trend we will naturally have to recalculate the forecasts we have made. This, then, is adaptive forecasting. As each new item of information becomes available, we adapt our forecast in the light of the new data we now have. Thus forecasting is constantly revised or adapted as we obtain more data on which it can be based.

As you know, one of the first problems we had to consider when we calculated the trend was that of deciding how many time periods to use. The choice is a critical one for the forecaster. Suppose we decided to use a three year moving average. As you can see from the table, the value of the trend fluctuates and adjusts very quickly to fluctuations in the data. Thus our forecasts based on the trend will also tend to fluctuate as data fluctuates in value. Our forecast will be highly *sensitive* to any variation in the data which may occur. Is this such a bad thing? Well, we want our forecast to take account of variations in data, yet, at the same time we must remember that data may vary for purely random reasons rather than as part of an underlying trend. If our forecast is too sensitive it will be reflecting these random influences rather than trend.

Data	3 Period total	trend	11 period total	trend
60				
52	163	54.33		
51	152	50.67		
49	144	48.00		
44	120	40.00		
27	96	32.00	498	45.27
25	93	31.00	502	45.64
41	108	36.00	499	45.36
42	127	42.33	498	45.27
44	149	49.67	495	45.00
63	171	57.00	506	46.00
64	176	58.67	535	48.64
49	163	54.33	554	50.36
50	145	48.33	569	51.73
46	151	50.33	578	52.55
55	157	52.33	590	53.64
56	155	51.67		
44	156	52.00		
56	151	50.33		
51	163	54.33		
56				

Suppose now we go to the other extreme and consider the calculation of trend using an eleven year moving average. There are still fluctuations in the value of the trend, but you can see from the table that fluctuations in the original data do not affect the value of this trend to anything like the same extent as they affected the three year moving average. There is little sensitivity in the trend but it is far more *stable* than the previous one. This difference is well brought out in the graph of the two trends drawn in Diagram 7.03.

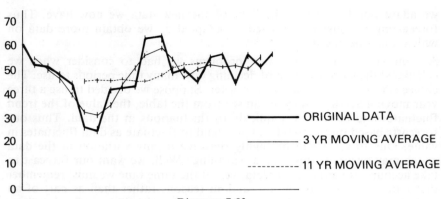

Diagram 7.03.

This, then, is the problem facing the forecaster. He must not provide a forecast which is so sensitive that it reflects every random variation; yet the forecast must not have such a high degree of stability that genuine changes in the underlying trend are not reflected in the forecast. This question of striking a balance between sensitivity and stability arises in any adaptive forecasting technique. If we are using moving averages, as we have seen, it is a question of choosing the right time period. Too short a period and we have extreme sensitivity but no stability; too long a period and we get a great deal of stability with no sensitivity. The balance the forecaster ultimately adopts will depend on the relative importance he himself attaches to sensitivity and stability.

Exponential Smoothing

In spite of its high sounding title *exponential smoothing* is merely an adaptive forecasting technique of great simplicity. Let us suppose that we had made a forecast that our December sales would be £35,000 but that in fact they were £38,000. All that exponential smoothing demands is that our forecast for the January sales be adjusted in the light of the difference of £3,000 between the actual and the forecast figures for the previous month. But what adjustment should we make? It seems logical that it should be some fraction of the difference — a half, or a quarter, or seven eighths and so on. Suppose we represent the chosen fraction by the Greek letter α (alpha), we can say:

New Forecast = old forecast + α (Actual Data − Old Forecast)

where α is a figure lying between 0 and 1.

Let us see how this would work in practice taking a value for α of 0.5.

When initially we begin to forecast our weekly sales, the only information we have is that this week's sales (week 1) were £3,500. The best forecast we can realistically make without any further information is that next week's sales will also be £3,500. In the event we find at the end of week 2 that our sales were in fact £3700. We will now adjust our forecast for week 3 in the light of this additional information. Our adjustment factor, α, is 0.5, so our forecast for week 3 is produced as follows:

New Forecast = Old Forecast + α (Actual Data − Old Forecast)
= 3500 + 0.5 (3700 − 3500) = 3600

So we will adjust our forecast for week 3 by an amount equal to half the error we made in week 2, that is to £3600. It is easy to see that this forecast takes into account all the information in our possession, both for week 1 and week 2.

Suppose now that actual sales in week 3 were £4100 or £500 above the forecast figure. For week 4 we would adjust our forecast further by an amount equal to this difference multiplied by α, that is, by £500 × 0.5. Thus our forecast for week 4 would be:

$$£3600 + 0.5 \times £500 = £3850$$

In tabular form the information would be presented as follows:

Week	Forecast £	Actual Sales £	Difference	Adjustment $\alpha = 0.5$	New Forecast
1		3500			3500
2	3500	3700	+ 200	+ 100	3600
3	3600	4100	+ 500	+ 250	3850
4	3850				

It is apparent that the new forecast we make will depend critically on the value we give to α. Let us consider an extreme case when the value we give to α is 0.0. Is it not obvious that once we have made our initial forecast this forecast will never be adjusted no matter how large the difference between the actual and the forecast figures?

Thus, having forecast sales of £3,500 in week 2 and achieved sales of £3,700, our forecast for week 3 will be:

$$£3,500 + 0.0 (£3,700 − £3,500) = £3,500 + 0 = £3,500.$$

We will never adjust our forecast because an alpha value of 0.0 means that we are taking no account whatsoever of any new information we may receive. We have a forecast which is absolutely stable, but which is completely insensitive to fluctuations in the data.

There is no golden rule which will tell you precisely what value to take for alpha, but in the tables below we have exponentially smoothed a series using a number of different alpha values, and you can see for yourself how sensitivity is gained at the expense of stability.

Actual Data	Forecast	Difference	Adjustment	New Forecast
When $\alpha = 0.1$				
103	72.5	+ 30.5	+ 3.05	75.55
112	75.55	+ 36.45	+ 3.645	79.195
110	79.195	+ 30.805	+ 3.0805	82.2755
135	82.2755	+ 52.7245	+ 5.2724	87.5479
143	87.5479	+ 55.4521	+ 5.5452	93.0931
152	93.0931	+ 58.9069	+ 5.8907	98.9838
150	98.9838	+ 51.0162	+ 5.1016	104.0854
175	104.0854	+ 70.9146	+ 7.0915	111.1769
183	111.1769	+ 71.8231	+ 7.1823	118.3925
192	118.3592	+ 73.6408	+ 7.3641	125.7593

Now as you can see in this table, an alpha value of 0.1 is not very satisfactory. Not only is the forecast consistently less than the actual data, the deficit is getting larger with each successive period. Beginning with a difference of only 30.5, we end up only nine periods later with a deficit of over 73. The trouble is that an alpha value as low as 0.1 has insufficient sensitivity. The forecast will constantly fall further and further below the actual data so long as this trend continues. Diagram 7.04, where both the original data and the forecast are plotted, shows this forcibly.

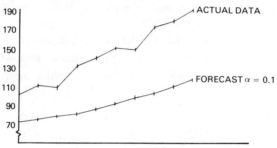

Diagram 7.04.

If an alpha value of 0.1 results in a forecast pattern which is not only bad but which may be misleading, can we improve the situation by increasing the value of alpha and so increasing sensitivity? Remember, though, that in judging the tables which follow that we can only increase sensitivity at the cost of stability.

Data	Forecast	Difference	Adjustment	New Forecast
α = 0.3				
103	77.5	+25.5	+ 7.65	85.15
112	85.15	+26.85	+ 8.06	93.21
110	93.21	+16.79	+ 5.03	98.24
135	98.24	+36.76	+11.03	109.27
143	109.27	+33.73	+10.12	119.39
152	119.39	+32.61	+ 9.78	129.17
150	129.17	+20.83	+ 6.25	135.42
175	135.42	+39.58	+11.87	147.29
183	147.29	+35.71	+10.71	158.01
192	158.01	+33.99	+10.20	168.20
α = 0.5				
103	82.5	+20.5	+10.25	92.75
112	92.75	+19.25	+ 9.63	102.38
110	102.38	+ 7.63	+ 3.81	106.19
135	106.19	+28.81	+14.41	120.59
143	120.59	+22.41	+11.20	131.80
152	131.80	+20.20	+10.10	141.90
150	141.90	+ 8.10	+ 4.05	145.95
175	145.95	+29.05	+14.53	160.47
183	160.47	+22.53	+11.26	171.74
192	171.74	+20.26	+10.13	181.87

$\alpha = 0.7$

103	87.5	+ 15.5	+ 10.85	98.35
112	98.35	+ 13.65	+ 9.56	107.91
110	107.91	+ 2.10	+ 1.47	109.37
135	109.37	+ 25.63	+ 17.94	127.31
143	127.31	+ 15.69	+ 10.98	138.29
152	138.29	+ 13.71	+ 9.59	147.89
150	147.89	+ 2.11	+ 1.48	149.37
175	149.37	+ 25.63	+ 17.94	167.31
183	167.31	+ 15.69	+ 10.98	178.30
192	178.30	+ 13.71	+ 9.59	187.89

$\alpha = 0.9$

103	92.5	+ 10.5	+ 9.45	101.95
112	101.95	+ 10.05	+ 9.05	111.00
110	111.00	− 1.00	− 0.90	110.10
135	110.1	+ 24.90	+ 22.41	132.51
143	132.51	+ 10.49	+ 9.44	141.95
152	141.95	+ 10.05	+ 9.05	151.00
150	151.00	− 1.00	− 0.90	150.10
175	150.10	+ 24.90	+ 22.41	172.51
183	172.51	+ 10.49	+ 9.44	181.95
192	181.95	+ 10.05	+ 9.04	191.00

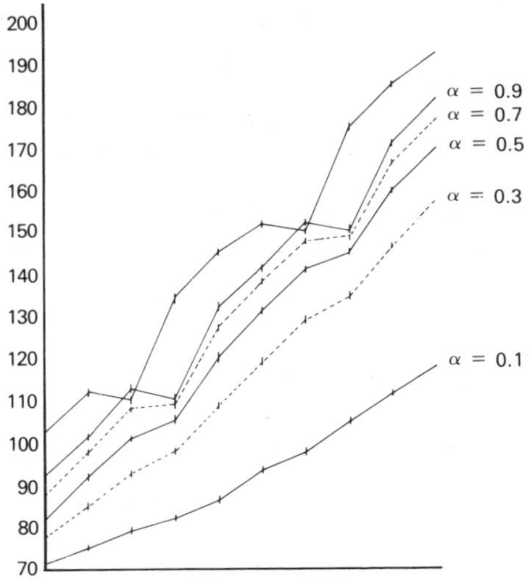

Diagram 7.05.

Now, as you can see from the graphs, if we take an alpha value of 0.9, the forecast we make reflects almost exactly what has been happening to the data in the previous year. Our forecast is highly sensitive, but in fact it reflects every minor variation in the actual data whether that fluctuation is purely random or not. At the other extreme an alpha value of 0.1 does not reflect any of the fluctuations in the original data. So true is this that an alpha value as low as this does not seem to give us a forecast reflecting even the trend. Our forecast figures are falling further and further below the actual data. We have a great deal of stability but no sensitivity. It would seem, from the graphs that we have drawn, that an alpha value of 0.3 would be the best to use. Our forecast will reflect adequately the generally rising trend of the figures, but is does not seem to be influenced by short term random fluctuations in the data. This conclusion, in fact, seems to reflect practice. The normal value of alpha used in this type of forecasting is very very rarely greater than 0.5 while a value of 0.3 is quite usual.

One final point is important. You will doubtless be worried by the fact that the forecasts we have made are *all* well below the actual data in every time period. This has enabled us to separate the graphs we have drawn for the sake of clarity. If you look carefully you will see that the low forecast springs in every case from the fact that our first forecast figure has been assumed to be well below the actual data. So long as there is an upward trend no value of alpha will ever enable us to catch up. In practice it is crucial, then, that if your forecast is to have any credibility the initial forecast must be compatible with the actual data. If it is not, every other forecast you make will reflect the same inaccuracy.

We must not end this chapter leaving you with the impression that forecasting is merely a simple arithmetic calculation. It is an art. Always look at your forecast and ask if it seems logical; if it fits the pattern of the series. If you feel it does not, obey your instinct and look at the trend again. The best forecasters are often those who get a gut feeling as to how the trend will behave a year or two ahead. Above all never place too much reliance on figures forecast for more than a short period ahead. You should realise that there are so many factors affecting the pattern of the time series, that it is unlikely that they will all remain constant or behave normally for any longer than the immediate future. Fortunately it is only rarely that you will be interested in many years ahead. In preparing budgets, in forecasting sales, in assessing how costs will behave, the most important period is — the very next year.

Exercises to Chapter 7
(Use additive model unless stated otherwise)

7.1 The number of prescriptions dispensed by chemists under the National Health Service in England and Wales during a five year period is shown in the table below.

N.H.S. Prescriptions (millions)

Year	Quarters			
	1	2.	3	4
1			60	71
2	69	67	62	69
3	73	66	62	68
4	72	66	65	67
5	75			

(a) Plot the data on a graph.
(b) Calculate and plot on the same graph a suitable moving average.
(c) Explain what additional calculations would be necessary to obtain the average seasonal variation. (Do *not* make these calculations).

7.2

Percentage Turnover of Labour Force

Year	Quarter			
	1	2	3	4
19-0	11	23	16	8
19-1	14	29	16	9
19-2	18	34	17	9
19-3	19	42	23	12

(a) Smooth this time series by means of a centred four-quarterly moving average.

(b) Calculate the average seasonal variations. (Use the multiplicative model).

(c) Explain how your calculations could be used as a basis for forecasting percentage turnover figures for the four quarters of 19-4.

7.3

Quarterly Sales (£000)

Year	Quarter			
	1	2	3	4
19-0	19	30	63	8
19-1	20	32	67	16
19-2	23	35	74	14

Calculate the average seasonal variation for each quarter of the year and, using the second quarter as an example, state what is meant by your calculated figure.

7.4

Number of Houses completed in England and Wales

Year	Quarter			
	1	2	3	4
19-0	32458	32881	35049	39462
19-1	33310	31953	33628	35983
19-2	30808	27935	27526	30946
19-3	28674			

(a) Rewrite the number of houses completed to the nearest hundred.

(b) Using the rounded data:
   (i) calculate, by means of four quarterly moving averages, the trend of the series;
   (ii) calculate the seasonal variations.

(c) State the main factors affecting a time series.

7.5 The following data gives the index numbers for cost of labour for a small firm during the period 19-5–19-8. By calculating an appropriate moving average obtain the trend and hence the seasonal factors for each quarter. Explain how to obtain a forecast of the index for the fourth quarter of 19-8.

Year	Quarters			
	1	2	3	4
19-5				103.5
19-6	102.2	101.2	95.1	103.5
19-7	99.1	103.8	96.8	109.0
19-8	112.1	110.5	105.2	

7.6

Supplies and Deliveries of Crude Steel
(Thousand Tons)

Year	Quarters			
	1	2	3	4
19-5	480	520	466	526
19-6	541	541	440	510
19-7	538	544	493	531

(a) Calculate the average seasonal variation for this data.
(b) Deseasonalise the series.
(c) Briefly comment on your findings.

7.7

Year	Quarters			
	1	2	3	4
19-0		110	120	140
19-1	110	124	130	154
19-2	123	134	143	176
19-3	139			

The above data gives the index numbers for the total value of sales in catering in Great Britain for the year 19-0 to 19-3. By calculating an appropriate moving average obtain the trend and hence the seasonal factors for each quarter. Deseasonalise the quarters of 19-2.

7.8

Days lost due to Industrial Accidents

	Quarters			
	1	2	3	4
19-4	152	204	134	213
19-5	187	198	311	299
19-6	210	257	231	323
19-7	201	225	236	318
19-8	180			

From the data given in the table calculate the average seasonal variation in days lost in the years 19-4 to 19-7 inclusive. Hence predict the number of days lost in the first quarter of 19-8. For what reasons might your prediction differ from the actual number of days lost?

C.I.P.F.A.

7.9

	Quarters			
	1	2	3	4
19-2	58	85	97	73
19-3	64	96	107	89
19-4	76	102	115	94

The above table shows the number of visitors (in hundreds) to an hotel during a period of three years. By using the method of moving averages rewrite the series without seasonal variations. Estimate as accurately as you can the quarterly figures for 19-5 and say how reliable you think your estimates are. (Use the multiplicative model).

7.10 The following table gives the sales (in hundreds) of components manufactured by Company A.

Quarterly Sales in Hundreds

	Quarter			
	1	2	3	4
19-2				13
19-3	14	16	9	14
19-4	16	17	12	17
19-5	18	20	13	

Calculate the seasonal fluctuations and rewrite the series with the seasonal variations removed. Draw a graph of the trend values and project it into 19-6. Hence, or otherwise, calculate the expected sales for the next four quarters. (Use the multiplicative model).

7.11 Quarterly production at a paper making plant was reported as follows:

Quarterly Production (000 tonnes)

	1	2	3	4
19-2	38.8	41.3	39.0	45.6
19-3	44.7	45.2	42.0	49.9
19-4	46.7	48.2	44.5	51.3
19-5	50.1	54.6		

Using the method of moving averages, find the average seasonal deviations, and thus estimate the production figures for the last two quarters of 19-5.

7.12 What do you understand by adaptive forecasting? Give an example of its use.

7.13 Use the exponential smoothing technique on the data in 11.10. Take alpha value of 0.3, 0.5 and 0.7. Which value of alpha would you recommend?

7.14 Suppose the trend line for the data in 11.9 is given by the equation Trend = 70 + 3t. Use this equation to predict the quarterly figures for 19-5 using both the additive and multiplicative models.

Chapter Eight

Index Numbers

In Chapter 5, we explored at some depth the meaning of the word average, and we were at pains to point out that an average is not necessarily representative of the data it describes. In this chapter, we are going to concern ourselves with averages at work — especially in an economic context. Make no mistake about it, averages do play a very important part in our lives — especially those that try to describe economic data. Now statisticians have constructed a device which attempts to measure the magnitude of economic changes over time — a device called an *index number*. This device is also used for international comparisons of economic data. Almost certainly, the Index of Retail Prices is the most widely quoted of all index numbers. Let us now examine just how important this index is to all of us.

You will realise that the Index of Retail Prices attempts to measure the change in the price of a whole range of goods and services that we regularly buy. So you can see that it is attempting to measure the cost of living — something that vitally concerns us all. In these times of inflation, the retail price index is probably more important than at any other time in its existence. The United Kingdom government has introduced measures in an attempt to reduce our rate of inflation to single figures, and this index will be used as a yardstick of its success. However, its importance does not stop here. Increases in the cost of living have been accepted as justification for an increase in pay, and the Index of Retail Prices has often been used as a basis for wage negotiations. In fact, a few years ago many million workers obtained an automatic 40p rise per week for each one-point rise in the cost of living index — the so-called threshold payments. Due to government pressure, it is no longer fashionable to link wage rates to the cost of living directly, but any incomes policy that ignores the index altogether seems doomed to failure. As further examples of the importance of this index we could note that the government have introduced an index-linked issue of savings certificates, the value of which will rise with the cost of living. Also, it is government policy to make pensions index-linked. If we are willing to accept index linking as desirable, we should be sure that we know what an index is, how it is calculated and what its limitations are.

Without doubt, an index is an extremely fashionable statistical tool. Open any copy of the Annual Abstract of Statistics and you are sure to be impressed by the number of indexes (or indices, if you prefer this form of the plural) that are calculated by the government statisticians. In fact, no self-respecting government department would fail to produce at least one index! We do not wish you to get the impression that all index number are generated by government departments — the Financial Times Ordinary Share Index, and the Economist's Key Indicators are notable indexes generated by the private sector. With the great number of indexes available, and the fact that they are used as a basis for taking decisions of national importance, it is worrying to be told that some statisticians doubt their validity. Now we cannot subscribe to this view. Index numbers are an invaluable tool in the decision-taking process — provided that they are treated with caution!

An Expenditure Index

An index number is merely a device to measure the change in some economic variable over time. One particular time period is chosen (called the *base* period) and the variable for that period is given an arbitrary value of 100. An *index* is then calculated for the remaining periods on the assumption that the base period has a value of 100. An index, then, gives the percentage change that has occured since that base period.

Example 1:

Al Coholic throws a (rather unusual) party each Christmas for his friends. Details of his expenditure on food and drink are as follows

	1978		1979	
	price	quantity	price	quantity
	p_0	q_0	p_n	q_n
Wine (per bottle)	£1.00	40	£1.15	50
Pork pies (each)	£0.20	100	£0.27	90
Christmas Cake (each)	£2.00	1	£2.20	1

In 1978, Al spent £1.00 × 40 + £0.20 × 100 + £2.00 × 1 = £62.00 and in 1979 he spent £1.15 × 50 + £0.27 × 90 + £2.20 × 1 = £84.00. If we call expenditure in 1978 100, then expenditure in 1979 would be

$$\frac{84}{62} \times 100 = 135.4$$

which shows that his expenditure in 1979 was 35.4% higher than in 1978. As you might expect, the index we have calculated is called an *expenditure index*. If we call base year prices and quantities p_0 and q_0, and current year prices and quantities p_n and q_n, then we can calculate an expenditure index like this:

$$\frac{\Sigma p_n q_n}{\Sigma p_0 q_0} \times 100$$

A Price Index

Now one reason for the change in expenditure incurred by Al Coholic is the change in the price of the goods he has bought. It would be interesting to construct an index number to measure the extent of the change in price. Now expenditure is made up of two components, price and quantities bought, and any change in either (or both) of these components will cause a change in expenditure. So it must follow that if we hold quantities constant, then any change in expenditure must be due to price changes.

We know that the total expenditure incurred by Al in 1978 was
$$1.00 \times 40 + 0.20 \times 100 + 2.00 \times 1 = £62$$
If Al had bought the same quantities in 1979, then his total expenditure would have been:
$$1.15 \times 40 + 0.27 \times 100 + 2.20 \times 1 = £75.20$$
As we have held the quantities constant, the change in outlay must be entirely due to the change in price. If we call the price index for 1978 100, then the index for 1979 is

$$\frac{75.20}{62} \times 100 = 121.3,$$

i.e. prices have increased by 21.3%.

Using the same symbols as before, we can calculate this index by taking

$$\frac{\Sigma p_n q_0}{\Sigma p_0 q_0} \times 100$$

It shows what the cost of goods in the nth year would be, assuming that we bought the same quantities as in the base year, and assuming that we call the base year price 100. An index calculated in this way is called *Laspeyre's Index*.

Now you may well complain that we have made a bit of a meal of calculating a price index. We could argue that the cost of 'one of each' in 1978 was £1.00 + £0.20 + £2 = £3.20, and the cost of 'one of each' in 1979 was £1.15 + £0.27 + £2.20 = £3.62. With 1978 as base year, the price index for 1979 is

$$\frac{3.62}{3.20} \times 100 = 113.1$$

An index calculated in this way is called a *simple aggregative index* and it has two major faults. Firstly, the value of the index will depend upon the pricing unit used — if we had quoted the price of wine as '£12 per case of twelve' for 1978 and '£13.80 per case of twelve' for 1979, we would have obtained a very different answer — even though we have not changed the price per bottle. The second drawback of this method is that it assumes that the items bought have the same significance to us, as it takes no account of the quantities bought. Surely, a 50% increase in the price of wine will have more significance to Al than a 50% increase in the price of Christmas cake? After all, he buys 40 bottles of wine, but only one Christmas cake! Fortunately, Laspeyre's Index suffers from neither of these defects. By taking into account the quantities bought, it recognises that some items bought have more significance to us than others. Using the statistician's jargon, Laspeyre's Index gives a *weighting* to the individual items to show their relative importance, using the quantities bought in the base year as weights.

Price Relatives

Before we judge the efficiency of Laspeyre's Index we should note that in practice it is not usually calculated in the way we have done — it is more usual to use *price relatives*. Price relatives refer to individual items in the index, and simply show the change in price per item since the base year. So the price relatives are calculated like this

$$\frac{p_n}{p_0} \times 100$$

The price relatives for the items bought by Al Coholic would be

	Price in 1978 p_0	Price in 1979 p_n	Price Relative
Wine	£1.00	£1.15	115.0
Pork Pies	£0.20	£0.27	135.0
Christmas Cake	£2.00	£2.20	110.0

To calculate the price index, we must now weight the price relatives, and the appropriate weights for Laspreyre's Index is expenditure in the base year.

	1978 price p_0	quantity q_0	expenditure (weight)	Relative	Weight × PR
Wine	£1.00	40	40	115.0	4600
Pork Pies	£0.20	100	20	135.0	2700
Christmas Cake	£2.00	1	2	110.0	220
			62		7520

The price index is now found by taking

$$\frac{\Sigma(\text{price relative} \times \text{weight})}{\Sigma(\text{weights})} \times 100$$

$$= \frac{7520}{62} = 121.3$$

So we have a choice of two methods for calculating Laspeyre's Index. If we wish to use base year quantities as weights, then we must use actual prices to calculate our price index. However, if we use base year expenditures as weights, then we use price relatives to calculate the index. As stated earlier, it is more usual in practice to use expenditure as weights. The reason is quite simply that it is easier to obtain data on expenditure (cost of living weights are obtained by the Survey of Household Expenditure — that is by sampling). Also, we can consider occasions when 'quantities' wouldn't make sense — how can one define 'quantities' of public transport? However, it would be quite possible to obtain details of personal expenditure on public transport.

Let us summarise what we have done so far. We have calculated an index called Laspeyre's Index, which tells us what we would have paid in year n for a collection of goods assuming we bought base year quantities. This description is true whether we use expenditure or quantities as weights. The point is that Laspeyre's Index uses base year statistics as weights. Now many statisticians doubt the validity of taking base year weights, as it implies that the quantities we buy do not vary over time. In many cases this will not be true. Suppose we have cost of living index based on 1970, and one of the items included is potatoes. The index for 1976 would assume that we bought the same quantity of potatoes as in 1970. But the period 1975-76 saw a dramatic increase in the price of potatoes, and one consequence of this was a fall in demand. Without doubt, people consumed considerably fewer potatoes and turned to substitutes, so because of this fall in quantity purchased the Laspeyre index tends to *overstate* increases in price. Another problem is that tastes tend to change fairly markedly over time, so weights that were appropriate for 1970 will not be appropriate for today. For example, few colour television sets were bought in 1970 in preference to monochrome, but today this situation is completely reversed.

In an attempt to overcome these problems, many statisticians have suggested that current year quantities should be used as weights. Our index now becomes

$$\frac{\Sigma p_n q_n}{\Sigma p_0 q_n} \times 100$$

and such an index is called a *Paasche index*. The approach here is quite different: we ask what would be the total outlay in the base year if we bought current year quantities, and compare this with current year outlay. Paasche's Price Index for the data supplied by Al Coholic is

	1978	1979			
	p_0	p_n	q_n	$p_0 q_n$	$p_n q_n$
Wine	£1.00	£1.15	50	50	57.5
Pork Pies	£0.20	£0.27	90	18	24.3
Christmas Cake	£2.00	£2.20	1	2	2.2
				70	84.0

$$\text{Paasche's Price Index} = \frac{84}{70} \times 100 = 120$$

This certainly overcomes the main objections to the Laspeyre Index, but does tend to raise problems of its own. Firstly, the Paasche index is not a pure price index as it also takes into account changes in quantities bought. Just, then, as the Laspeyre index tends to overstate the effect of rising prices, the Paasche index tends to understate the effects. Secondly, it can be a long and expensive job calculating the values for weights with a Paasche index. This must be done for every period, while with a Laspeyre index this is done for the base year only. The current cost of living index is a compromise between the two: the index is published monthly and the weights are adjusted annually by the Survey of Consumer Expenditure.

Volume Index

So far, we have assumed that the change in expenditure incurred by Al Coholic has been due to price changes, and we have attempted to measure the extent of the price change with a price index. However, the change in expenditure could be due to changes in quantities bought, and we can measure the extent of these changes by a *volume index*. We have been comparing Al Coholic's expenditure at two moments of time. If we keep the quantities bought the same, then any variation in outlay must be due to changes in price, so we can calculate a price index. If, however, we keep the prices the same, then any variation in outlay must be due to changes in quantities bought, so we can calculate a volume index. If, then, we use quantities as weights we obtain a price index, but if we use prices as weights then we obtain a volume index. Again, we have the choice between current year and base year weighting, so we have

$$\text{Laspeyre's Volume Index} \quad \frac{\Sigma p_0 q_n}{\Sigma p_0 q_0} \times 100$$

$$\text{Paasche's Volume Index} \quad \frac{\Sigma p_n q_n}{\Sigma p_n q_0} \times 100$$

Let us now calculate both indexes for the data supplied by Al Coholic.

	1978 Price	Quantity	1979 Price	Quantity				
	p_0	q_0	p_n	q_n	$p_0 q_0$	$p_0 q_n$	$p_n q_0$	$p_n q_n$
Wine	1.00	40	1.15	50	40	50	46.0	57.5
Pork Pies	0.20	100	0.27	90	20	18	27.0	24.3
Christmas Cake	2.00	1	2.20	1	2	2	2.2	2.2
					62	70	75.2	84.0

$$\text{Laspeyre's Volume Index} = \frac{70}{62} \times 100 = 112.9$$

$$\text{Paasche's Volume Index} = \frac{84.0}{75.2} \times 100 = 111.7$$

Selecting a Suitable Base Period

One of the main problems in constructing index numbers is the choice of a suitable base period. We wish really to choose a base when the price is as 'normal' as possible, i.e. when the price is not unduly high or unduly low. Otherwise, the index will move away from the base figure too quickly and show very large deviations from it. Suppose, for example, the price of a particular stock at certain periods of time was

April	125p	(takeover rumour)
July	69p	(takeover unsuccessful, poor dividends announced)
Dec.	95p	(quite good dividends forecast)

then the price index for this particular stock could be

April	=	100.	July	=	100	Dec.	=	100
April		100			181.2			131.6
July		55.2			100			72.6
Dec.		76.			137.7			100.

The best base to choose is probably December, because this minimises the greatest deviation from the base (31.6). Does this matter? Well, there is evidence to show that people are more likely to understand and appreciate smaller percentage changes than larger ones. This is one of the reasons that statisticians tend to update the bases that they use. Another reason is that bases in the not-too-distant past tend to be much more meaningful to the users of the index. It would be much more reasonable to compare prices now with prices ten years ago than it would be to compare them with prices in 1949.

This problem of choosing a suitable base can also be illustrated if we are constructing an index to measure the volume of production. If we take as our base a month in which there is a major strike, then the index in the following months would be bound to show a substantial increase and give a misleading impression of the prosperity of the industry. If we were constructing an index of motor-car sales, which do you think would be the *worst* month to choose as a base?

An alternative to up-dating the base period is to use instead a *chain-based system*, where the base used is the previous period. This method has found particular favour in the United States. To illustrate the difference between the chain-based and fixed base system, consider the following example which refers to the average price of a certain security over six months.

Month	Price(p)	Price index (chain-based, previous month = 100)	Price index (fixed base, August = 100)
August	155		$\frac{155}{155} \times 100 = 100$
September	143	$\frac{143}{155} \times 100 = 92.3$	$\frac{143}{155} \times 100 = 92.3$
October	144	$\frac{144}{143} \times 100 = 100.7$	$\frac{144}{155} \times 100 = 92.9$
November	139	$\frac{139}{144} \times 100 = 96.5$	$\frac{139}{155} \times 100 = 89.7$
December	140	$\frac{140}{139} \times 100 = 100.7$	$\frac{140}{155} \times 100 = 90.3$
January	131	$\frac{131}{140} \times 100 = 93.6$	$\frac{131}{155} \times 100 = 84.5$

Consider the price in (say) January. Using the chain base numbers, we can see that the price was 6.4% lower than the previous month and using the fixed base numbers we can see that the price was 15.5% lower than in August. We can see, then, that chain based index numbers are particularly suited for period-by-period comparisons, but if we are to compare the movement of prices over time then the fixed based indexes are much easier to interpret.

An 'Ideal' Index

Earlier, we stated that Laspeyre's Index tends to overstate and Paasche's Index tends to understate changes in prices or quantities. So we would expect Laspeyre's Index to exceed Paasche's Index, and if we examine the indexes of Al Coholic's expenditure, we find that this is indeed the case.

	Volume	Price	Expenditure
Laspeyre's	112.9	121.3	135.4
Paasche's	111.7	120.0	

However, you must not assume that Laspeyre's Index will always be greater than Paasche's Index. (In fact, you should consider what the implications would be if Paasche's Index is greater than Laspeyre's Index). What is reasonable, however, is to conclude that the 'true' index lies somewhere between Laspeyre's Index and Paasche's Index, and Fisher has suggested that it is equivalent to the geometric mean of the two i.e.

Fisher's Index = $\sqrt{\text{(Laspeyre's Index)} \times \text{(Paasche's Index)}}$.

Fisher's Price Index = $\sqrt{121.3 \times 120} = 120.6$

Fisher's Volume Index = $\sqrt{112.9 \times 111.7} = 112.3$

It is claimed that Fisher's is an 'ideal' index and it now remains to state in what sense it is ideal. At the begining of this chapter, we stated that expenditure was made up of prices and quantities, so if we have a price index and a quantity index we should be able to correctly predict the expenditure index. Remembering that expenditure is price times quantities, the prediction for the expenditure index is

(a) Using Laspeyre's data,

$$\frac{112.9 \times 121.3}{100} = 136.9$$

(which overstates the expenditure index)

(b) Using Paasche's data,

$$\frac{111.7 \times 120.0}{100} = 134.0$$

(which understates the expenditure index)

(c) Using Fisher's data,

$$\frac{120.6 \times 112.3}{100} = 135.4$$

Fisher's Index numbers are ideal in the sense that they are the only ones that correctly predict the expenditure index.

Using an Index for 'Deflating' a Series

It is a well-known fact that wages have risen dramatically since 1960, and standards of living have risen too. However, as we all know, price rises can erode increases in earnings, and economists have coined the phrase 'real wages' which shows how incomes have changed, taking price changes into account. Let us examine the relevant statistics.

Year	Average weekly earnings of Manual Workers in Manufacturing Industries (£)	Retail Price Index
1960	15.16	114.5
1961	15.89	117.5
1962	16.34	100.0
1963	17.29	103.6
1964	18.66	107.0
1965	20.16	112.1
1966	20.78	116.5
1967	21.89	119.4
1968	23.62	125.0
1969	25.54	131.8
1970	28.91	140.2
1971	31.37	153.4
1972	36.20	164.3
1973	41.52	179.4
1974	49.12	191.8

Source: Annual Abstract of Statistics.

The first problem that we notice is that the price index has undergone a change of base. Earlier figures have 1956 as base, while later figures have 1962 as base. We will need to obtain a price index related to a single base, and it is usual to take the most recent base as the base for the entire series. So we will want to recalculate price indexes for 1960 and 1961 with 1962 as base. Now when a base changes, it is usual to calculate a price index for a few periods using both the new and the old base. Consulting the Annual Abstract, we see that for 1962 the price index using the old base was 119.3, so we have:

$$\text{Price index for 1960} = 114.5 \times \frac{100}{119.3} = 95.98$$

$$\text{Price index for 1961} = 117.5 \times \frac{100}{119.3} = 98.49$$

We can now use the retail price index to deflate actual earnings and obtain real wages (taking into account price rises since 1962) For example, average wage for 1970 at 1962 prices is

$$\frac{28.91}{140.2} \times 100 = £20.62$$

So actual and real wages over the period would look like this:

Year	Average Weekly Earnings (£)	Real Income
1960	£15.16	£15.79
1961	£15.89	£16.13
1962	£16.34	£16.34
1963	£17.29	£16.69
1964	£18.66	£17.44
1965	£20.16	£17.98
1966	£20.78	£17.84
1967	£21.89	£18.33
1968	£23.62	£18.90
1969	£25.54	£19.38
1970	£28.91	£20.62
1971	£31.37	£20.45
1972	£36.20	£22.03
1973	£41.52	£23.14
1974	£49.12	£25.61

When actual weekly earnings are deflated, a very different picture emerges; though without doubt living standards have risen. Notice that in all years except 1971 wages were rising faster than prices. We could if we wish calculate an index of real wages with 1962 (or any other year for that matter) as base by dividing current year real wages by 16.34 and multiplying by 100. So the index for 1970 would be:

$$\frac{20.62}{16.34} \times 100 = 126.2$$

Exercises to Chapter 8

8.1 What considerations must be borne in mind when an index number is compiled? You should illustrate your answer by reference to any index number with which you are familiar.

8.2 What is the purpose of an index number? Describe the main methods for constructing index numbers, indicating the advantages and disadvantages of each. Illustrate one of the methods by reference to any index number in current use.   C.I.P.F.A.

8.3 Why are weights used in the construction of index numbers? What is meant by:
(a) base weighting;
(b) current weighting;
and what are the advantages and disadvantages of each?   I.C.M.A.

8.4 Define the following kinds of index numbers:
(a) fixed base;
(b) chain base;
(c) base weighted;
(d) current weighted.
Give an example of any published index with which you are familiar, briefly indicating how it is compiled and what the information it gives is intended to show.

8.5 A company has reached an agreement with representatives of its employees that wages and salaries will in future be tied to a local cost of living index which the company will compile. Advise the company how they should gather information and compile this local cost of living index. What are the problems the company will encounter in completing this exercise?

8.6 (a) Explain briefly the principles of index number construction.

(b) A certain company uses approximately 6,2,5,3 thousand units of raw materials, A,B,C,D respectively. The average price (£) per unit of the raw materials in 1970 and 1973 is given in the following table:

	Average Price in 1970	Average Price in 1973
Raw material A	10	15
B	21	39
C	2	3
D	46	101

Calculate the Simple Aggregative Index for the year 1973 with 1970 as base year. Calculate also a Weighted Aggregative Index for the period 1970-3 and explain why this may be considered preferable to the Simple Aggregative Index.

8.7 The average prices of four commodities for the years 1979, 1980 and 1981 are shown in the following table:

Commodity	Average Price per Unit (£)		
	1979	1980	1981
A	8	15	14
B	54	70	72
C	15	22	29
D	75	84	78

(i) Calculate the *Simple Aggregative Index* for each of the years 1980 and 1981 with 1979 as base year.

(ii) Write down *two* main disadvantages in the use of the Simple Aggregative Index.

(iii) The number of units used annually by a certain company is approximately 5000, 1000, 3000 and 8000 for commodities. A, B, C and D respectively. Calculate a Weighted Aggregative Index for the period 1979-81 and explain to what extent this Index overcomes the disadvantages you mentioned in part (ii) of this question.

8.8 *Index Numbers of Wholesale and Retail Prices*

	Food Manufacturing Industries Price Indices		Food Index of Retail Prices (16/1/62 = 100)
	Materials and Fuel Used (1970 = 100)	Output (1970 = 100)	
1972			
January	111.4	112.2	163.9
February	110.1	111.8	165.1
March	110.3	112.0	166.0
April	108.9	111.4	164.6
May	109.8	111.9	166.3
June	109.4	112.5	169.2
July	110.1	113.2	169.2
August	113.0	115.4	172.3
September	113.1	115.9	172.4
October	114.8	116.8	172.8
November	118.6	117.6	174.3
December	125.4	118.5	176.9

Source: Monthly Digest of Statistics.

(a) Explain what is meant by (i) *index number* and (ii) *base year*.

(b) Using January 1972 as the base period reduce the above figures to a comparable basis.

(c) Comment briefly on the results in (b).

8.9 (a) Describe the main features of the Index of Industrial Production.
(b) The following table shows the U.S. consumption (millions of pounds) and price (dollars per pound) of vegetable oil products.

Type of Oil	1963		1964		1965	
	Quantity	Price	Quantity	Price	Quantity	Price
Soybean	322	0.13	368	0.12	367	0.13
Cotton seed	96	0.15	114	0.14	123	0.14
Linseed	32	0.13	31	0.13	19	0.12

Required: Taking 1963 as the base year, calculate Laspeyre's base weighted index numbers for the general level of prices of these products for 1964 and 1965.   A.C.A.

8.10 (a) In relation to index numbers, explain what is meant by
  (i) base year,
  (ii) weights.
(b) Calculate the missing value x, the weighting for housing, and y, the weighted arithmetic mean, the index for all items in the table below.

Index of Retail Prices for 1962 = 100

Group	Weights (1972)	Index (Sept. 1972)
Food	251	172
Alcoholic drink and tobacco	119	153
Housing	x	192
Fuel and light	60	173
Durable household goods	58	141
Clothing and footwear	89	144
Transport and vehicles	139	159
Other	163	177
All items	1000	y

Source: Monthly Digest of Statistics.

8.11 (a) What is meant by the term 'index number'?
(b) *Index Numbers of Retail Prices in the U.K.*

Group	Index	Weight
Food	104	319
Housing	110	104
Durable household goods	100	64
Miscellaneous goods	108	63

(i) Using a weighted arithmetic mean calculate an index of retail prices for the group combining all four of these items.
(ii) If the index of retail prices for all household items (total weight = 1000) was 104, calculate the index for all household goods.

8.12 Recalculate Laspeyre's index numbers for the data in question 8.9 but this time use price relatives and make 1964 the base year.

8.13 A steel stockist notices that prices and values of sales for the main units of steel supply were:

	19-4		19-9	
	Price per tonne £	Sales £m	Price per tonne £	Sales £m
Ingots	162	324	200	600
Steel bars	188	564	190	760
Steel strip	220	880	275	1100

Calculate a Paasche index number for 19-9 prices, taking 19-4 = 100

Recalculate the Paasche price index using price relatives and the appropriate weights.

I.C.S.A.

8.14 A company employs three grades of male direct operators, M1, M2 and M3, and three grades of direct female operators, F1, F2, and F3. The following table gives the number of operators employed and the rates paid.

Labour Grade	19-1		19-3	
	rate per hour	No. of Ops.	rate per hour	No. of Ops.
M1	0.66	32	0.80	35
M2	0.62	14	0.74	10
M3	0.56	16	0.66	12
F1	0.44	40	0.64	45
F2	0.41	18	0.58	16
F3	0.36	25	0.52	30

With 19-1 as base calculate an index number for the average wage in 19-3 using

(a) base weighting, and
(b) current weighting

Comment on the relationship between the two indices.

8.15 The prices and sales by value of four commodities are shown below for 19-8 and 19-2. Calculate Laspeyre's and Paasche's index numbers for 19-2 taking 19-8 = 100. Would you expect the two answers to have the same value?

Commodity	19-8		19-2	
	Price £	Sales by value £m	Price £	Sales by value £m
A	2.00	8.0	3.00	12.0
B	1.00	6.5	1.50	9.0
C	4.50	9.0	5.50	11.0
D	7.00	14.0	8.50	17.0

I.C.S.A.

8.16 The following data shows the price index numbers of three groups of commodities for five years. Groups 1 and 2 have been prepared on the fixed base principle, but group 3 numbers are chain based and unlinked.

Year	1	2	3	4	5
Group 1	100	106	113	122	128
Group 2	84	94	100	108	114
Group 3	100	102	104	101	103

(a) Recalculate the Group 2 numbers to make them more easily comparable with the group 1 numbers.
(b) Convert the group 3 numbers into a series linked to the index number of 100 for year 1.
(c) Tabulate the figures for group 1 and the revised figures for Groups 2 and 3 and briefly comment on the situation revealed.

8.17 Imports Through Certain U.K. ports

	Declared Value £ million	Value at 19-0 prices £ million
19-0	522	522
19-1	430	533
19-2	351	469

Construct index numbers for price and volume with 19-0 as base year.

8.18

	Laspeyre's	Paasche's	Fisher's
Price Index	—	—	107.4
Volume Index	—	108.5	—

Find the missing data given that the Expenditure Index is 115.7, and given that base year expenditure at base year prices is £58,430 million.

8.19

Year	Profit (£000)	Wholesale price index
1970	1132	112.3
1971	1245	120.7
1972	1345	129.3
1973	1412	100.0
1974	1582	110.4
1975	1898	127.1
1976	2182	147.4
1977	2488	162.1
1978	2736	175.1
1979	2955	183.9

Before 1973, the wholesale price index had 1965 as its base year. Index for 1973 with 1965 = 100 is 136.5.

(a) Use the index of wholesale prices to deflate profits, and so obtain profits at 1973 prices.
(b) Find the index numbers of 'real profits' for 1970-1979 with 1973 = 100.

8.20 (a) From the information stated below construct a quantity index for the products made by Multiproducts Ltd. for the period 1966-9, weighted as to 1966 prices, and
(b) explain the purpose of preparing a quantity index, state what the indices calculated in (a) above indicate about the production of Multiproducts Ltd. for the years 1966-9, and discuss the influence of individual products on the index.

Multiproducts Ltd.

Product	1966 Average Price £	1966 Production (000's)	1967 Production (000's)	1968 Production (000's)	1969 Production (000's)
Pliers	2.00	62	65	66	90
Wrenches	3.00	138	120	110	80
Bolts	0.50	500	540	580	800
Drills	4.50	10	10	10	10

A.C.A.

8.21 Given the following data, calculate a measure of the increase in physical output in the U.K. of motor vehicles (passenger and commercial vehicles, taken together) in 1960 compared with 1958. For what reason may your measure be a poor guide to the actual change in the output level?

U.K. Motor Industry — Production and Deliveries

	Monthly Averages		Value of Deliveries	
	No. of Passenger Cars Produced 000s	No. of Commercial Vehicles Produced '000s	Cars £m	Commerical Vehicles £m
1958	87.6	26.1	34.4	16.3
1959	99.2	30.9	40.7	20.1
1960	112.7	38.2	44.3	23.7

Source: Monthly Digest of Statistics.

C.I.P.F.A.

8.22 *Index of Industrial Production (Average 1963 = 100)*

	Weight	Index 1969
Mining and quarrying	56	80.2
Total manufacturing	749	125.6
Construction	127	118.9
Gas, electricity, water	68	136.2
Total	1000	

Source: Annual Abstract of Statistics.

(a) Explain the purpose of weights.
(b) Calculate the index of industrial production for 1969.
(c) If the index for construction rises to 140 and the other data remains the same, find the change in the index of industrial production.
(d) Calculate to the nearest integer an index for 1963 taking the value from (b) as base.

8.23   *The A.B. Co. Ltd. Indices of Production (1960 = 100)*

		Indices	
Dept.	Weights	1962	1964
A	15	108	116
B	26	122	112
C	4	92	106
D	35	134	130

(a) Calculate for each of the two years 1962 and 1964 an index number of production for the whole firm.

(b) Calculate index numbers for each department and for the whole firm for 1964 taking 1962 as the base year.

8.24 (a) Define an Index Number.

(b)   *Exports of Selected Commodities*

	Quantities (tons)		Value (£)
	1955	1960	1955
Tiles	104,150	139,350	476,200
Sanitary ware	212,350	120,200	1,286,800
China	13,210	13,820	848,800
Electrical ware	23,740	17,720	302,400
Other earthenware	250,450	213,520	3,886,450

Calculate, for this group of exports, an index number of volume for 1960 (1955 = 100).

8.25 (a) What are the main problems to be faced in the construction of an index number?

(b) Compute a price and a quantity index number for 1974 from the following figures (1970 = 100).

Item	Quantity ('00 tons)		Cost (£'000s)	
	1970	1974	1970	1974
A	172	197	154	185
B	34	47	128	151
C	49	42	277	248
D	64	66	219	206

Chapter Nine

Dispersion

In Chapter 6 we found it desirable to summarise data with a measure of central tendency and the main reason we did this was to enable us to compare sets of data or frequency distributions. We spent quite some time deciding how such measures could be truly representative of the data, and discussed the suitability of using the arithmetic mean (or average), the median and the mode. However, it is not sufficient to summarise data with a measure of central tendency only, we also need a measure of dispersion. The problem with quoting a measure of central tendency only is that we can be led into drawing the wrong conclusions. Suppose, for example, we were examining income at two periods of time, and we find that there has been no change in average income. We might be tempted to conclude that economic welfare has not changed. However, although average income may not have changed, it is quite possible that welfare did increase through a more equal distribution of income. While our measure of central tendency will detect changes in average income, a measure of dispersion will detect changes in the spread of income. Clearly, both measures are desirable, so we shall now critically examine methods of measuring dispersion.

Measures of the Range

To measure dispersion, we could simply take the difference between the greatest and the least value in a series or distribution. We call this measure the *range*. To illustrate how the range is calculated, consider the data below which gives the average price of a certain security over a six month period.

	Average Price (p)
Jan.	155
Feb.	143
March	144
April	139
May	140
June	141

So the highest price during the period was 155p and the lowest price was 139p, giving a range of 16p. Nothing, then, could be more simple both in concept and calculation. However, the range does have two great faults. Firstly, it ignores the bulk of the data available to us, being concerned only with extreme values. The second disadvantage is that, being concerned with the extreme values only, the range may be quite unrepresentative of the spread of items, especially as the extreme values are probably quite untypical of the data. Suppose, for example, we were to use the range as a measure of dispersion of the heights of males in Liverpool. The range would be the difference between the tallest giant (say 6 foot 8 inches) and the smallest dwarf (say 4 foot 6 inches) — a range of 26 inches. However, we would probably find that by far the bulk of the population would be contained within the limits 5 foot 6 inches to 6 foot 2 inches — a range of 8 inches. Despite these criticisms, the range is used considerably — especially in statistical quality control. The reason is that in this field a measure of dispersion is often required quite quickly, and the range is eminently suited to this condition. However, it should be pointed out that, in these circumstances, the range will be found repeatedly from a large number of samples, and this does minimise the chances of obtaining untypical values at the extremes.

Because the range can be so misleading, statisticians have turned to alternative measures of dispersion. The one we shall now examine finds the range containing the central 50% of a distribution. Consider the data below which refers to the bonuses paid to the salesmen of a particular firm during a particular month.

Monthly bonus (£)	Number of salesmen	Cumulative Frequency
Under 60	8	8
60 and under 70	10	18
70 –	16	34
80 –	14	48
90 –	10	58
100 and under 110	5	63
110 and over	2	65
	65	

We want to find the range of bonuses earned by the central 50% of salesman. We can do this by finding the bonus earned by the $\frac{65+1}{4} = 16.5$th salesmen and the bonus earned by the $\frac{3(65+1)}{4} = 49.5$th salesman. Diagram 9.01 shows the relevant part of the ogive from which we can read that 16.5 salesmen earn a bonus of less than £68.50 and 49.5 salesmen earn a bonus of less than £91.50. So the range earned by the central 50% of salesmen is £91.50 — £68.50 = £23. We call this range the *interquartile range* (it is the range of bonuses earned by the central two quarters of the distribution). We call the lower end of this range (£68.50) the *lower quartile* (Q_1) and 25% of salesmen earn less than this figure. Likewise, the upper end of the range is called the *upper quartile* (Q_3) and 25% of salesmen earn more than this figure. Calling the lower quartile Q_1 and the upper quartile Q_3 would naturally lead us to call the median Q_2. Notice also that although the median *salesman* lies midway between the upper quartile salesman and the lower quartile salesman, the median *bonus* is not at the centre of the interquartile range. Beware of this — many students fall into this trap.

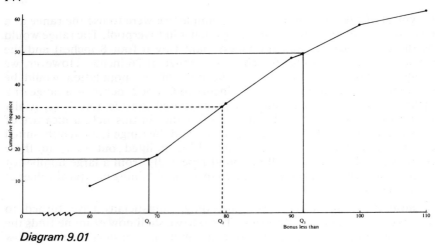
Diagram 9.01

Sometimes you will come across the *quartile deviation* — this is merely half the interquartile range, i.e.

$$\text{Quartile deviation} = \frac{Q_3 - Q_1}{2}$$

so the quartile deviation for monthly bonuses is $\frac{23}{2}$ = £11.50.

Notice that in this example, we could not have calculated the range for bonuses, as we have no way of knowing the highest bonus nor the lowest bonus earned. The range cannot be used for open-ended distributions. This disadvantage will probably not apply to the interquartile range, as it is unlikely that the central 50% of frequencies will penetrate the open ended class. However, the interquartile range does have the same disadvantage as the range — it uses only two values and ignores the rest. What we shall now examine are measures of dispersion that use all the data available to us.

Measures of Average Deviation

So far, we have examined two predetermined points in a series or distribution and calculated the difference between them. We have fixed the points either at the ends of the distribution (the range) or about the central 50% (the quartile deviation). We could of course, have fixed these two points anywhere we wished; for example, covering the central 40% or 60% of the items. But let us now consider dispersion in an entirely different way: we could choose some central value and calculate the deviation of the items from this value. What central value should we choose? Well, the arithmetic mean is one choice that automatically springs to mind. However, there is a snag with choosing the arithmetic mean — can you see what it is? Clearly, we are not going to be interested in the individual deviations themselves — we would still have a series of numbers rather than a single measure of dispersion. Surely, it would be more sensible to consider the average deviation from the mean. Now here's where the snag comes in. If you remember the definition of the arithmetic mean, then you will appreciate that the sum of the deviations from the mean is zero. Hence the average deviation from the mean must always be zero! Clearly, this will not do — we cannot use the average deviation from the mean as a measure of dispersion, because irrespective of the distribution under consideration our measure of dispersion will always be zero.

One way out of this problem is to find the *absolute deviations* from the mean, irrespective of whether they are positive or negative. Let us do this for the data on security prices considered earlier. The average price was

$$\frac{155 + 143 + 144 + 139 + 140 + 141}{6} = \frac{862}{6} = 143\tfrac{2}{3}\text{p}$$

and the deviations from this price are

Price (x)	Deviation x−x̄	Absolute Deviation
155	11⅓	11⅓
143	− ⅔	⅔
144	⅓	⅓
139	− 4⅔	4⅔
140	− 3⅔	3⅔
141	− 2⅔	2⅔
	0	23⅓

So the average absolute deviation from the mean is

$$\frac{23\tfrac{1}{3}}{6} = 3.89$$

The measure of dispersion we have just calculated is called the *mean absolute deviation*, and using the Σ notation we calculate it like this:

$$\text{Mean absolute deviation} = \frac{\Sigma |x - \bar{x}|}{N}$$

The vertical lines enclosing $x - \bar{x}$ simply means, "take the absolute deviations", whereas brackets would imply that we must take account of signs. Only a minor adjustment is needed to this formula if we are confronted with a frequency distribution.

$$\text{Mean absolute deviation} = \frac{\Sigma f |x - \bar{x}|}{\Sigma f}$$

where f represents the frequencies and x represents the centre point of the various classes.

The mean absolute deviation has a number of disadvantages. Because we have taken absolute deviations and ignored the signs, it is not capable of being manipulated mathematically in the ways that the statistician requires. A second disadvantage is that it will be quite cumbersome to calculate when the mean is not a whole number. Thirdly, we cannot combine a number of mean absolute deviations to obtain an overall measure of dispersion. Suppose, for example, a number of samples will be made available to us over a period of time, and we intend to use the samples to estimate dispersion in the population. We cannot calculate the mean deviation of each sample, and pool them, adjusting our estimate as more samples become available. We would have to pool all the samples, making one single sample. We do not wish, however, to leave you with the impression that the mean absolute deviation is of little use — far from it. One thing it cannot be criticised for is its ability to be a *representative* measure of dispersion. Probably no other measure of dispersion does this quite so well. The advantage of the mean absolute deviation is that conceptually it is so easy to understand. So what advice can we give you about when to use it? If you are interested merely in representing dispersion, then you can do no better than to use the mean absolute deviation, but if further statistical analysis is required we must look to other measures.

The main problems surrounding the mean absolute deviation arise because we took all the deviations as being positive in order to prevent the sum of the deviations being zero. Is there any other course of action open to us? Well, we could square all the deviations — the square of a negative number is a positive number, so the pluses and minuses will no longer cancel each other out. If we find the average of the square of the deviations, then we have a measure called the *variance*.

$$\text{Variance} = \frac{\Sigma(x - \bar{x})^2}{n}$$

So we would calculate the variance of security prices as follows:

Price (x)	$(x - \bar{x})$	$(x - \bar{x})^2$
155	11.33	128.44
143	−0.67	0.44
144	0.33	0.11
139	−4.67	21.78
140	−3.67	13.44
141	−2.67	7.11
		171.32

$$\text{Variance} = \frac{171.32}{6} = 28.55\text{p}$$

The trouble with the variance is that (in this example) it is measured in units of the square of the price deviations. Strictly speaking, the variance of security prices is not 28.55p but 28.55 (pence)². It would seem more sensible, then, to take the square root of the variance. We give the name *standard deviation* (σ)[1] to this measure of dispersion, and of all the measures this is the one that is most commonly used.

$$\sigma = \sqrt{\frac{\Sigma(x - \bar{x})^2}{n}}$$

So the standard deviation of the price of securities is $\sqrt{28.55} = 5.34\text{p}$.

1. This is the Greek small letter sigma. Do not confuse it with the capital letter sigma (Σ) which we met earlier.

It is quite easy so show that an alternative form of the standard deviation is

$$\sigma = \sqrt{\frac{\Sigma x^2}{n} - \left(\frac{\Sigma x}{n}\right)^2}$$

Notice that this expression does not ask us to calculate the deviations from the arithmetic mean — and this will be very useful when the mean is an awkward number. We do not intend to prove that the two expressions are the same (you can do this for yourself) but we shall demonstrate that both expressions give the same result by calculating the standard deviation of the price of securities this time using the second expression.

x	x^2
155	24,025
143	20,449
144	20,736
139	19,321
140	19,600
141	19,881
862	124,012

$$\sigma = \sqrt{\frac{124012}{6} - \left(\frac{862}{6}\right)^2} = 5.34p$$

If we wish to find the standard deviation of a frequency distribution, we simply replace x with fx and n with f.

$$\sigma = \sqrt{\frac{\Sigma fx^2}{\Sigma f} - \left(\frac{\Sigma fx}{\Sigma f}\right)^2}$$

where x is the centre point of the class and f is the frequency. Let us now look at an example.

ABC Ltd. Debtors' Balances at 31st December 19-9

Balance Outstanding (£)	Number of Accounts
20 and under 40	1
40 and under 60	3
60 and under 80	6
80 and under 100	10
100 and under 120	5
120 and under 140	3
140 and under 160	2
	30

We need first to find the centre point of each class.

Centre point (x)	f	fx	fx² = fx×x
30	1	30	900
50	3	150	7500
70	6	420	29400
90	10	900	81000
110	5	550	60500
130	3	390	50700
150	2	300	45000
	$\Sigma f = 30$	$\Sigma fx = 2740$	$\Sigma fx^2 = 275000$

$$\sigma = \sqrt{\frac{275{,}000}{30} - \left(\frac{2740}{30}\right)^2}$$

$$= £28.72$$

Notice that in constructing the table for calculating the standard deviation, we have also obtained all the information we need to calculate the arithmetic mean. It would make sense, then to calculate the mean as well as the standard deviation.

$$\bar{x} = \frac{\Sigma fx}{\Sigma f} = \frac{2740}{30} = £91.33$$

Unless you have a calculator at your disposal, it can be very tedious to calculate the standard deviation using this method. Fortunately we can simplify the calculations (you will remember that we did this when calculating the arithmetic mean) by taking an assumed mean — in this case we will take £90. We then calculate the deviation of the centre points x from the assumed mean (column 2). As all the deviations are divisible by 20, this is done in the column headed d. We can now use the column d to calculate the standard deviation.

x	(x – 90)	d	f	fd	fd²
30	– 60	– 3	1	– 3	9
50	– 40	– 2	3	– 6	12
70	– 20	– 1	6	– 6	6
90	0	0	10	0	0
110	20	1	5	5	5
130	40	2	3	6	12
150	60	3	2	6	18
			$\Sigma f = 30$	$\Sigma fd = 2$	$\Sigma fd^2 = 62$

Calculating the standard deviation as before, we have

$$\sigma = \sqrt{\frac{62}{30} - \left(\frac{2}{30}\right)^2}$$

$$= £1.436$$

However, we divided all the deviations by 20, so we must multiply the standard deviation above by 20

$$\sigma = £1.436 \times 20 = £28.72$$

which agrees exactly with our previous result. Let us see if we can write a formula which describes the above calculation. If d is the deviation from an assumed mean x_0, and if d is divided by a constant c, then

$$\sigma = c \times \sqrt{\frac{\Sigma fd^2}{\Sigma f} - \left(\frac{\Sigma fd}{\Sigma f}\right)^2}$$

Relative Dispersion

So far, we have been trying to measure dispersion within a series or distribution. We shall now attempt to measure dispersion between distributions, and we shall find that the standard deviation cannot do this. Consider the two series below:

A = 8,9,10,11,12,13,14   Mean 11
B = 1008,1009,1010,1011,1012,1013,1014   Mean 1011

Both A and B have the same standard deviation — can you see why? The deviations from the mean are identical in both series ($-3, -2, -1$ etc.) and both series have the same number of observations, so $\Sigma(x - \bar{x})^2$, and n would be the same whichever series we examined. You should verify that the standard deviation is, in fact, 2 for both series. Now does this mean that both series show the same degree of spread? Surely not! An increase from the smallest to the largest value in A is an increase of

$$\frac{14 - 8}{8} \times 100 = 75\%$$

but the same increase for series B is only

$$\frac{1014 - 1008}{1008} \times 100 = 0.595\%!$$

Clearly, A has a much greater percentage spread than B, and if we are going to compare dispersion between series we must take into account not only dispersion within each series (i.e. the standard deviation) but also the mean of the series. We can do this by calculating the *coefficient of variation*

$$V = \frac{100 \times \sigma}{\bar{x}} \%$$

For the series above, the coefficients of variation are

$$\frac{100 \times 2}{11} = 18.18\% \text{ (for A)}$$

and

$$\frac{100 \times 2}{1011} = 0.198\% \text{ (for B)}$$

confirming our suspicion that A has a greater relative spread than B. Notice that the coefficient of variation is a percentage, and not a particular measurement (centimetre, dollar, kilogram, etc.) and so it can be used for comparing distributions which have different units.

Now we would like to issue a word of warning. If you attempt to quantify the differences in dispersion using the coefficient of variation, ask yourself whether the result you have obtained 'feels right'. Compare your result with a 'gut feeling' you have obtained by visually examining the histogram of each distribution. If your results do not feel right then you would be well advised to compare the distributions using the quartiles. Earlier, we defined the quartile deviation as $½(Q_3-Q_1)$, and we could take the median Q_2 as a measure of central tendency, so an alternative measure of relative dispersion is

$$\frac{100 \times ½(Q_3-Q_1)\%}{Q_2} = \frac{50(Q_3-Q_1)\%}{Q_2}$$

and we shall call this measure the *quartile coefficient of variation*. We shall say no more about it now, but in the exercises at the end of this chapter you will find an example where its use is obviously required.

Finally, do you remember that in Chapter Five we introduced the concept of skewness? Now that we know how to measure dispersion, we can also measure skewness with the coefficient of skewness

$$\text{Coefficient of Skewness} = \frac{3\,(\text{mean} - \text{median})}{\text{Standard deviation}}$$

If the distribution is symmetrical, then the mean and median are the same, so the coefficient would be zero. With negative skewness, the median exceeds the mean, and the coefficient is negative. With positive skewness, the mean exceeds the median and the coefficient is positive. By itself, the coefficient is of little use, but it is useful when we are trying to decide which of a number of distributions shows the greater degree of skewness.

Exercises to Chapter 9

9.1   The monthly production in 2 factories in thousands of units is as follows:
A: 23, 30, 28, 31, 29, 26, 34, 36, 28, 32, 25, 26
B: 53, 65, 70, 50, 62, 58, 52, 63, 69, 72, 64, 54
(a) Compare the variability of production in the 2 factories by calculating the standard deviation in each case.
(b) Calculate the coefficient of variation in each case.

9.2   From the figures below state what is
(a)   the range
(b)   the arithmetic mean
(c)   the median
(d)   the lower quartile
(e)   the upper quartile
(f)   the quartile deviation
(g)   the mean deviation

Explain what is meant by the mean deviation and quartile deviation.

8	35	45	50	60	68
13	37	46	52	61	70
26	40	47	55	65	71
29	41	48	58	67	75
33					

I.C.M.A.

9.3 The table shows the age distribution of the estimated U.K. population.

Age in Years	Total U.K. Population (millions)
0 and under 5	4.4
5 and under 10	4.7
10 and under 15	4.3
15 and under 20	3.9
20 and under 25	4.1
25 and under 30	4.0
30 and under 35	3.3
35 and under 40	3.2
40 and under 45	3.3
45 and under 50	3.4
50 and under 55	3.4
55 and under 60	3.3
60 and under 65	3.2
65 and over	7.5

Source: Monthly Digest.

(a) Draw an appropriate graph and from it estimate the median and the upper and lower qualities. Hence obtain the semi-interquartile range.

(b) Detail how the quartiles could be obtained other than by using the graph.

9.4 Details of the annual bonuses paid to sales staff employed by a large wallpaper manufacturer have recently been made available.

Annual Bonus (£)	Number of Bonuses
under 60	8
60 and under 70	10
70 and under 80	16
80 and under 90	14
90 and under 100	10
100 and under 110	5
over 110	2

Source: Company Records.

(a) Using graphical methods, estimate the median bonus paid and comment on the suitability of the median in interpreting this type of data. How many sales staff receive an annual bonus of
  (i) Less than £88?
  (ii) At least £63 but less than £75?
  (iii) £96 or more?

(b) From the graph, estimate a measure of dispersion, using the quartiles. What is meant by *dispersion*?

9.5 *Estimated Wealth of Individuals in Gt. Britain*

Ranges of Net Wealth (£)	No. of Cases (thousands)
Under 1,000	5,190
1,000 – under 3,000	5,415
3,000 – under 5,000	2,918
5,000 – under 10,000	2,191
10,000 – under 15,000	598
15,000 – under 20,000	288
20,000 – under 25,000	148
25,000 – under 50,000	326
50,000 – under 100,000	121
100,000 – under 200,000	40
200,000 and over	20
Total	17,255

Source: Board of Inland Revenue.

(a) Calculate: (i) the median net wealth,
 (ii) the quartile deviation.
(b) How appropriate are these measures to this distribution?
(c) State briefly the advantages and disadvantages of using quartile deviation and standard deviation as measures of dispersion.

9.6 Profits made by a sample of firms in a certain industry in a given year are as follows:

(£) Profits	% of Firms
Up to 2,000	6.8
2,001 –	11.4
4,001 –	14.1
6,001 –	13.5
8,001 –	11.5
10,001 –	24.1
12,001 –	6.4
13,001 or more	5.8
Showing a loss	6.4
All firms	100%
Number of firms	5231

Calculate the median and quartile deviation of the profits made by the 5231 firms included in the sample. Construct a block diagram to represent the distribution and discuss whether the mean is likely to be high or lower than the median. C.I.P.F.A.

9.7 The following data gives the monthly expenditure on advertising in 1969 of the branches of Cosmetics Ltd.:

Cosmetics Ltd. Monthly Advertising Expenditure

£	Number of Branches
800 and less than 1000	50
1000 and less than 1200	200
1200 and less than 1400	350
1400 and less than 1600	150
1600 and less than 1800	100
1800 and less than 2000	75
2000 and less than 2200	50
2200 and less than 2400	25

(a) From the above data calculate the median monthly expenditure, and explain what it indicates about the branches' advertising expenditure.
(b) Calculate the semi-interquartile range from the data and explain the purpose of this calculation. A.C.A.

9.8

Deliveries of Gravel from a Quarry

Weight of Load (cwt)	No. of Loads Week 1	Week 2
5 and under 10	84	45
10 and under 20	86	63
20 and under 30	122	140
30 and under 40	86	77
40 and under 60	16	21
60 and under 80	6	4
	400	350

(a) Express the frequencies as percentages of their respective totals and use those figures to draw, on the one graph, cumulative frequency curves for the loads of gravel delivered in Week 1 and Week 2.
(b) Use your graph to find the median weight and the semi-inter-quartile range for the two weeks.
(c) Comment briefly on your results.

9.9

Local Authorities with Central Purchasing Organisations

Average % Discount Obtained	Number of Authorities
8 and under 12	0
12 and under 16	0
16 and under 20	2
20 and under 24	0
24 and under 28	3
28 and under 32	0
32 and under 36	6
36 and under 40	7
40 and under 44	2
44 and under 48	3

Local Authorities without Central Purchasing Organisations

Average % Discount Obtained	Number of Authorities
8 and under 12	5
12 and under 16	16
16 and under 20	12
20 and under 24	8
24 and under 28	13
28 and under 32	1
32 and under 36	0
36 and under 40	1
40 and under 44	1
44 and under 48	0

Source: L.G.O.R.U.

For each of these distributions calculate

(a) the arithmetic mean of the average % discount obtained;
(b) the standard deviation of the average % discount obtained.

Compare the arithmetic means and suggest what conclusions might be drawn from these statistics.

9.10 The table below gives an analysis of the debtors' balances of Fuel Suppliers Ltd.

Fuel Suppliers Ltd. Debtors' Balances

Balance Outstanding (£)	Number of Accounts
20 – 39.9	1
40 – 59.9	3
60 – 79.9	6
80 – 99.9	10
100 – 119.9	5
120 – 139.9	3
140 – 159.9	2

You are required to:
(a) calculate the mean balance and the standard deviation, and
*(b) explain the value of knowing the standard deviation in respect of the outstanding balances due to Fuel Suppliers Ltd. A.C.A.

9.11 The distribution shown below is the output of the factories of Quality Clothing Ltd., for the month of May 1972. You are required to:
(a) calculate the standard deviation from these figures, and
*(b) contrast the mean deviation and the standard deviation as measures of dispersion and indicate briefly what the standard deviation calculated in (a) means for the monthly output of Quality Clothing Ltd.

Quality Clothing Ltd.

Monthly Output Men's Suits (000's)	Number of Factories
23 and under 28	10
28 and under 33	20
33 and under 38	20
38 and under 43	24
43 and under 48	20
48 and under 53	16
53 and under 58	8
58 and under 63	2

A.C.A.

9.12 From the following frequency distribution of the weekly overtime earnings of the staff of Catering Ltd., you are required to:
(a) calculate (i) the mean and (ii) the standard deviation, and
*(b) indicate what the value of the standard deviation tells you about the weekly overtime earnings of the staff of Catering Ltd.

Catering Ltd.

Weekly Overtime Earnings (pence)	Number of Staff
50 – 74.9	2
75 – 99.9	5
100 – 124.9	10
125 – 149.9	17
150 – 174.9	10
175 – 199.9	4
200 – 224.9	2

A.C.A.

9.13 • A furnace whose size is nominally 200 tons is used to cast steel ingots of 10 tons weight. The amount of steel in the furnace cannot be controlled accurately, so that an incomplete ingot is normally produced when ingots are cast. For example, a furnace load of 198 tons will produce 19 full ingots and one 8-ton ingot. The following data show the weight of 100 furnace loads of steel.

Weight of Furnace Load (tons)	Frequency
190.0 and under 192.5	1
192.5 and under 195.0	4
195.0 and under 197.5	8
197.5 and under 200.0	19
200.0 and under 202.5	36
202.5 and under 205.0	20
205.0 and under 207.5	8
207.5 and under 210.0	4

Required: Compile a frequency distribution of the weight of incomplete ingots, and calculate its mean and standard deviation. A.C.A.

9.14

Distribution of Personal Incomes before Tax

Lower Limit of Range of Income (£)	Number of Incomes (thousands)
400	883
500	1,161
600	1,270
700	1,212
800	2,322
1,000	5,720
1,500	4,296
2,000	3,341
3,000	836
5,000	328
Total	21,369

Source: Board of Inland Revenue.

(a) Rewrite the numbers of incomes to the nearest hundred thousand.
(b) Using the rounded data calculate the standard deviation.
*(c) Explain the importance of this measure of dispersion.
(d) The source gives the number of all incomes as 21,368 (thousands). Why is this number different from the total in the above table?

9.15 A survey of house prices in a local newspaper yielded the following information:

Price of House (£'000s)	Number of Houses for Sale
below 10	16
10 and under 12	41
12 and under 14	39
14 and under 16	22
16 and under 18	10
18 and under 20	11
20 and under 22	4
22 and under 24	5
24 and under 26	1
26 and under 28	5
28 and under 30	4
30 and over	2
	160

(a) Calculate the mean price of houses in the area and the standard deviation, stating clearly any assumptions you may make.

(b) For what reason may the mean and standard deviation be inappropriate for such a distribution? Suggest better alternatives.

9.16 At the General Election in February 1974, the sizes of constituencies were distributed as shown in the table below:

Number of Voters ('000s)	Number of Constituencies
20 and under 30	7
30 and under 40	17
40 and under 50	74
50 and under 60	176
60 and under 70	183
70 and under 80	115
80 and under 90	48
90 and under 100	13
100 and over	2
Total	635

(a) Calculate the average number of voters per constituency and also the standard deviation.

*(b) In order that constituencies may be kept to a roughly similar size, constituency boundaries are from time to time revised. From your results in (a) suggest a range of sizes outside which changes might be made, and give reasons for your suggestions.

9.17   *Electric Lamps*

Hours of Life (hundreds)	Number of Lamps
0 – 5	5
5 – 10	10
10 – 20	38
20 – 40	36
40 and over	15

(a) Calculate the mean and standard deviation of the life of the whole batch of lamps whose pattern of duration is given in the table above.
(b) Compare and contrast the standard deviation and the quartile deviation as measures of dispersion. C.I.P.F.A.

9.18 (a) Explain the meaning of absolute and relative measures of dispersion and compare their use.
(b) In Department A of a firm the average weekly earnings are £366 with a standard deviation of £28.2. Calculate the relative dispersion.
(c) In Department B the earnings in pounds in a certain week of the 10 members are as follows:

237   245   283   296   253   249   236   254   305   242

Calculate the mean and the standard deviation of this department. Compare this result with Department A and comment.

9.19 Explain the use of the coefficient of variation.
A company has two factories A and B, situated in different parts of England. Labour turnover in factory B is much higher than in factory A. The consultant engaged by the firm to investigate the causes of labour turnover suggests that a possible cause of the higher turnover in factory B is a wider variation in the annual wages of workers in factory B than in factory A: he found that the standard deviation of wages of workers in factory B was £600 compared with a standard deviation of £500 for factory A. Given the information that the mean annual wage of workers in factory B is £2000 and in factory A £1500. Show how the consultant's suggestion may be tested by use of the coefficient of variation, and say whether he is right. A.C.A.

9.20

Projected Population in Scotland
(tens of thousands)

Age Group	Males	Females
0 – 14	75	72
15 – 29	59	58
30 – 44	46	47
45 – 59	43	48
60 – 74	31	41
75 and over	6	14
All ages	260	280

(a) Calculate the standard deviation for each distribution.
(b) Calculate the arithmetic mean of the male distribution and hence obtain a relative measure of dispersion.

9.21

Monthly Salary £	Number of Trainee Draughtsmen
under 72	4
72 and under 76	9
76 and under 80	16
80 and under 84	28
84 and under 88	45
88 and under 92	66
92 and under 96	85
96 and under 100	72
100 and under 104	54
104 and under 108	38
108 and under 112	27
112 and under 116	18
116 and under 120	11
120 and under 124	5
124 and over	2

(a) The above figures relate to the monthly salaries of the trainee draughtsmen of a local company. Calculate the arithmetic mean and the standard deviation.

(b) Information relating to the trainees employed by a rival firm gives a coefficient of variation of 18%. Are they, or the local company's trainees, more variable with respect to salaries?

9.22 A company selling a consumer product directly to retail outlets has collected the following information:

Average Number of Orders Taken per Month by Individual Salesmen	Number of Salesmen
20 and under	3
21 – 30	7
31 – 40	16
41 – 50	22
51 – 60	19
61 – 70	8
71 and over	2

Calculate (a) the approximate range;
(b) the standard deviation;
(c) the coefficient of variation.

All of these are measures of dispersion. Describe, with additional examples, when the use of each would be appropriate.

9.23 The following table shows the distribution of weekly income for two towns

Weekly Income (£)	Town A No.	Town A Income (£)	Town B No.	Town B Income (£)
0 and under 10	0	0	1	5
10 and under 20	4	60	5	75
20 and under 30	9	225	9	225
30 and under 40	17	595	13	455
40 and under 50	23	1035	19	855
50 and under 60	19	1045	23	1265
60 and under 70	13	845	17	1105
70 and under 80	9	675	9	675
80 and under 90	5	425	4	340
90 and under 100	1	95	0	0
Totals	100	5000	100	5000

(i) Show that the arithmetic mean income is the same in both towns. In the light of this result, do you feel that the arithmetic mean is a useful measure to use for comparison of the two town's 'average' income? If not, derive an alternative measure that you consider better, giving reasons for your choice.

(ii) Show that the standard deviation of incomes is the same in both towns. In the light of this result do you feel that the standard deviation is a useful measure to use for comparison of the two towns' dispersion of income? If not, derive an alternative measure that you consider better, giving reasons for your choice.

(iii) Given your discussion of (i) and (ii) above, what do you consider to be the essential differences between the two towns' distributions of income? How *might* you express this by a statistical measure?

C.I.P.F.A.

9.24 The table below show the salaries of the 600 manual employees of a certain company.

No. of Employees	Range of Salary
0	Under £600
80	£600 but under £800
170	£800 but under £1000
90	£1000 but under £1200
80	£1200 but under £1400
70	£1400 but under £1600
50	£1600 but under £1800
40	£1800 but under £2000
20	£2000 but under £2200
0	£2200 and over

(a) Calculate the mean and standard deviation salary of the employees.

(b) Obtain and comment on an arithmetical measure of the skewness of the salary distribution.

(c) State the main reason for normally using the median salary as a measure of position rather than the arithmetic mean or average; illustrate your answer with reference to the above data.

9.25 The table shows the age distribution of those males receiving retirement pensions at 31st December 1972. Calculate the arithmetic mean and standard deviation for this distribution. If the median is 71.8 years, obtain a measure of skewness and comment on what this tells you about the shape of the frequency curve of the distribution.

Retirement Pensions (Male) at 31st December 1972

Age (years)	Number Receiving Pensions (thousands)
65 – 69	1030
70 – 74	820
75 – 79	460
80 – 84	240
85 – 89	80
90 and over	20

Source: Dept. of Health and Social Security.

*These questions can be completed after reading Chapter 16.

Chapter Ten

The Normal Distribution

We have considered in detail measures of central tendency and measures of dispersion, and we have found that many such measures are available to us. So far, we have used such measures for descriptive purposes, and the time has now come to consider their analytical implications. To do this, we shall confine ourselves to two measures: we shall use the arithmetic mean as a measure of central tendency, and the standard deviation as a measure of dispersion. We shall start by analysing a simple problem involving comparison.

Acme Ltd. is considering the profits that it made last year. Its Office Equipment Division earned a profit of £65,000 and its Domestic Appliance Division earned a profit of £55,000. At first sight, it seems that the Office Equipment Division had the more successful year. Suppose, however, that we were told that the average profit earned in the office equipment industry was £50,000, and that the average profit earned in the domestic appliance industry was £45,000. Seemingly, the profit opportunities in office equipment are greater. How, then, can we compare the performance of the two divisions? If we were told that the standard deviation of profits in the office equipment industry was £10,000, then a profit of £65,000 is

$$\frac{65,000 - 50,000}{10,000}$$

= 1.5 standard deviations above the mean. Suppose also that we know the standard deviation of profit earned in the domestic appliance industry was £5,000, then a profit of £55,000 is

$$\frac{55,000 - 45,000}{5,000}$$

= 2 standard deviations above the mean. So we see that in the domestic appliance industry the relative performance (i.e. the performance with respect to competitors) was better than in the office equipment industry. Data calculated in this way — in terms of standard deviations measured from the mean — are called *standard scores or Z scores,* and for any variate x drawn from a population with a mean μ and a standard deviation σ, we can calculate the Z score like this

$$Z = \frac{x - \mu}{\sigma}$$

Notice that the Z score is in relative, not absolute units; so they can be used to compare values in distributions that do not use the same unit of measurement.

Now Z scores are very important to statisticians: under certain conditions Z scores can be used to predict the proportion of a distribution that is more than, or less than a certain measurement. Let us first examine the conditions necessary for us to be able to do this. The distribution must be continuous; it must be symmetrical and it must be bell-shaped. In other words, the distribution must be shaped like the one in diagram 10.01. We call such a distribution a *Normal Distribution,* and call its shape *mesokurtic.* Now you may (with some justification) think that these conditions are very stringent and will not often be met in practice. We will have more to say about this in later chapters and for the moment we will concentrate on learning how to make predictions based on the Normal Distribution.

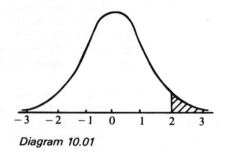

Diagram 10.01

If you consult the tables at the end of this book you will find a table headed the Normal Distribution Function. In the first column there is a list of Z scores ranging from 0.0 to 4.0. The second column gives the proportion of the Normal Distribution with a Z score greater than the corresponding figure in the first column. Suppose that we wanted to find the proportion in a Normal Distribution with a Z score greater than 2.0 (i.e. the proportion greater than two standard deviations above the mean). The shaded area in diagram 10.01 represents the proportion diagrammatically. Finding 2.0 in the first column, we read that the corresponding figure in the second column is 0.02275 (or 2.275%), so 2.275% of a Normal Distribution has a Z score greater than 2.0. This being so, it must follow that 100 − 2.275 = 97.725% of a Normal Distribution has a Z score less than 2.0 (we imply here that a negligible proportion will have a Z score *exactly* equal to 2.0).

The remaining columns in the table enable us to take an extra place of decimals when calculating our Z score. Suppose we wanted to find the proportion in a Normal Distribution with a Z score greater than 2.12. We find the required proportion at the junction of the row with a Z score 2.1, and the column headed .02 ie. the proportion required is 0.01321. The Normal Distribution tables, then, follow the standard format followed by other mathematical tables.

How can we apply this knowledge? Suppose we know that electric light bulbs have an average life of 2000 hours and a standard deviation of 60 hours. Furthermore, we know that the lives of bulbs are Normally Distributed (this would be a reasonable assumption) and we wish to know the proportion of bulbs failing after 2120 hours. We could represent the problem diagrammatically like this:

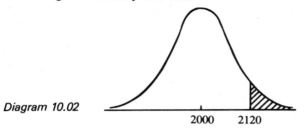

Diagram 10.02

Notice that as the distribution is symmetrical, the mean bisects the distribution, and the mean, median and mode coincide. The shaded area will, of course, represent the proportion of bulbs failing after 2120 hours, and only a small proportion will do so. The first thing we do is calculate the Z score

$$Z = \frac{2120 - 2000}{60} = 2 \text{ standard deviations.}$$

Now we already know that 2.275% of items in a Normal Distribution have a Z score more than 2, and in this particular Normal Distribution any bulb with a life above 2120 hours has a Z score more than 2. So it must follow that 2.275% of bulbs will fail after 2120 hours.

So far, we have considered cases where the Z score is positive, but it is perfectly possible for a Z score to be negative. This will occur when an item under consideration has a value less than the arithmetic mean. Suppose, for example, we wished to find the proportion of bulbs with lives more than 1850 hours. This involves finding the proportion of items with Z scores more than

$$\frac{1850 - 2000}{60} = -2.5$$

If we consult the table, we notice that only positive Z scores are given — but the table can also be used for negative Z scores. A glance at diagram 10.03a and 10.03b will confirm that as the distribution is symmetrical, the proportion of items with Z scores more than -2.5 is the same as the proportion with Z scores less than 2.5. Reading from the table, we see that 0.621%. of items have Z scores more than 2.5, so 99.379% have Z scores less than 2.5, and so 99.379% must have Z scores greater than -2.5. So we can see that 99.379% of bulbs can be expected to fail after 1850 hours.

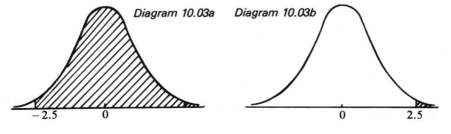

Diagram 10.03a Diagram 10.03b

Now you may find this all a bit confusing — sometimes the Z score is positive and sometimes it is negative; sometimes we read the percentage directly from the table, and sometimes we subtract it from 100. Can we offer some advice? When considering problems involving the Normal Distribution, *always* draw a sketch of the problem and shade the area under consideration. You can then see at a glance whether the area required is greater or less than 50%. If the area is less than 50%, then the proportion in the table is the one required. If the area required is greater than 50%, subtract from 100 the percentage obtained from the table.

Suppose we know that the weights of bags of flour are Normally Distributed with a standard deviation of 0.01 kilograms. The bags are marked 'weight not less than 1 kilogram', and the machine is set to fill bags to an average weight of 1.015 kilograms. We require to know the proportion of bags that satisfy the claim printed on the bag. Firstly, let us draw a sketch of the problem (see diagram 10.04).

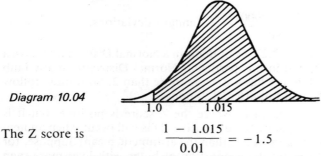

Diagram 10.04

The Z score is $\dfrac{1 - 1.015}{0.01} = -1.5$

and looking up 1.5 in the table we obtain a proportion of 6.68%. Now the diagram tells us immediately that this cannot be the percentage we require — it obviously should be more than 50%. So the proportion of bags meeting the claim is $100 - 6.68 = 93.32\%$.

Sometimes it is necessary to use the Normal Distribution in reverse, ie. given a proportion, we enter the tables to find the corresponding Z score, and so find a corresponding value in a given Normal Distribution. Suppose that in order to satisfy legal requirements, a pork pie manufacturer may only produce 0.2% of pies below a weight of 75 grams. The pie-producing machine operates with a standard deviation of 0.5 grams. If the weights of pies are Normally Distributed, to what (mean) weight should the machine be set? Diagram 10.05 illustrates the situation, and μ is the machine setting required.

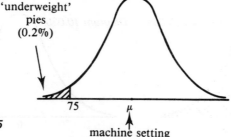

Diagram 10.05

Consulting the table we see that the nearest Z score to a proportion 0.2% is 2.88. Notice that the weight we are considering (75 grams) is *less than* the mean, so its Z score must be negative. In other words

$$\frac{75 - \mu}{0.5} = -2.88,$$

so the machine setting (μ) is 76.44 grams. (ie. $2.88 \times 0.5 + 75$).

Let us take this example a stage further. The pie manufacturer's weekly output is 200,000 pies, and the pie contents cost 5p per 100 grams. Pies with weights in excess of 77 grams require additional packing at a cost of 0.5p per pie. Find the firms weekly cost.

Firstly, we are going to have to find the proportion of 'overweight' pies. This is illustrated in diagram 10.06. The Z score is

$$\frac{77 - 76.44}{0.5} = 1.12,$$

so 13.14% of output will need extra packaging.

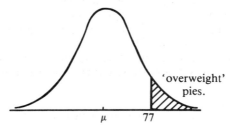

Diagram 10.06

We can now calculate the weekly cost like this:
Weekly pie content requirement is $200,000 \times 76.44 = 15,288,000$ grams.
Weekly cost of contents is $15,288,000 \times 0.05\text{p} = £7,644$.
Each week 13.14% of $200,000 = 26,280$ pies need extra packing.
Weekly cost of extra packing is $26,280 \times 0.5\text{p} = £131.40$.

Finally, let us suppose that a new pie producing machine is available; operating with a standard deviation of 0.2 grams, but costs an additional £225 per week to operate. Would you recommend that the pie manufacturer purchases the new machine? To answer this question, we must first determine the machine setting using the new machine, and determine the percentage of 'overweight' pies. To obtain the machine setting, we know that

$$\frac{75 - \mu}{0.2} = -2.88,$$

so the machine setting (μ) must be 75.576 gms. (Notice that the lower standard deviation enables the machine setting to be lowered). To obtain the percentage of 'overweight' pies, we know that

$$Z = \frac{77 - 75.576}{0.2} = 7.12.$$

Now if you consult the Normal Distribution tables, you will see that the Z scores only go up to 4.0, and the proportion of a Normal Distribution with a Z score of more than 4 is only 0.003%. Clearly, then, it follows that as 77 grams has a Z score of 7.12, we would not expect to have any 'overweight' pies if the new machine is used.

We can now calculate the cost of using the new machine.
Weekly pie content requirement is $200,000 \times 75.576 = 15,115,200$ gms.
Weekly cost of contents is $15,115,200 \times 0.05p = £7,557.60$
and as there are no 'overweight' pies, this is the total weekly cost of using the new machine. So the new machine involves a weekly cost saving of £7,775.40 − £7,557.60 = £217.80. However, as the new machine costs an extra £225 per week to operate, the manufacturer would be well advised to retain the use of the old machine.

Before ending this chapter, two points deserve attention. Firstly, the Normal Distribution is of paramount importance in statistics, and you would be well advised to master it before moving on to the following chapters. Secondly, you may be rather worried by the stringent conditions imposed before the Normal Distribution may be used. Well, there are many cases in which a population shows a very good approximation to the Normal Distribution, and any errors resulting from its application should not be too serious. More importantly, although populations may not approximate to the Normal Distribution, there are sound theoretical reasons for assuming that sampling statistics will do so. It is to this important truth that we shall shortly turn our attention.

Exercises to Chapter 10

10.1 (a) What are the more important characteristics of the normal distribution?
 (b) Discuss the importance of the mean and standard deviation in the use of the normal distribution.

10.2 The length of rods in a large batch is normally distributed with mean 120 mm and standard deviation 1.5 mm. What percentage of the rods would you expect to measure
 (i) over 122.5 mm;
 (ii) between 116 mm and 124 mm?

10.3 A random sample was taken of 200 candidates who had submitted scripts in a certain question paper at a public examination. The results obtained by the 200 candidates in the examination were as follows:

Marks Obtained out of a Possible 100	Number of Candidates
Under 20	0
20 – 29	6
30 – 39	18
40 – 49	41
50 – 59	68
60 – 69	49
70 – 79	14
80 – 89	4
90 and over	0
Total	200

(a) Calculate:
   (i) the mean;
   (ii) the standard deviation.
(b) (i) What does the standard deviation measure?
   (ii) How does the standard deviation you have calculated differ from what might be expected if results were normally distributed?

I.C.M.A.

10.4 (a) What are the principal features of the normal distribution?
(b) Suppose that the distribution of women by size of feet is normal, with an arithmetic mean length of 8 in. and a standard deviation of 1 in. A manufacturer of women's shoes has an output of 10,000 pairs per week and he wants to know how many pairs of shoes of each length to produce. How could a table of the area under a normal curve help him to plan his output? How many pairs per week should be (a) above 8½ in.; (b) below 6½ in.; (c) above 7½ in.; (d) below 9 in.?

C.I.P.F.A.

10.5 (a) Sketch a normal distribution curve. On this sketch indicate the approximate proportions of the area under the curve which are contained within the following limits:

arithmetic mean $\pm$ 1 standard deviation,
arithmetic mean $\pm$ 2 standard deviations, and
arithmetic mean $\pm$ 3 standard deviations.

(b) A company packing biscuits knows that the weights of 500 packets form a normal distribution with a mean weight of 16 ounces and a standard deviation of 0.2 ounces. How many of these 500 packets can be expected to weigh:
   (i) less than 15.6 ounces;
   (ii) between 15.8 and 16.4 ounces;
   (iii) at least 16.2 ounces?

10.6 Describe a normal distribution, paying particular attention to precise definitions of its most important properties. Give examples of statistical data which might be expected to conform closely to a normal distribution.

An automatic machine produces packages whose weights are normally distributed with a standard deviation of 0.07 ounces. To what weight should be machine be set so that at least 97½% of the packages are over 1 lb in weight?

10.7 (a) Illustrate graphically the relationship between the standard deviation and areas under the normal curve of distribution.
(b) Assuming that the hub thickness of a certain type of gear is normally distributed around a mean thickness of 2.00 inches, with a standard deviation of 0.04 inches, say:
   (i) approximately how many gears will have a thickness between 1.96 and 2.04 inches in a production run of 5000 gears;
   (ii) in a random selection of one gear from this production run of 5000, with what degree of confidence could it be predicted that its thickness would be between 1.92 and 2.08 inches?

A.C.A.

10.8 A manufacturer makes chocolate bars with a mean weight of 110 grams and a standard deviation of 2 grams. The weights are normally distributed. What proportion of the bars are likely to be less in weight than 106 grams? The manufacturer decides to make 'bigger' bars with a mean weight of 115 grams, with the same standard deviation as before. What proportion of the new bars is likely to be less in weight than the old ones? (The weights of the 'bigger' bars are also normally distributed.)

It is decided that the covers of these 'bigger' bars will be marked 'Minimum weight 115 grams'. What mean weight will have to be aimed at if no more than 1 bar in 100 is to be less than 115 grams in weight?

I.C.S.A.

10.9 Components made by a certain process have a thickness which is normally distributed about a mean of 3.00 cm and a standard deviation of 0.03 cm. A component is classified as defective if its thickness lies outside the range 2.95 cm to 3.05 cm.
Required:
(1) What is the proportion of defective components?
(2) Find the change in the proportion of defective components if the mean thickness is increased to 3.01 cm, the variability remaining the same.

A.C.A.

10.10 (a) Explain the meaning of the term normal distribution and illustrate with a simple diagram.
(b) Records kept by the goods inwards department of a large factory show that the average number of lorries arriving each week is 248. It is known that the distribution approximates to the normal with a standard deviation of 26. If this pattern of arrivals continues, what percentage of weeks can be expected to have a number of arrivals of:
(i) less than 229 per week?
(ii) more than 280 per week?

10.11 The length of rods in a large batch is normally distributed with mean 120 mm and standard deviation 1.5 mm. What percentage of the rods would you expect to measure
(i) over 122.5 mm;
(ii) between 116 mm and 124 mm?

10.12 The mean weight of a consignment of 500 sacks of sugar is 151 lb and the standard deviation 15 lb. Assuming that the weights are normally distributed, find how many sacks weigh:
(a) between 120 and 155 lb,
(b) more than 185 lb, and
(c) less than 128 lb.

10.13 The average number of newspapers purchased by a population of urban households in 1972 was 400, and the standard deviation was 100. Assuming that newspaper purchases were normally distributed, what is the probability that households bought:
(a) between 250 and 500 papers;
(b) less than 250 papers;
(c) between 500 and 600 papers;
(d) more than 500 papers?

10.14 The thermal efficiency of electricity generating stations in the U.K. in 1969-70 varied from below 10% to over 34% as shown in the table:

Electricity: Output and Efficiency
Thermal Efficiency of Steam Stations

Thermal Efficiency of Stations (%)	No. of Stations with given Efficiency
34 and over	7
32 – 34	17
30 – 32	12
28 – 30	17
26 – 28	21
24 – 26	22
22 – 24	27
20 – 22	27
18 – 20	15
16 – 18	15
14 – 16	12
12 – 14	4
10 – 12	4
10 and under	3
Total	203

Source: Department of Trade and Industry, Digest of Energy Statistics, 1971.

(a) Calculate
  (i) the average thermal efficiency,
  (ii) the standard deviations.
(b) Assuming the distribution to be nearly normal, what would you expect the range to be, observing that the end classes are 34 *and over* and 10 *and under?*

10.15 A firm is considering the purchase of a machine to turn ball-bearings. A machine is borrowed for testing purposes, and it is found that 6.68% of ball-bearing have mean diameters greater than 5.03 mm. Assuming the machine to be set at 5 mm, find the standard deviation to which the machines operates. The firm wishes to produce ball-bearings to a design dimension of 5 ± 0.05 mm, and any ball-bearings outside this range would be rejected. Assuming the machines produces 1 million units per month, and each unit rejected costs the firm 1p, find the monthly cost of rejects.
A more accurate machine is available, which costs an additional £100 per month to purchase. If this machine is known to operate with a standard deviation of 0.016mm, which machine should be purchased?   I.C.A.

10.16 A manufacturer finds that although he promises delivery of a certain item in 7 weeks, the time he takes to deliver to customers is approximately normally distributed with a mean of 6 weeks and a standard deviation of 2 weeks.
Required:
  (1) what proportion of customers receive their deliveries late?
  (2) what proportion of customers receive deliveries within 4 to 7 weeks?
  (3) to what figure should his delivery promise be amended if it is required that only 20% of deliveries should be late?
  (4) what proportion of customers will receive deliveries within 5 weeks if the manufacturer reduces the standard deviation of delivery time to 1 week, keeping the mean time at 6 weeks?

10.17 The heights of 8585 adult males born in the British Isles were measured. The standard deviation and arithmetic mean for this sample were calculated and found to be 2.57 inches and 67.52 inches respectively. Use areas under the normal curve to find the proportion of men, during the year in which this operation was carried out, who would have been (i) over 71 inches, (ii) less than 61 inches, (iii) either taller than 71 inches or shorter than 61 inches, (iv) between 71 inches and 61 inches. I.C.A.

10.18 A maintenance engineer has established that component XYZ has an operating life of 2500 hours and a standard deviation of 250 hours. Assuming a normal distribution, calculate the probability that any one component XYZ chosen at random would have an operating life of:
(a) less than 2200 hours;
(b) between 2300 and 2700 hours;
(c) between 2300 and 2400 hours;
(d) more than 3000 hours;
(e) between 2000 and 3000 hours.

10.19 A machines is packaging nominal 8 oz packets and it has been found that over a long period the actual weight put in the packet has been normally distributed with a standard deviation of 0.05 oz. The company wishes to ensure that no more than 1% of the packets have a weight of less than 8 oz; at what mean weight should the machine be set?
The weekly output of packages averages 2,400,000 and the cost, in pence, of producing a packet of weight w ounces is given by the relationship.

$$c = 6 + 0.5 w$$

If, by installation of new machinery, the standard deviation could be reduced to 0.02 oz, and the mean allowed to fall just far enought to give the same percentage below 8 oz as before, determine the average saving in pounds per week. I.C.A.

Chapter Eleven

Probability

If you have read the previous chapters carefully, you will have realised by now that much of a statistician's time is spent measuring data and drawing conclusions based on his measurements. Sometimes, all the data is available to the statistician, and the measurements are bound to be accurate. In such circumstances, we can say that he has perfect knowledge of the population he is investigating. Unfortunately, this will not be the usual situation. In most cases, the statistician will not have the details he want about the entire population, and will be unable to collect all the information he wants because of the cost and labour involved. To take an example; suppose it is required to find the average height of adult males in a particular country; it would not be possible to measure the height of every male. Instead, the statistician would have to make do with the average height of a sample. Now if he is careful about how his sample is drawn, and if the sample is not too small, then the sample average will give a good approximation to the population average. However, because the entire population has not been examined, the statistician can never be completely sure of his results, so when quoting conclusions based on sample evidence it is usual to state just how confident we are of our results.

So you will often see the estimates which are based on sample results quoted "with 95% confidence". Now what, exactly, does this statement mean? Suppose that we are selecting a sample and calculating its average in order to provide an approximation of the population average. To say we are 95% confident cannot mean that the sample we have selected has a 95% probability of being a good or a satisfactory approximation. After all, once we have selected a sample it either gives a good approximation of the population mean or it does not. No, what we mean is that if we were to select a large number of all the possible samples, 95% of them would give us a good approximation. The 95% confidence refers to the probability of our selecting a sample which does this, not the probability that a sample which has already been selected will do so. You may argue, of course, that we are splitting hairs. The difference is not so great as all that. But it is important for sampling theory that you get the distinction clear in your own mind.

In order to understand the principles of sampling, then, we must first understand the meaning and theory of probability. Furthermore, the emphasis of this book will now undergo a change. Previously, we have been concerned with statistical measurement and representation, but from now on we will concentrate on *statistical analysis*. The basis for all statistical analysis is the theory of probability — something that must be mastered if we are to appreciate fully the work of the statistician.

Some Definitions

Let us suppose that we toss a coin. This experiment can have two outcomes: the coin can land with either heads showing or tails showing. The result of such an experiment is called an *event*. Notice that in this case the events are *mutually exclusive*, that is, if a head occurs, then a tail cannot occur at the same time. Now not all events are mutually exclusive, and we must be careful to recognise when events are mutually exclusive and when they are not. Suppose, for example, we are selecting a card from a pack; and the first event is that the card is red and the second event is that the card is an ace. If we draw the ace of hearts or the ace of diamonds, then both events have occurred simultaneously. Clearly, the events are *not* mutually exclusive. Usually, we use a capital E with a suitable subscript to identify the events in an experiment. We could write the events in the coin-spinning experiment like this:

E_1 = the coin shows a head
E_2 = the coin shows a tail

Notice too that E_1 and E_2 are *collectively exhaustive:* that is, they account for all the logical possibilities (we treat with contempt the suggestion that the coin lands on its edge!).

Sometimes we will find it convenient to consider the situations when the event would not occur, and we can do this by placing a dash after the symbol for the event. So we could write

E_1' = the coin does not show a head

Can you see that in this case E_1' and E_2 are equivalent events? If the coin does not show a head then it must show a tail. We can conclude that if $E_1' = E_2$, then E_1 and E_2 must be both mutually exclusive and collectively exhaustive.

How we Measure Probability

Let us consider a certain experiment, and list all the possible events; that is, we will ensure that the events are collectively exhaustive. Events that are equally likely will be assigned the same 'weighting'. Suppose, for example, we again consider tossing a coin. We have

E_1 = the coin shows a head
E_2 = the coin shows a tail

If we make the assumption that the coin is unbiassed, then E_1 and E_2 must be equally likely. What we will now do is to assign 'weights' to the events in proportion to the likelihood that they will occur. So we have

	Weight
E_1	1
E_2	1

Now the probability that the event E occurs is

$$P(E) = \frac{\text{Weighting for event E}}{\text{Sum of the weights}}$$

So $\qquad P(E_1) = \frac{1}{2}$ (or 0.5)

and $\qquad P(E_2) = \frac{1}{2}$

So we can see that the probability of obtaining a head with a single throw of a coin is one half. Just what do we mean by this? Well, it is obvious that we cannot demonstrate probability with a single toss of a coin. In fact, the outcome of a single toss depends on the force we exert together with the way the coin was originally facing. It has nothing to do with probability! When we state that the probability of a head is a half, we are surely making some prediction as to the proportion of heads occurring if we repeat this experiment many times. The outcome of each *individual* toss is determined by the forces mentioned earlier, but the outcome of many tosses obeys a law of 'mass behaviour', and it is this 'mass behaviour' that our probability measure is trying to predict. So when we state that the probability of a head is a half, we mean that if the coin is tossed a large number of times, then we would expect heads to occur on 50% of occasions.

Let us now see if we can restate our measure of probability. If an experiment has N *equally likely* outcomes, n of which constitute event E, then we can state that

$$P(E) = \frac{n}{N}$$

Suppose, then, we wished to find the probability of drawing an ace from a well-shuffled pack of cards. Here we have $N = 52$ (there are 52 cards in a pack, all of them having an equal chance of being drawn) and $n = 4$, so

$$P(\text{ace}) = \tfrac{4}{52} = \tfrac{1}{13}$$

Easy, isn't it?

Now suppose that $n = N$. This implies that each and every outcome must constitute event E. In other words, if we perform the experiment we are absolutely certain that event E will occur. Moreover, if $n = N$, then $P(E) = 1$, so we assign a probability measure of 1 to events that are absolutely certain to occur. If the event E cannot possibly occur, then $n = 0$ and $P(E) = 0$, so we assign a zero probability measure to events that are absolutely impossible. As absolute certainty and absolute impossibility are at opposite ends of the spectrum, we have now obtained limits to our measure of probability – $P(E)$ must lie between zero and one. We can write this statement mathematically like this:

$$0 \leq P(E) \leq 1$$

So let this be a warning to you — if you are calculating the probability of an event, and your result is either negative, or greater than one, then you have made a mistake somewhere. We, as examiners, have frequently met solutions to probability problems in the form of (say) $P(E) = 1.5$. Now not only has the candidate obviously performed the calculations incorrectly, but also he has demonstrated that he does not know that $P(E)$ cannot exceed one — you cannot be more certain than absolute certainty! So if you obtain an answer like this in an examination, and you cannot discover where you have gone wrong, please do state that your answer *is* wrong, and state *why* it is wrong!

Earlier, we stated that an experiment has N equally likely outcomes, n of which constitute event E. Hence, it must follow that $N - n$ of the outcomes would *not* constitute event E. We can now formulate that

$$P(E') = \frac{N - n}{N}$$

$$= 1 - \frac{n}{N}$$

So $$P(E') = 1 - P(E)$$

We have already discovered that the probability of drawing an ace from a pack of cards is $\frac{1}{13}$, so it must follow that the probability of not drawing an ace is $1 - \frac{1}{13} = \frac{12}{13}$. Later, we will find this formula extremely useful.

The Three Approaches to Probability

Well, we have now seen how to measure probability. However, we have so far been making an assumption without actually spelling it out. We have assumed that we not only know all the possible outcomes of an experiment, but also that we can weight the probability of each outcome in proportion to its likelihood. More importantly, we have assumed we can do both of these things *before the experiment is performed*. In other words, we assume a prior knowledge of the outcomes — we have been using the so-called *a priori* approach to probability. Now although it is true that in many cases we will have the necessary information to use an a priori approach (it is true, for example, when considering games of chance), there are many cases in which an a priori approach cannot be used.

Suppose we have a large case of wood screws, and we wish to find the probability that one screw chosen at random is defective. Clearly, it is possible here to define all the events (the screw is either defective or it isn't) but it is not possible to weight the events in proportion to their likelihood. The only way we can determine probability in this case is to draw a sample of N screws, and count the number of defectives (call this n). We can *estimate* the probability that a randomly chosen screw is defective is $\frac{n}{N}$.

This is the so-called *empirical approach* — there is just no way of estimating the probability without drawing that sample!

An appreciation of these two approaches to probability helps to explain a problem that confuses so many students. The problem runs something like this — if I spin a penny 100 times, and on 99 occasions the coin shows heads, what is the probability that it will show heads on the next spin? Some people would argue that the outcome can be either a head or a tail, and the coin has no memory of the 100 previous tosses. So the probability that the coin will show heads on the next spin must be $\frac{1}{2}$. Others would argue that the coin is more likely to show heads than tails, and would estimate the probability of obtaining a head on the next spin to be $\frac{99}{100}$. Well, which approach is the correct one? Surprisingly, the answer is both! In the first case, we are using an a priori approach, reasoning that the experiment has produced a fluke result which does not detract from the fact that the coin is unbiassed. In the second case, we are using an empirical approach, stating that the experimental evidence indicates that the coin is biassed in favour of heads. Now ask yourself this — if you were a gambler, which approach would you prefer to use?

There is a third approach to probability that we must now examine. Suppose we wished to find the probability that a particular horse wins the Derby — clearly we cannot use an a priori approach. Nor can we use an empirical approach, as this would demand that the same race be repeated many times under identical conditions! The only way we can obtain this probability is to give a personal, 'gut feeling' of the horse's chances. This is the so-called *subjective* approach to probability, and it is the method used by bookmakers when fixing odds for a particular horse to win a race. Initially, the odds will be determined by the personal view of the bookmaker, and will be modified as the race approaches according to the collective, subjective views of the punters.

These three distinct approaches to probability raise an interesting philosophical problem. We know that *P(E)* can never exceed one, but does *P(E)* = 1 imply absolute certainty? It all depends on the approach used. If we use an a priori approach then *P(E)* = 1 means that *E must always* occur. However, using an empirical approach *P(E)* = 1 means that *E has always* occurred — which does not imply that it *must* occur in the future. Likewise, using a subjective approach *P(E)* = 1 means that *we think that E will occur* — which again does not imply that it must occur.

Without doubt, the empirical and subjective approaches are more interesting and more useful than the a priori approach. However, in an introductory book such as this, it is preferable to concentrate our attention on a priori probability, and, unless we state to the contrary, you should assume that an a priori approach is being used.

The Laws of Probability

Let us suppose that we cast two dice, and add the scores of the dice. We could represent all the outcomes in a table like this:

	Second Die[1]					
	1	2	3	4	5	6
1	2	3	4	5	6	7
2	3	4	5	6	7	8
First 3	4	5	6	7	8	9
Die 4	5	6	7	8	9	10
5	6	7	8	9	10	11
6	7	8	9	10	11	12

The number in front of each row represents the possible scores of the first die, and the number at the head of each column represents the possible scores of the second die. The numbers in the main body of the table represent the sums of the two possible scores. So we see that if we score a total of 11, we must have thrown either a six with the first die and a five with the second, or a five with the first die and a six with the second. We see, then, that there are 36 equally likely total scores ($N = 36$). Let us now define two events.

E_1 = the sum of the scores is 7
E_2 = the sum of the scores is 9

Now there are 6 ways of scoring a total of 7, so $P(E_1) = \frac{6}{36} = \frac{1}{6}$. Again, there are 4 ways of scoring a total of 9, so $P(E_2) = \frac{4}{36} = \frac{1}{9}$. Now suppose we wished to find the probability that the sum of the scores is *either* 7 or 9. We can write the probability symbolically like this:

$$P(E_1 \cup E_2)$$

where $\cup$ is a shorthand way of writing 'either – or'. Consulting the table, we see that there are 10 ways of obtaining a total score of either 7 or 9, so $P(E_1 \cup E_2) = \frac{10}{36} = \frac{5}{18}$. Notice that the 10 ways are obtained by adding the number of ways for E_1 and E_2. This gives us our first law of probability: the so-called addition law.

$$P(E_1 \cup E_2) = P(E_1) + P(E_2)$$

a law which is true only if E_1 and E_2 are mutually exclusive. If the events are not mutually exclusive, then using this law will not yield the correct probability. Suppose we draw a card from a pack, and E_1 is that the card is an ace. So $P(E_1) = \frac{4}{52}$. If E_2 is that the card is a heart, then $P(E_2)$ is $\frac{13}{52}$. If we want to find the probability that the card is either a heart or an ace, we notice that there are 52 equally likely outcomes, 16 of which would be either a heart or an ace (i.e. 13 hearts plus the three other aces). So the probability that the card is either a heart or an ace is $\frac{16}{52}$. Notice that if we had applied the addition law, we would have obtained $\frac{13}{52} + \frac{4}{52} = \frac{17}{52}$ – the wrong probability! So beware – before using this law make absolutely sure that the events are mutually exclusive!

1. Die is the singular, and dice the plural.

Look again at the table we obtained earlier which refers to the sum of the possible scores from casting two dice. Notice that there are 36 equally likely total scores. We could deduce this as follows: there are six ways the first die can fall, each of which can combine with any one of the six ways that the second die can fall. So there are $6 \times 6 = 36$ ways that both dice can fall. Now suppose we have two piles of cards. The first pile contains the two red aces and the four kings, and the second pile contains the two red aces and the ace of spades and the four kings. Suppose we draw a card from each pile – as there are six cards in the first pile and seven cards in the second, there will be $6 \times 7 = 42$ ways of drawing a pair of cards, one from each pile. We could represent the situation in a table like this:

		Second Card						
		A♥	A♦	A♠	K♥	K♦	K♠	K♣
	A♥	1	2	3	4	5	6	7
	A♦	8	9	10	11	12	13	14
First	K♥	15	16	17	18	19	20	21
Card	K♦	22	23	24	25	26	27	28
	K♠	29	30	31	32	33	34	35
	K♣	36	37	38	39	40	41	42

In this table we have numbered all the possible 42 combinations of events 1, 2, 3, ... etc., so combination 25, for example, means drawing the king of diamonds with the first card and the king of hearts with the second. Let us now define two events.

E_1 = the first card is an ace, so $P(E_1) = \frac{2}{6}$
E_2 = the second card is an ace, so $P(E_2) = \frac{3}{7}$

Suppose we wanted to find the probability that both cards were aces: We could write it symbolically like this

$$P(E_1 \cap E_2)$$

where the symbol ∩ is a shorthand form for 'both ... and'. Notice that there are six ways of obtaining two aces, so $P(E_1 \cap E_2) = \frac{6}{42}$. Now we could have obtained this result by multiplying $P(E_1)$ and $P(E_2)$ together ($\frac{2}{6} \times \frac{3}{7} = \frac{6}{42}$). This gives us the second law of probability: the so-called multiplication law.

$$P(E_1 \cap E_2) = P(E_1).P(E_2)$$

a law which applies only if E_1 and E_2 are *independent* (i.e. as long as the outcomes in no way affect each other). We will discuss this point more fully later, but it is worth noting now that independent events cannot be mutually exclusive, and mutually exclusive events cannot be independent.

The second law of probability enables us to modify the first law to take account of events that are not mutually exclusive. A little earlier, we considered the case of drawing a card from a pack, calling E_1 that the event was a heart (so $P(E_1) = \frac{4}{52}$), and calling E_2 that the card is an ace (so $P(E_2) = \frac{13}{52}$). We stated that the probability that the card is either a heart or an ace $P(E_1 \cup E_2)$ is *not* $\frac{4}{52} + \frac{13}{52} = \frac{17}{52}$. Now why doesn't the addition law work? Surely the fault here is that the card we draw could be the ace of hearts *and we have counted this card twice* — once as a heart and once as an ace. So the probability that the card is either an ace or a heart is $\frac{17}{52}$ minus the probability that the card is the ace of hearts. Using the second law, the probability that the card is the ace of hearts is $P(E_1 \cap E_2) = \frac{4}{52} \times \frac{13}{52} = \frac{1}{52}$, so the probability that the card is either an ace or a heart is $\frac{17}{52} - \frac{1}{52} = \frac{16}{52}$ — which agrees precisely with the result we obtained from first principles. We can now restate the addition law to take account of situations when the events are not mutually exclusive:

$$P(E_1 \cup E_2) = P(E_1) + P(E_2) - P(E_1 \cap E_2)$$

This is the so-called *general law of addition,* and it works whether the events are mutually exclusive or not (if the events are mutually exclusive, then they cannot both occur, so $P(E_1 \cap E_2)$ will be zero.

Applications of the Laws of Probability
EXAMPLE 1
Two types of metal A and B, which have been treated with a special coating of paint have probabilities of $\frac{1}{4}$ and $\frac{1}{3}$ respectively of lasting four years without rusting.

If both types of metal are given the special coating on the same day, what is the probability that

(i) both last 4 years without rusting
(ii) at least one of them lasts 4 years without rusting

For part (i), we can use the multiplication law to obtain a probability of $\frac{1}{4} \times \frac{1}{3} = \frac{1}{12}$ that both last for four years. Turning to the second part, we should first notice that four distinct outcomes are possible

(a) both last 4 years   (probability is $\frac{1}{4} \times \frac{1}{3} = \frac{1}{12}$)
(b) A lasts, B doesn't   (probability is $\frac{1}{4} \times \frac{2}{3} = \frac{2}{12}$)
(c) B lasts, A doesn't   (probability is $\frac{3}{4} \times \frac{1}{3} = \frac{3}{12}$)
(d) neither lasts   (probability is $\frac{3}{4} \times \frac{2}{3} = \frac{6}{12}$)

Now any of the outcomes (a), (b) or (c) satisfies the condition that at least one lasts four years, so that probability we require is $\frac{1}{12} + \frac{2}{12} + \frac{3}{12} = \frac{6}{12} = \frac{1}{2}$. Notice that we could have calculated the probability more directly by using the fact that

$$P \text{ (at least one lasts)} = 1 - P \text{ (neither lasts)}$$

This second method can save quite a lot of time, and you should always use it in preference to writing out all the possible outcomes.

EXAMPLE 2

An item is made in three stages. At the first stage, it is formed on one of four machines, A, B, C or D, with equal probability. At the second stage it is trimmed on one of three machines, E, F or G, with equal probability. Finally, it is polished on one of two polishers, H and I, and is twice as likely to be polished on the former as this machine works twice as quickly as the other. Required:

(1) what is the probability that an item is:
 - (i) polished on H?
 - (ii) trimmed on either F or G?
 - (iii) formed on either A or B, trimmed on F and polished on H?
 - (iv) either formed on A and polished on I, or formed on B and polished on H?
 - (v) either formed on A or trimmed on F.

(2) Suppose that items trimmed on E or F are susceptible to a particular defect. The defect rates on these machines are 10% and 20% respectively. What is the probability that an item found to have this defect was trimmed on F? (A.C.A.)

First, we shall determined the probabilities for each machine. As the formation stage is equally likely to occur on any one of the four machines, we have

$$P(A) = P(B) = P(C) = P(D) = \tfrac{1}{4}$$

Again, trimming is equally likely to occur on any one of the three machines, so

$$P(E) = P(F) = P(G) = \tfrac{1}{3}$$

Polishing is twice as likely to occur on machine H as machine I, so

$$P(H) = \tfrac{2}{3}, \; P(I) = \tfrac{1}{3}$$

(i) The probability that an item is polished on $H = P(H) = \tfrac{2}{3}$.

(ii) The probability that an item is trimmed on either F or $G = P(F \cup G) = P(F) + P(G) = \tfrac{2}{3}$.

(iii) Formed on either A or B, trimmed on F and polished on H

$$= P(A \cup B) \cap F \cap H$$
$$= [P(A) + P(B)] \cdot P(F) \cdot P(H)$$
$$= (\tfrac{1}{4} + \tfrac{1}{4}) \times \tfrac{1}{3} \times \tfrac{2}{3} = \tfrac{1}{9}$$

(iv) Either formed on A and polished on I or formed on B and polished on H

$$= P[(A \cap I) \cup (B \cap H)]$$
$$= [P(A) \cdot P(I)] + [P(B) \cdot P(H)]$$
$$= (\tfrac{1}{4} \times \tfrac{1}{3}) + (\tfrac{1}{4} \times \tfrac{2}{3})$$
$$= \tfrac{1}{4}$$

(v) Either formed on *A* or trimmed on *F*. These events are *not* mutually exclusive as it is possible to form on *A and* trim on *F*, so we need the general rule of addition, i.e.

$$P(A \cup F) = P(A) + P(F) - P(A \cap F)$$
$$= \tfrac{1}{4} + \tfrac{1}{3} - (\tfrac{1}{4} \times \tfrac{1}{3})$$
$$= \tfrac{1}{2}$$

Dealing with the second part of this question, we notice that *F* is twice as likely to produce a defective item as is machine *E*. So, given that the item is defective, the probability that it was trimmed on *F* must be $\tfrac{2}{3}$.

EXAMPLE 3
In the past, two building contractors, *A* and *B*, have competed for twenty building contracts of which ten were awarded to *A* and six were awarded to *B*. The remaining four contracts were not awarded to either *A* or *B*. Three contracts for buildings of the kind in which they both specialise have been offered for tender.

Assuming that the market has not changed, find the probability that
(a) *A* will obtain all three contracts;
(b) *B* will obtain at least one contract;
(c) Two contracts will not be awarded to either *A* or *B*;
(d) *A* will be awarded the first contract, *B* the second, and *A* will be awarded the third contract.   (I.C.M.A.)

The probability that *A* gets the contract $P(A) = \tfrac{10}{20} = \tfrac{1}{2}$, the probability that *B* gets the contract is $P(B) = \tfrac{6}{20} = \tfrac{3}{10}$, and the probability that neither *A* nor *B* gets the contract is $P(A \cup B)' = \tfrac{4}{20} = \tfrac{1}{5}$.

(a) The probability that *A* will obtain all three contracts is $\tfrac{1}{2} \times \tfrac{1}{2} \times \tfrac{1}{2} = \tfrac{1}{8}$

(b) The probability that *B* will obtain at least one contract
$$= 1 - P(B \text{ obtains no contracts})$$
$$= 1 - (\tfrac{7}{10} \times \tfrac{7}{10} \times \tfrac{7}{10}) = \tfrac{657}{1000}$$

(c) The probability that a contract is awarded to *A* or $B = P(A \cup B) = \tfrac{4}{5}$. We require to know the probability that two contracts will not be awarded to either *A* or *B*: this is equivalent to finding the probability that one of the contracts is awarded to either *A* or *B*. But the contract awarded to either *A* or *B* could be either the first, second or third contract. So the probability we require is

$\tfrac{4}{5} \times \tfrac{1}{5} \times \tfrac{1}{5}$ (*A* or *B* wins the first contract)
plus $\tfrac{1}{5} \times \tfrac{4}{5} \times \tfrac{1}{5}$ (*A* or *B* wins the second contract)
plus $\tfrac{1}{5} \times \tfrac{1}{5} \times \tfrac{4}{5}$ (*A* or *B* wins the third contract)
$$= 3 \times \tfrac{4}{5} \times \tfrac{1}{5} \times \tfrac{1}{5} = \tfrac{12}{125}$$

(d) The probability that *A* is awarded the first contract, *B* the second and *A* the third is $\tfrac{1}{2} \times \tfrac{3}{10} \times \tfrac{1}{2} = \tfrac{3}{40}$.

Tree Diagrams

One of the major problems that occur when dealing with probability is that of ensuring that all logical possibilities are considered. In fact, this is probably the most common student error when dealing with this topic. A tree diagram is a device to help us avoid such errors: it is a diagram which looks something like a tree, each branch of which represents one logical possibility. We shall illustrate tree diagrams by means of an example.

Buggsy Flynn owns 60% of protection rackets and 80% of illegal gambling in Chicago, and he is informed that the police are about to investigate both these activities. Police records show that 70% of investigations into the protection racket and 90% of investigations into gambling lead to court action. What is the probability that Buggsy will end up in court as a result of the investigations? (You may care to attempt to solve this problem before reading on.)

This problem involves an examination of whether Buggsy is investigated and whether he is charged. We can begin the tree diagram by calculating the probability of an investigation into Buggsy's protection rackets. Since he owns only 60% of the protection in Chicago there is a 0.6 probability that the police investigation will concern him, and the position can be illustrated as:

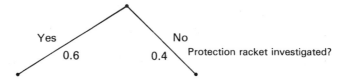

Notice that the probabilities have been inserted on to the diagram. We can now add the possibility of an investigation into illegal gambling on to the diagram.

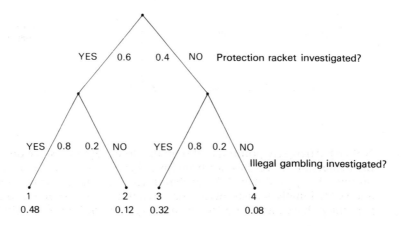

We now have four routes on our tree, and if we multiply the probabilities along each route, then we have deduced the probability of certain events occuring at each point. For example, point 1 represents the situation where Buggsy is investigated on both activities and we find the probability of this happening is 0.48; point 3 is the situation where Buggsy is not investigated on his protection racket but he is investigated on his illegal gambling, and the probability of this is 0.32. Notice that four points cover all the logical possibilities of investigations, so the probabilities must total 1 (0.48 + 0.12 + 0.32 + 0.08 = 1.0).

We will now insert on to the diagram the possibilities of court action resulting from a protection racket investigation.

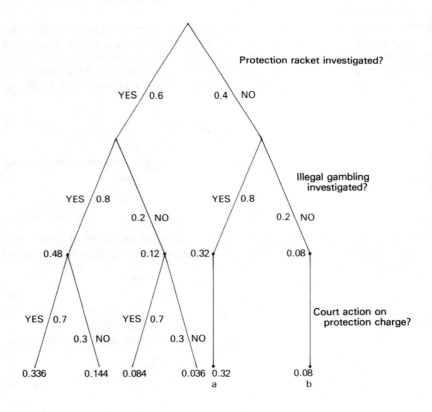

Notice particularly points a and b on this tree diagram. Since on this branch the protection racket has not been investigated there can be no court action on that charge and hence the probabilities at points a and b remain at 0.32 and 0.08. Finally we can insert the possibility of court action on an illegal gambling charge and find the probability of each logical possibility.

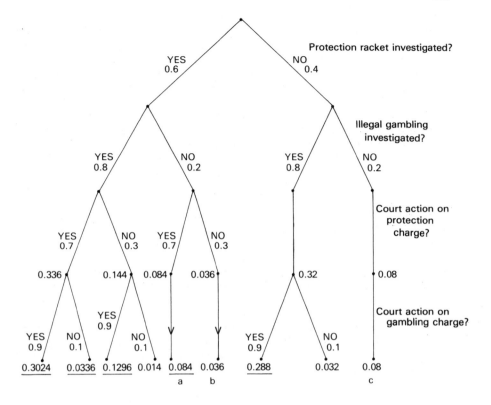

Notice again points a, b and c. In each of these branches there has been no investigation of illegal gambling. Hence there can be no court charge and the probabilities remain as they were at the previous junction.

In the diagram we have underlined every logical possibility that results in Buggsy landing up in court. The probability of his facing a court charge is the sum of the probabilities of all of these, i.e.

$$0.3024 + 0.0336 + 0.1296 + 0.084 + 0.288 = 0.8376$$

Conditionally Probability

In the first section, we stated that events cannot be both mutually exclusive and independent. This should not be taken to mean that if events are not mutually exclusive then they must be independent: there is a third category that we must now examine. When we state that events are independent, we mean that the outcome of one event in no way affects the outcome of the other. Now there are many cases where this is not true: the outcome of the second event is *conditional* on the outcome of the first event. Two examples may clarify this point.

Suppose we draw a card from a well-shuffled pack, and this card happens to be an ace. We now draw a second card — what is the probability that this card is also an ace? Well, having removed one card already, there must be 51 equally likely outcomes left, three of which would yield an ace. So the probability of an ace with the second card is $\frac{3}{51}$. Suppose the first card was not an ace: there would be 51 equally likely outcomes of which four would yield an ace. In this case, the probability of an ace with the second card would be $\frac{4}{51}$. In the first case, we are drawing from a pack with a lower proportion of aces than originally, and in the second case the pack has a higher proportion. Whichever way you consider this problem, the outcome of the first trial affects the outcome of the second. Now this raises an interesting philosophical problem: we cannot predict the outcome of the second event until we know the outcome of the first. If I deal a card to you, and you do not reveal it to me, the probability that I deal myself an ace must be $\frac{4}{52}$. Why is this so when a card has been removed from the pack? It is the *information* from the first card that is important, not the fact that it has been dealt. In this case, as you do not reveal the card it has a zero information value to me: from the information viewpoint, it is irrelevant to me whether the card is in your hand or in the pack. However, if your card is revealed to me, then I can use this information to calculate the probability that I deal myself an ace — the probability will be $\frac{4}{51}$ or $\frac{3}{51}$, depending on whether you have an ace or not. We need a new notation to take into account the fact that events can be conditional upon each other. If we have two events E_1 and E_2 then $P(E_2|E_1)$ is the probability that E_2 occurs given that E_1 has occurred. So we can now modify our multiplication law to take account of conditional probability

$$P(E_1 \cap E_2) = P(E_1) \cdot P(E_2|E_1)$$

If E_1 is draw an ace with the first card, then $P(E_1) = \frac{1}{13}$. If E_2 is draw an ace with the second card, then $P(E_2|E_1) = \frac{3}{51}$. The probability of drawing an ace with both cards is $P(E_1 \cap E_2) = \frac{1}{13} \times \frac{3}{51} = \frac{1}{221}$.

Now let us consider a second example. Suppose we have a box of ten machine parts, three of which are defective. From this we draw a sample of two parts — what is the probability they are both defective? If E_1 is that the first part is defective, and E_2 is the second part is defective, then the events are conditional. $P(E_1) = \frac{3}{10}$, $P(E_2|E_1) = \frac{2}{9}$ and $P(E_1 \cap E_2) = \frac{3}{10} \times \frac{2}{9} = \frac{1}{15}$. Can you see that if we had replaced the first part before drawing the second then E_2 would not be conditional on E_1, and $P(E_2)$ would also be $\frac{3}{10}$? Using the statisticians' jargon, we would say that *sampling without replacement makes the events conditional.* Is this always true? We might have drawn two parts from a very large consignment indeed, and it would seem rather pedantic to state that the consignment is poorer in defectives if the first item drawn is defective. Moreover, the proportion of defectives in a very large consignment can only be an estimate. When sampling from a large a large population, then E_1 and E_2 can for all intents and purposes be considered independent.

The importance of conditional probability is that it enables us to modify our probability predictions in the light of any additional information that is made available to us.

EXAMPLE 4

When exploration for oil occurs a test hole is drilled. If, as a result of this test drilling it seems likely that really large quantities of oil exist, (a bonanza) then the well is said to have structure. Examination of past records reveals the following information:

Probability (structure and bonanza)	0.20
Probability (structure but no bonanza)	0.15
Probability (no structure but a bonanza)	0.05
Probability (no structure and no bonanza)	0.60

We can put this information into a table like this

	Structure	No structure	
Bonanza	0.20	0.05	0.25
No bonanza	0.15	0.60	0.75
	0.35	0.65	1.00

Calculating the row totals and the column totals (often called the marginal probabilities), we can deduce that

Probability (Bonanza) = 0.25

since a bonanza can occur either with or without structure. Similarly

Probability (No Bonanza) = 0.75
Probability (Structure) = 0.35
Probability (No structure) = 0.65

Such probabilities are known as prior probabilities – they are obtained from past records. But suppose we are given some additional information – in particular suppose we know that a well has been sunk and structure is revealed. The number of possible outcomes has been reduced to two. We can ignore the "no structure" column of the table and conclude that the weighting for a "bonanza" is 0.2 and the weighting for "no bonanza" is 0.15. So Probability (Bonanza|structure) = $\frac{0.2}{0.2+0.15}$ = 0.571 and

Probability (No bonanza|structure) = $\frac{0.15}{0.2+0.15}$ = 0.429

Now suppose we were given the information that the hole does not have structure. We can deduce that

Probability (Bonanza|no structure) = $\frac{0.05}{0.65}$ = 0.077

and Probability (No bonanza|no structure) = $\frac{0.6}{0.65}$ = 0.923

The probabilities we have just calculated are called posterior probabilities — they cannot be established until after something has happened (i.e. the test hole drilled). The knowledge whether the test hole has structure or not is valuable. Without this knowledge we would estimate the probability of obtaining a bonanza at 0.25, but once the test hole is drilled we can revise our estimates. If the test hole reveals structure then the probability of a bonanza rises to 0.571; if it does not reveal structure the probability of a bonanza falls to 0.077.

Sometimes it is more useful to use tree diagrams to solve conditional probability problems.

EXAMPLE 5

Suppose we have 100 urns. Type 1 urn (of which there are 70) each contains 5 black and 5 white balls. Type 2 urn (which there are 30) each contain 8 black and 2 white balls. An urn is randomly selected and a ball is drawn from that urn. If the ball chosen was black, what is the probability that the ball came from a type 1 urn?

Firstly we notice that

P (urn type 1)   =   0.7
P (urn type 2)   =   0.3
P (Black|type 1)   =   0.5
P (White|type 1)   =   0.5
P (Black|type 2)   =   0.8
P (White|type 2)   =   0.2.

we can now draw the tree diagrams and find the prior probabilities.

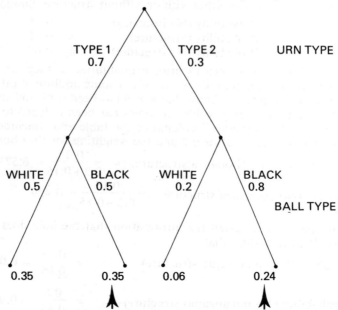

Now given that a black ball was drawn, the logical possibilities of the tree are reduced from four to the two arrowed. So the probability that the ball came from a type 1 urn is $\frac{0.35}{0.35 + 0.24} = 0.593$

One of the great difficulties of statistics examiners is the problem of ensuring that the questions they set are not ambiguous. Consider this case.

EXAMPLE 6

Firm A is one of many firms competing for a contract to build a bridge. The probability that Firm A is the first choice is $\frac{1}{9}$; the probability that it is the second choice is $\frac{1}{3}$, and the probability that it is the third choice is $\frac{1}{2}$. What is the probability that Firm A will not be the first, second or third choice?

The ambiguity in this question arises from the precise meaning of the expression "second choice" and "third choice". Are these probabilities joint probabilities, or are they conditional? Suppose we say that the probabilities are conditional — then the $\frac{1}{3}$ probability means that Firm A is the second choice, given that it is not the first choice. The question can then be approached like this.

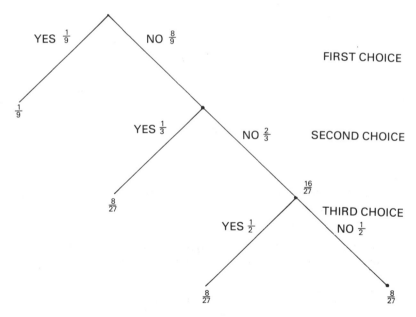

So, assuming the probabilities are conditional, the probability that the firm is neither the first, second nor third choice is $\frac{8}{27}$.

On the other hand, if the probabilities are joint probabilities, the probability of $\frac{1}{3}$ is the probability that the firm is not the first choice but is the second. Think carefully about this distinction. If the probabilities are conditional we say that the probability that the firm is the second choice given that it is not the first is P(not the first) × P(it is the second) = $\frac{8}{9}$ × $\frac{1}{3}$ = $\frac{8}{27}$. But if the probabilities are joint we are given the probability that the firm is not the first choice and the probability that it is (both not the first and is the second). That is we now say $\frac{8}{9}$ × P(it is the second) = $\frac{1}{3}$. The question would be approached like this.

We will calculate the conditional probability that Firm A is the government's second choice, given that it is not the first as follows:

P(it is not the first) × P(it is the second|not the first) =
P(it is both not the first and is the second)

that is $\frac{8}{9}$ × Conditional probability = $\frac{1}{3}$
Conditional probability = $\frac{1}{3}$ × $\frac{9}{8}$ = $\frac{3}{8}$

Thus we can build up our tree diagram.

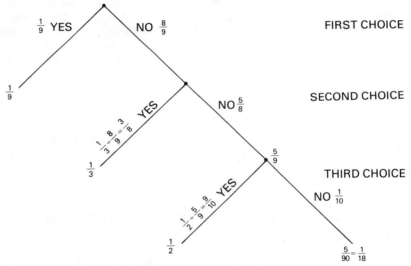

We could deduce the probability that the firm is not the first, second nor third choice by subtraction

$$1 - (\tfrac{1}{9} + \tfrac{1}{3} + \tfrac{1}{2}) = \tfrac{1}{18}$$

but it is more informative to work through conditional probabilities that we can calculate. Thus the probability that the firm is not the first choice is $\frac{8}{9}$. So

P (Firm A is second choice|not first choice) = $\frac{1}{3} \div \frac{8}{9} = \frac{3}{8}$
P (Firm A is not second choice|not first choice) = $1 - \frac{3}{8} = \frac{5}{8}$
P (Firm A is neither first nor second choice) = $\frac{8}{9} \times \frac{5}{8} = \frac{5}{9}$
P (Firm A is third choice|not first or second choice) = $\frac{1}{2} \div \frac{5}{9} = \frac{9}{10}$
P (A is not third choice|neither first not second) = $1 - \frac{9}{10} = \frac{1}{10}$
P (Firm A is not first nor second nor third choice) = $\frac{1}{10} \times \frac{5}{9} = \frac{1}{18}$
the same answer as before.

Now a few words to conclude this chapter. The concepts underlying probability are extremely easy to understand, and there are very few rules that must be learnt. However, putting these rules (or combinations of these rules) into practice can be very tricky indeed. You would be well advised to re-read the section on the laws of probability to make absolutely sure that you understand them. Also, make certain that you understand the logic employed in the previous examples before you attempt the following exercises. Before you attempt any calculations on the exercises, you should be certain you know the way or ways the desired event can occur (for example, 'at least one' means one or more). Also, you should decide on the relationship between the events – whether they are mutually exclusive, independent, or conditional. Suppose that, after all this, you still cannot cope with probability – what then? Well, at least you can console yourself with the knowledge that although some students find probability a highly stimulating intellectual exercise, by far the bulk of students find probability problems the nearest thing to purgatory! Frequently you will find that although you can see that your answer is wrong, it is extremely difficult to see just where you have gone wrong! Fortunately, a lack of capability at solving probability problems need not hold you up in a statistics course, as a knowledge of what probability means and how it is measured are the important things.

Exercises to Chapter 11

11.1 Explain carefully what is meant by 'the probability of an event'. Outline the main differences in philosophy of the Classical as opposed to the Beyesian School. I.C.A.

11.2 A bag contains nineteen balls, each of which is painted in two colours. Four are red and white, seven are white and black, and eight are black and red. A ball is chosen at random and seen to be partly white. What is it the probability that its other colour is red?

11.3 (a) Explain what is meant by the terms
  (i) mutually exclusive
  (ii) independent
 (b) A manufacturer purchases two machines A and B. The probability that A will last 5 years is $\frac{4}{5}$ and the probability that B will last 5 years is $\frac{3}{4}$. Find the probability that
  (i) both machines will last 5 years
  (ii) only machine A will last 5 years
  (iii) at least one machine will last 5 years

11.4 (a) Explain what you understand by
  (i) independent events
  (ii) conditional probability
 (b) 8% of the items produced in a manufacturing process are known to be faulty. A sample of 5 items is drawn at random. Calculate the probability that the sample
  (i) does not contain any faulty items
  (ii) only contains one faulty item
 (c) Two independent events A and B are such that $P(A) = 0.2$ and $P(B) = 0.4$. Find the values of (i) $P(AB)$, (ii) $P(A+B)$ and (iii) the probability of A occurring given that B has already occurred.

11.5 (a) State the theorems of addition and multiplication of probabilities.
(b) An analysis of the national origins of the employees of a manufacturing company produced the following result:

English	488
Scots	230
Irish	112
Welsh	108
American	30
Other European	16

If two employees of this company are chosen at random, state the probability that
(i) both will be other European
(ii) both will be English
(iii) both will be either Irish or Welsh
(iv) at least one of the two will be a Scot.

11.6 (a) Explain what is meant by (i) mutually exclusive events and (ii) independent events.
(b) (i) What is the chance of throwing a number greater than four with an ordinary die?
(ii) If the probability of throwing a seven with a pair of dice is $\frac{1}{6}$, what is the probability of rolling two sevens in a row?
(iii) If three dice are thrown together, what is the probability of obtaining at least one five?
(iv) Out of 10 steel valve springs 3 are defective. Two springs are chosen at random for testing. What is the probability that both test specimens are: (a) not defective;
(b) defective?

11.7 The following results were obtained from interviews with 500 people who changed their cars recently, cars being classified as large, medium or small in size, depending on their length.

Previous Car Size	Present Car Size			Total
	Large ($>15'$)	Medium ($13'-15'$)	Small ($>13'$)	
Large	75	47	22	144
Medium	36	75	69	180
Small	11	63	102	176
Total	122	185	193	500

Required:
(1) What proportion of people in the survey changed to:
(i) a smaller car?
(ii) a larger car?
(2) What effect would such car-changing habits have generally on the size of car owned in the future?
(3) What is the probability that a person from the survey selected at random who bought a large car, previously had a small or medium car?
(4) Estimate the average length of car owned at present by the 500 people surveyed, if no owner has a car less than 9 feet or greater than 19 feet.

A.C.A.

11.8 The probability that machine A will be performing a useful function in five years' time is $\frac{1}{4}$ while the probability that machine B will still be operating usefully at the end of the same period is $\frac{1}{3}$.
Find the probability that in five years' time:
    (a) both machines will be performing a useful function;
    (b) neither will be operating;
    (c) only machine B will be operating;
    (d) at least one of the machines will be operating.   I.C.M.A.

11.9 The probability that a man now aged 55 years will be alive in 2000 is $\frac{5}{8}$ while the probability that his wife now aged 53 years will be alive in 2000 is $\frac{5}{6}$.
Determine the probability that in 2000:
    (a) both will be alive;
    (b) at least one of them will be alive;
    (c) only the wife will be alive.   I.C.M.A.

11.10 A sub-assembly consists of three components A, B and C. Tests have shown that failures of the sub-assembly were caused by faults in one, two and sometimes all three components. Analysis of 100 sub-assembly failures showed that there were 70 faulty components A, 50 faulty components B, and 30 faulty components C. Of the failures 44 were caused by faults in two components only (i.e. A and B, or A and C, or B and C) and 10 of the 44 faults were in B and C.
If a faulty component is randomly selected, find the probability that it has faults in:
    (a) all three components;
    (b) component A on its own.   I.C.M.A.

11.11 (a) One bag contains 4 white balls and 2 black balls; another contains 3 white balls and 5 black balls. If one ball is drawn from each bag, find the probability that (i) both are white, (ii) both are black, (iii) one is white and one is black.
    (b) A purse contains 2 silver coins and 4 coppers coins, and a second purse contains 4 silver coins and 3 copper coins. A coin is selected at random from one of the two purses. What is the probability that it is a silver coin?
    (c) A box contains M components of which N are defective. Selection occurs without replacement. Find the probability that:
        (i) the first selection is defective;
        (ii) the first two selections consist of one defective and one non-defective;
        (iii) the second selection is defective.   I.C.A.

11.12 An amplifier circuit is made up of three valves. The probabilities that the three valves are defective are $\frac{1}{20}$, $\frac{1}{25}$ and $\frac{1}{50}$ respectively. Calculate the probability that (a) the amplifier works, (b) that the amplifier has one defective valve.

A bag contains r red balls and w white balls, $r > w$. If they are drawn one by one, show that the chance of drawing first a red, then a white and so on alternately until only red balls are left is

$$\frac{r!\,w!}{(r+w)!}$$

    I.C.A.

N.B. Do not attempt the second part of this question until you have read Chapter 13.

11.13 In a particular factory, an automatic process identifies defective items produced. Defectives can be classified as lacking strength, incorrect weight or incorrect diameter. A random sample of 1000 items were checked and the following results recorded:

120 have a strength defect;
80 have a weight defect;
60 have a diameter defect;
22 have strength and weight defects;
16 have strength and diameter defects;
20 have weight and diameter defects;
8 have all three defects.

Find the probability that a randomly chosen item
(i) is not defectve;
(ii) has exactly two defects.

I.C.A.

11.14 A manufacturer supplies transistors in boxes of 100. A buyer takes a random sample of 5 transistors and if one of them is faulty he rejects the box; otherwise he accepts the box.
(i) If there are 10 faulty transistors in the box what is the probability that the buyer will accept it?
(ii) What is the probability that he will reject a box which in fact only contains one faulty transistor?

11.15 The table given below shows a frequency distribution of the lifetimes of 500 light bulbs made and tested by ABC Limited:

Lifetime (hours)	Number of light bulbs
400 and less than 500	10
500 and less than 600	16
600 and less than 700	38
700 and less than 800	56
800 and less than 900	63
900 and less than 1000	67
1000 and less than 1100	92
1100 and less than 1200	68
1200 and less than 1300	57
1300 and less than 1400	33
	500

(a) Using the information in this table construct an ogive.
(b) Determine the percentage of light bulbs whose lifetimes are at least 700 hours but less than 1200 hours.
(c) What risk is ABC Limited taking if it guarantees to replace any light bulb which lasts less than 1000 hours?
(d) Instead of guaranteeing the life of the light bulb for 1000 hours, ABC Limited suggests introducing a 100-day money-back guarantee. What is the probability that refunds will be made, assuming the light bulb is in use:
   (i) 7 hours per day;
   (ii) 11 hours per day?

A.C.A.

11.16 A businessman estimates that the probability of gaining an important contract to build a factory is 0.65. If this contract is obtained he will certainly have a probability of gaining a further contract of building an associated computer block – he estimates that this probability is 0.8. If he fails to obtain the contract to build the factory, he will not be asked to deal with the computer block construction. As an alternative to the factory/block contract, there is a probability of 0.35 that the businessman could obtain the contract to build an office block – he could only deal with this if the first set of contracts was not obtained. What is the probability that both factory and computer block contracts will be gained? Also, what is the probability that the businessman will obtain either the computer block contract or the office block contract? I.C.S.A.

11.17 A company in which the training period for apprentices is five years is considering its intake for the coming year. Information concerning apprentices recruited in previous years is given below:

Year of intake	1963	1964	1965	1966
Number of apprentices recruited	700	500	150	250
Number of apprentices leaving in First year	28	18	8	14
Second year	29	18	6	10
Third year	17	8	2	1
Fourth year	13	8	1	2
Fifth year	2	2	1	1

(a) What is the probability that an apprentice will qualify?
(b) What is the probability that an apprentice will stay for longer than two years?
(c) How many apprentices should be recruited in 1971 to provide the company with 300 qualified men in 1976? I.C.M.A.

11.18 (a) Explain briefly the value of conditional probability calculations to the businessman.
(b)

Table: Number of Boxes

Firm	Defective Electron Tubes per box of 100 units			
	0	1	2	3 or more
Supplier A	500	200	200	100
Supplier B	320	160	80	40
Supplier C	600	100	50	50

From the data given in the above table, calculate the conditional probabilities for the following questions:
(i) If one box had been selected at random from this universe what are the probabilities that the box would have come from Supplier A; from Supplier B; from Supper G?
(ii) If a box had been selected at random, what is the probability that it would contain two defective tubes?
(iii) If a box had been selected at random, what is the probabiulity that it would have no defectives and would have come from Supplier A?
(iv) Given that a box selected at random came from Supplier B, what is the probability that it contained one or two defective tubes?
(v) If a box came from Supplier A, what is the probability that the box would have two or less defectives?
(vi) It is known that a box selected at random has two defective tubes. What is the probability that it came from Supplier A; from Supplier B; from Supplier C? A.C.C.A.

11.19 Two companies A and B regularly bid against each other for building contracts. A has a probability of $\frac{2}{3}$ of winning any given contract while B has a probability of $\frac{1}{3}$. Assuming independence, calculate the probabilities that, of the next two contracts, (i) A wins both (ii) B wins both, (iii) each company wins one.

11.20 (a) If an 'event' is defined as an outcome of an experiment, explain what is meant by
  (i) mutually exclusive events,
  (ii) independent events, and
  (iii) conditional events.
  Give examples of events that would fall into each category.
  (b) To control the quality of output, a sample of ten items is examined each hour and if no defective items are found the process continues, otherwise the process is stopped and adjustments are made. What is the probability that the process will be stopped after a sample is drawn if 10% of the items produced by this process is defective? How large a sample should be drawn to ensure that if 10% of items is defective, then the probability that the process is stopped is at least 99%?
  I.C.A.

11.21 A circuit is protected by two fuses, F and G, so arranged that the correct operation of either of them is sufficient to stop the circuit being damaged. The reliability (i.e. the probability that the device operates successfully when required) of the fuses, F and G, is 0.95 and 0.90 respectively. What is the probability that on a given occasion the circuit will not be protected?

11.22 In an experiment the probability of success is $\frac{1}{3}$. If it is performed 6 times, what is the probability of occurrence of (i) 5 successes, (ii) more than 4 successes?

11.23 (a) Explain what are meant by mutually exclusive events, independent events and dependent events.
  (b) A public company holds two accounts: an account A with a government department and an account B with a merchant bank. It has been established that on any given day there is a finite and distinct probability that each account will exceed its overdraft facilities. It may be assumed that the probability that either or both accounts exceed their overdraft facilities is 0.55 whilst the probability that account B exceeds its overdraft facility is 0.25.
  (i) If the fluctuations of the accounts are independent what is the probability that account A exceeds its overdraft facilities on any given day?
  (ii) Had the fluctuations of the accounts been dependent what would have been the probability that account A exceeded its overdraft facilities on any day in which account B also exceeded its overdraft facilities if the probability that both accounts exceeded their overdraft facilities had been 0.12?
  I.C.A.

11.24 If three cards are withdrawn from a pack (without replacement) find the probability that they are
  (a) Jack, Queen and King (in that order)
  (b) A Jack a Queen and a King.

11.25 Look again at example 4. Suppose E_1 is that oil is found in commercial quantities and E_2 is that the well has structure. Given that $P(E_1 \cap E_2) = 0.42$, $P(E_1 \cap E'_2) = 0.18$, $P(E'_1 \cap E_2) = 0.28$ and $P(E'_1 \cap E'_2) = 0.12$ what would you conclude?

11.26 A certain mass produced article sometimes has a dimension defect and sometimes a surface defect. Let
$\quad E_1$ = an article has a dimension defect
$\quad E_2$ = an article has a surface defect
$\quad E_3$ = an article is non-defective
If $P(E_1) = 0.06$, $P(E_2) = 0.07$ and $P(E_3) = 0.9$, what would you conclude?

11.27 Tom, Dick and Harry are candidates for the post of works manager. The managing Director will recommend a candidate and the recommendation must be ratified by the board of directors. The probability that the managing director recommends Harry is 60%, with 25% for Dick and 15% for Tom. The probabilities that the board ratifies are 40% for Harry, 35% for Tom, and 25% for Dick
(a) Find the probability that none of the candidates are appointed.
(b) Find for each candidate the probability that he is appointed given that the board has ratified the managing directors recommendation.
N.B. Use a tree diagram to solve this problem.

Chapter Twelve

Decision Criteria

We are all experts in taking decisions. Does this surprise you? Well throughout your life you have had many decisions to take, and the fact that you are alive today, reading this book, indicates that you have taken many decisions correctly. Take a simple problem — the need to get to the other side of a busy road. You have to decide when it is safe to cross. The fact that you have so far taken the correct decision proves that you have carefully weighed up the evidence, and interpreted the evidence correctly. Sometimes the decision will be easy to take; the road might be completely empty and you can cross with safety; and at other times the road will be so busy that to attempt to cross would be suicidal. In other words, we have all the necessary information available to make the correct decision every time. A cautious (and according to Road Safety Officers a sensible) person would obey the following decision rules: Only cross the road when no traffic is in sight. If we obey this rule, then we will always cross the road safely (unless we fall down an uncovered manhole!) Unfortunately, however, it is not always possible to obey this rule, as there are many roads so busy that we could never cross. With normal city centre roads we will seldom find a situation where they are completely empty of traffic, and we must use our judgement to decide when it is safe to cross.

Now of course, the decision taking problems faced by businessmen are much more complex than the decision whether to cross the road. Not only must the businessman decide between many alternatives, but also he often cannot be sure of the consequences of any decision he may make. In this chapter, we shall attempt to discover criteria that may aid the decision taking process.

Decision Criteria

Guy Rope owns a camp site in the Dordogne, and he wishes to develop the site in order to increase his profits. He realises that three options are open to him: he could build a swimming pool and charge the campers for its use, he could build a tennis court (and charge for its use), or he could build a restaurant to supply the campers with simple hot meals. He has sufficient funds to undertake just one of the options — what should he do? Obviously, the decision taken by Guy will depend on the profitability of the options, and he realises that the profitability will depend upon the weather. If the summer is too hot, then tennis may prove to be too exhausting; also evening barbecues in the open air will be more popular than restaurant meals. Given a hot summer, then, the swimming pool would be the most profitable option. If the summer is poor, then Guy reckons that his campers would welcome somewhere where they could buy hot meals, and so the restaurant would be the most profitable. People seem to prefer playing tennis when the weather is neither too hot nor too cold, so the tennis court would prove to be the most profitable given an 'average' summer. Armed with these beliefs, Guy estimates the annual profitability of each course of action as follows (all figures in thousand francs).

Table 11.01

		State of Nature		
		Summer is		
		Cool	Average	Hot
	Swimming pool	50	100	150
Strategies	Tennis court	30	180	90
	Restaurant	170	100	40

Notice that the options facing Guy are called *strategies:* a list of strategies is a list of courses of action facing the decision taker who has direct control over which course is chosen. However, Guy has no control whatsoever over the weather (if he had then there would be no problem!). The different types of summer that can occur are called *states of nature;* a list of events outside the control of the decision taker. The table above is usually called a *payoff matrix.* For each strategy it shows the payoff (in this case, the profit) that would result from each state of nature.

What should we advise Guy to do? Unfortunately, statistics alone cannot help us – we need to know something about Guy's character. What sort of person is he? Is he a gambler, a risk taker who is supremely confident that lady luck is always on his side? If Guy is this type of person, then he would reason as follows – "Lucky me – I am always right. If I build a swimming pool, then the summer will be hot and I will earn 150,000FF. If I build a tennis court, then the summer will be average and I will earn 180,000FF. If I open a restaurant, the summer will be cool and I will earn 170,000FF. The most logical thing for me to do is to go for the greatest profit, so I will build a tennis court".

Table 11.02 summarises Guy's reasoning. For each strategy, he notes the maximum payoff that can result. He then chooses the strategy with the greatest maximum payoff. If Guy acts in this way, then Guy is applying the *maximax criterion:* he is maximising his maximum possible payoff by gambling that the summer will be average.

Table 11.02

	Cool	Average	Hot	Maximum Payoff
Swimming Pool	50	100	150	150
Tennis Court	30	180	90	180 ←
Restaurant	170	100	40	170

Decision takers who practice the maximax criterion must be very rare specimens. If maximax was widely practised, then although we would have a few more millionaires, we would certainly have many more bankruptcies! Lady luck smiles on very few of us indeed! We shall now move to the other end of the spectrum and suppose that Guy is a born pessimist, who always assumes that the states of nature work against him. He would argue something like this: "Suppose I build a swimming pool. Will the summer be hot or even average? No chance! You can bet your bottom dollar that the summer will be cool! Likewise, one sure fire way to ensure a cool summer is for me to build a tennis court. Of course, if I open a restaurant then the summer will be the hottest for years and only a few campers will want feeding. The big money always avoids me. I might as well build a swimming pool as this at least guarantees me 50,000FF. profit. If I build a tennis court I could only be sure of 30,000FF. and opening a restaurant guarantees me no more than 40,000FF. A swimming pool it is, then". Table 11.03 summarises Guy's reasoning.

Table 11.03

	Cool	Average	Hot	Minimum Payoff
Swimming Pool	50	100	150	50
Tennis Court	30	180	90	30
Restaurant	170	100	40	40

For each strategy, he notes the minimum payoff that can result. He then selects the strategy that maximises his minimum payoff. Guy is applying the *maximin criterion;* he is maximising his minimum possible payoff.

So we see that the two criteria are fundamentally different, reflecting quite different states of mind. If Guy adopts the maximax criterion then he is pushing the ceiling on his profits to the highest level (180,000FF.) by building a tennis court. No other strategy could earn this much. However, he is taking a risk: his profit would only be 30,000FF if the summer was cool. On the other hand, if Guy adopts the maximin criterion, then he is raising the floor on his profits to the highest level (50,000FF) by building a swimming pool. No other strategy could guarantee as much as this. However, by building a swimming pool he foregoes the possibility of really large profits (180,000FF from a tennis court or 170,000FF from a restaurant).

There is a third way of analysing the problem facing Guy – a way that neither assumes that Guy is ultra optimistic nor assumes that he is ultra pessimistic. The reasoning is something like this: suppose the summer turns out to be cool – if Guy had opened a restaurant then he would have made the right decision. He would have no 'regret' at all, and earn the maximum possible profit under the circumstances (170,000FF). But suppose he had built a swimming pool — he has made the wrong decision and would certainly regret it. His profit would be 50,000FF, 120,000FF less than it would have been had he made the correct decision. It would seem reasonable, then, to use the 120,000FF as a measure of the extent of his regret. In a similar fashion, if the summer was cool and Guy had built a tennis court, then he would regret the 170,000 − 30,000 = 140,000FF he had foregone. If we use similar reasoning and assume an average summer; then assume a hot summer; we can calculate Guy's regret under all possible circumstances. Our results are summarised in a *regret matrix*. (see table 11.04).

Table 11.04

	Cool	Average	Hot	Maximum Regret
Swimming Pool	120	80	0	120
Tennis Court	140	0	60	140
Restaurant	0	80	110	110

For each strategy, the maximum regret has been identified (for example, had he built a tennis court, then Guy's maximum regret would have been the 140,000FF in profit foregone during a cool summer). Examining the matrix we conclude that Guy should open a restaurant, as this is the strategy that minimises the maximum regret he could experience. If he uses this reasoning as a basis for decision taking, then Guy is applying the *minimax criterion*.

In this section, we have examined three quite distinct criteria for decision taking, and we shall now state a few simple rules to summarise them. The rules will assume that in the payoff matrix the rows refer to the strategies and the columns refer to states of nature.

Rule 1

For each row in the payoff matrix, find the maximum value. If we select the row with the greatest maximum value then we are applying the maximax criterion

	Cool	Average	Hot	Maximum
Swimming Pool	50	100	150	150
Tennis Court	30	180	90	180
Restaurant	170	100	40	170

Rule 2

For each row in the payoff matrix, find the minimum value. If we select the row with the greatest minimum value then we are applying the maximin criterion.

	Cool	Average	Hot	Minimum
Swimming Pool	50	100	150	50 ◄
Tennis Court	30	180	90	30
Restaurant	170	100	40	40

Rule 3

Find the maximum value for each column in the payoff matrix, and subtract all the payoffs in each column from their corresponding maximum value. This gives the regret matrix.

Rule 4

For each row in the regret matrix, find the maximum value. If we select the row with the least maximum value, then we are applying the minimax criterion.

	Cool	Average	Hot
Swimming Pool	50	100	150
Tennis Court	30	180	90
Restaurant	170	100	40
Maximum	170	180	150

	Cool	Average	Hot	Maximum Regret
Swimming Pool	120	80	0	120
Tennis Court	140	0	60	140
Restaurant	0	80	110	110 ◄

Which Criteria?

We have applied three different criteria to the problem facing Guy. The Maximax criterion suggests he should build a tennis court. The maximin criterion suggests he should build a swimming pool. The minimax criterion suggests he should open a restaurant. Poor Guy must be utterly confused! Which criteria is appropriate to this problem? As we suggested earlier, this is an impossible question to answer as the strategy selected will depend upon Guy's character. However, it is possible to comment generally on the criteria.

1. For the problem facing Guy, each criterion suggested a different strategy, but it frequently happens that different criteria would suggest that we should select the same strategy. For example, if the payoff matrix was

	Cool	Average	Hot
Swimming Pool	30	100	170
Tennis Court	50	180	90
Restaurant	170	100	40

then whatever criterion is applied the strategy selected would be the same — build a tennis court (you should verify for yourself that this is true). This strategy is said to be *dominant,* and no problem of strategy selection exists.

2. The minimax criterion minimises the decision taker's maximum regret, and so it attempts to minimise the consequences of taking the wrong decision. This is an intrinsically satisfying criterion — it seems to be a highly logical method of decision taking. Moreover, this criterion fits in well with economists' ideas of opportunity cost (if you do not know what opportunity cost is, talk to an economist).

3. If we examine the payoff matrix facing Guy, then we notice that whatever strategy he selects and whatever the state of nature, Guy always makes a profit. Because of this, Guy may be tempted to apply the maximax criterion — especially if Guy is what economists call a profit maximiser. He would be following the well quoted 'law' that you must speculate to accumulate. But suppose it was possible for losses to occur: this could well be the case if he opened a restaurant and the summer was hot, or he built a swimming pool and the summer was cool. Surely, Guy would then be strongly tempted to apply the maximin criterion, especially if the losses could threaten his survival as a camp site owner.

4. Guy has a problem selecting the appropriate strategy because he has no control over the states of nature, but if he has *more information* about the states of nature then his decision can be more soundly based. What additional information would help? Well, the states of nature will not be equally likely to occur, so if Guy can assign probabilities to the states of nature this should increase his insight into the problem.

Expected Monetary Value

Suppose that Guy examines the Dordogne weather records over the last 100 years. On 20 occasions the summer was cool, on 70 occasions the summer was average and on 10 occasions the summer was hot. This information enables Guy to deduce the following empirical probabilities

State of Nature	Probability
Cool	0.2
Average	0.7
Hot	0.1

Suppose Guy thinks ahead over the next ten years — using the probability distribution he would predict 2 cool, 7 average and 1 hot summer. So if he was to build a swimming pool he would estimate his earnings (in thousands francs) at

$$2 \times 50 + 7 \times 100 + 1 \times 150 = 950$$

which is an average of $950 \div 10 = 95$ per year. This average earning is called the *expected monetary value* (EMV) of building a swimming pool, and we could calculate it more directly like this:

$$0.2 \times 50 + 0.7 \times 100 + 0.1 \times 150 = 95.0$$

In a similar fashion we could deduce that
EMV (build tennis court) = 0.2 × 30 + 0.7 × 180 + 0.1 × 90 = 141
EMV (open restaurant) = 0.2 × 170 + 0.7 × 100 + 0.1 × 40 = 108
We now have another criterion for selecting the appropriate strategy — *select the strategy with the greatest EMV* as this maximises the long run average gains. Guy would be well advised to build a tennis court and earn an average of 141,000FF per year.

Of the four criteria we have examined, EMV has the soundest logical base. Maximising your average long run profits certainly seems a sensible thing to do. However, the criterion does have its drawbacks. We have stated that Guy should build a tennis court because this has the greatest EMV (141,000FF) — but on any particular year he cannot earn this amount. He will earn either 30,000FF or 180,000FF or 90,000FF. Suppose one of these options was a loss, then we would have to enquire whether Guy is capable of shouldering the loss. Sooner or later, losses would be bound to occur, and if they could result in bankruptcy then Guy is clearly using the wrong criterion. The maximin criterion would be more appropriate.

The second problem with using the EMV criterion arises when the term 'long run payoff' makes no sense. In particular, let us suppose that Guy intends to run the camp site for just one summer, obtain the maximum profit he can, then leave the industry (he would be what economists call a 'snatcher'). The EMV of building a tennis court cannot be his long run average profit as there will be no long run! His average profit will be the profit for the year — either 30,000FF or 180,000FF or 90,000FF. But if he stays in the industry then as the years elapse, the closer will his average annual profit move towards 141,000FF. If Guy is a snatcher, must we conclude that the EMV criterion is inappropriate? Not necessarily. If Guy applies the EMV criterion to all the decision problems facing him, however diverse they may be, then he would be maximising his returns over the entire range of problems.

The Value of Perfect Information

For the rest of this problem, we will assume that Guy is a snatcher, basing his decisions on the EMV criterion. He decides to build a tennis court as this has the greatest EMV (141,000FF). Now suppose additional information is available to Guy — in particular let us suppose that a peasant is prepared to predict the weather and that his predictions are always right. In the jargon, we would say that Guy has *perfect information* available to him. This will certainly help the decision taking process. If the peasant predicts a cool summer (and there is a 20% chance that he will) then Guy will open a restaurant. If he predicts an average summer (and there is a 70% chance that he will) then Guy will build a tennis court. If he predicts a hot summer (and there is a 10% chance that he will) Guy will build a swimming pool. Given perfect information, then, Guy will earn either 170,000FF or 180,000FF or 150,000FF. His expected earnings, then, are

$$0.2 \times 170 + 0.7 \times 180 + 0.1 \times 150 = 175,000\text{FF}$$

Without this information, Guy's expected earnings were 141,000FF. So the information has a value to Guy of 175,000 − 141,000 = 34,000FF. This 34,000FF is the *expected value of perfect information* (EVPI) Now it is highly likely that the peasant will charge for the information, but the information would certainly be worth buying if its cost was less than EVPI. In other words, if the peasant charges less than 34,000FF for his predictions, then the predictions are worth buying, but if the peasant charges more than 34,000FF then Guy would be advised not to buy a prediction from the peasant and simply apply the EMV criterion.

The value of Additional Information

What is more likely is if Guy could buy reasonably accurate, though not perfect, information about the likely state of the weather during the summer. In the jargon, we call this *additional information*. To be particular, let us suppose that for 25,000FF Guy can buy a prediction about the weather, and such predictions have proved in the past to be 90% reliable. Should Guy buy this information, or should he follow the EMV criterion and build a tennis court?

It is certainly most convenient to analyse this problem with a tree diagram

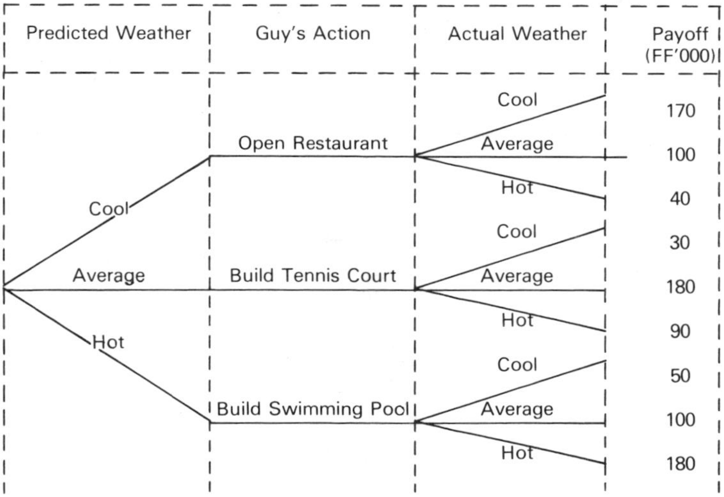

Predicted Weather	Guy's Action	Actual Weather	Payoff (FF'000)
	Open Restaurant	Cool	170
		Average	100
		Hot	40
Cool	Build Tennis Court	Cool	30
Average		Average	180
Hot		Hot	90
	Build Swimming Pool	Cool	50
		Average	100
		Hot	180

Diagram 12.1

Whether or not Guy should buy this information depends upon its expected value (EVAI) and in order to evaluate this we need the appropriate probabilities. Now we already know the probabilities for the predicted weather ie.

$$P(Cool) = 0.2, \; P(Average) = 0.7, \; P(Hot) = 0.1$$

However, the probabilities for the actual weather are conditional on the predictions made. Now we know that 90% of all predictions are accurate, so we can deduce that

P (cool summer|cool summer predicted) = 0.9
P (average summer|average summer predicted) = 0.9
P (hot summer|hot summer predicted) = 0.9

Now suppose that a cool summer is predicted — if the summer is average or hot then the prediction is wrong. Moreover, we know that the probability of a wrong prediction is 0.1, so

P (average summer|cool summer preducted) + P (hot summer|cool summer predicted) = 0.1,

We also know that the ratio of average summers to hot summers is 7:1, so

P (average summer|cool summer predicted) $= 0.1 \times \frac{7}{8} = 0.0875$
P (hot summer|average summer predicted) $= 0.1 \times \frac{1}{8} = 0.0125$

Using similar reasoning, we can deduce that

P (cool summer|average summer predicted) $= 0.1 \times \frac{2}{3} = 0.0666$
P (hot summer|average summer predicted) $= 0.1 \times \frac{1}{3} = 0.0333$
P (cool summer|hot summer predicted) $= 0.1 \times \frac{2}{9} = 0.0222$
P (average summer|hot summer predicted) $= 0.1 \times \frac{7}{9} = 0.0777$

We can now fill in the probabilities on to the tree diagram, calculate the joint probabilities and hence find the EVAI

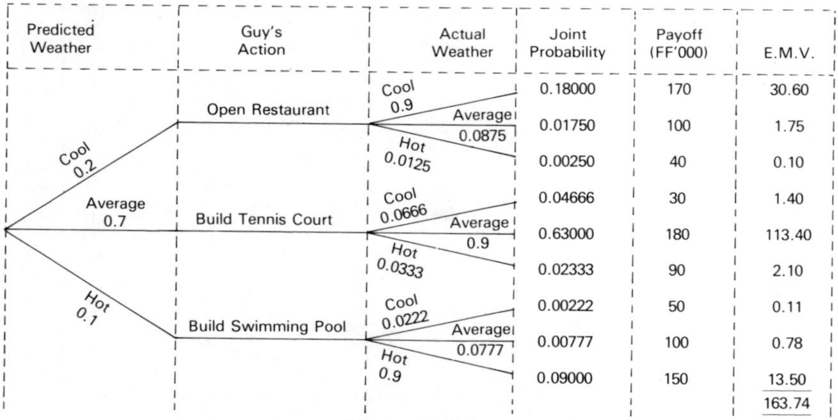

Predicted Weather	Guy's Action	Actual Weather	Joint Probability	Payoff (FF'000)	E.M.V.
Cool 0.2	Open Restaurant	Cool 0.9	0.18000	170	30.60
		Average 0.0875	0.01750	100	1.75
		Hot 0.0125	0.00250	40	0.10
Average 0.7	Build Tennis Court	Cool 0.0666	0.04666	30	1.40
		Average 0.9	0.63000	180	113.40
		Hot 0.0333	0.02333	90	2.10
Hot 0.1	Build Swimming Pool	Cool 0.0222	0.00222	50	0.11
		Average 0.0777	0.00777	100	0.78
		Hot 0.9	0.09000	150	13.50
					163.74

Diagram 12.2

If Guy buys the information, then his expected earnings are

163.74 − 25 = 138,740FF

and as this is less than the EMV for building a tennis court (141,000FF) we must advise Guy not to buy the information. Although the information does raise the EMV, the raise is more than offset by the cost of the information. We can calculate the Expected Value of Additional Information like this

EVAI = EMV (with information) − EMV (without information) − cost of information.

The information is only worth buying if EVAI is positive.
The EVAI for the problem facing Guy is

$$163.74 - 141 - 25 = -2.26$$

which again shows that the information is not worth buying.

Multi-stage Decision analysis

We shall now examine a more complex problem, in which the decision taking process is divided into stages. Flint McRae owns the oil prospecting and development rights for a plot of land in California, and a property development company has offered him $50,000 for the plot. Flint must now decide whether to sell the land or to exploit the land as an oilfield. Suppose that Flint decides to exploit the land — he sinks the well but finds no oil. In the jargon of the oil industry, the well is said to be 'dry', and Flint would lose $20,0000. However, the well might yield oil in commercial quantities — a so-called 'wet' well, and this would earn Flint $100,000 in total income. Exceptionally, Flint could obtain a real bonanza from his oilwell — a so-called 'soaking' well, and this would earn $500,000 in total net income. From past experience, Flint knows that there is a 70% chance that the well is 'dry', a 20% chance that it is 'wet' and a 10% chance that it is 'soaking'. If he wishes, Flint can engage a firm of geologists to undertake a seismic survey of the land, and this will cost him $30,000. The geologists supply the following information as to the reliability of such surveys (all figures are prior probabilities).

		Survey report is		
		bad	good	
	dry	0.44	0.26	0.7
true state of the well	wet	0.05	0.15	0.2
	soaking	0.01	0.09	0.1
		0.50	0.50	1.0

If Flint always uses the EMV criterion, what should he do?

In this problem, there are two sets of strategies facing Flint

1) Should he engage the geologist's services?
2) Should he drill or should he sell?

Likewise, there are two sets of states of nature facing Flint

1) The result of the survey (is the report good or bad?)
2) The true state of the well (is it dry, wet or soaking?)

Armed with this information, we can construct a tree diagram of the problem facing Flint. The boxes represent decision points and the circles represent states of nature. So we see that there are four distinct decision points facing Flint.

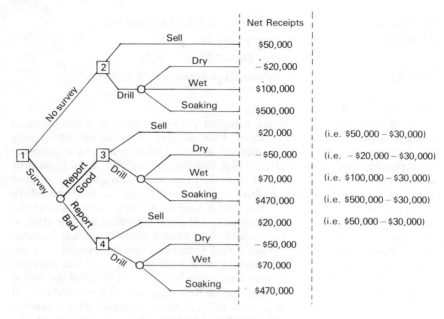

Diagram 12.3

Now before Flint can decide on decision point 1, he must decide on decision points 2, 3 and 4. Hence we will analyse the tree diagram from right to left — a method called the *rollback principle*.

Firstly we shall analyse decision point 2

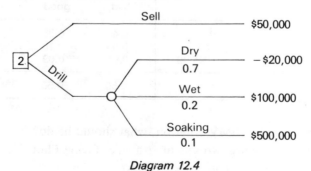

Diagram 12.4

Assuming that Flint reaches decision point 2
    EMV (Sell)   = $50,000
    EMV (Drill) = $0.7 \times -\$20,000 + 0.2 \times \$100,000 + 0.1 \times \$500,000$
                = $56,000

So we see that if decision point 2 is reached, then Flint will drill. If we now turn to decision point 3, we will need conditional probabilities, ie the probabilities that the well is dry, wet or soaking given a good report from the geologist

P (dry well|good report) = $0.26 \div 0.5 = 0.52$
P (wet well|good report) = $0.15 \div 0.5 = 0.30$
P (soaking well|good report) = $0.09 \div 0.5 = 0.18$

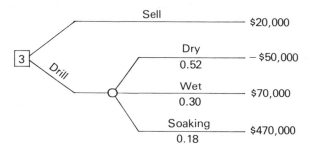

Diagram 12.5

Assuming that Flint reaches decision point 3
  EMV (Sell) = $20,000
  EMV (Drill) = $0.52 \times -\$50{,}000 + 0.3 \times \$70{,}000 + 0.18 \times \$470{,}000$
              = \$79,000

So if decision point 3 is reached, Flint should drill. Turning now to decision point 4

P (dry well|bad report) = $0.44 \div 0.5 = 0.88$
P (wet well|bad report) = $0.05 \div 0.5 = 0.10$
P (soaking well|bad report) = $0.01 \div 0.5 = 0.02$

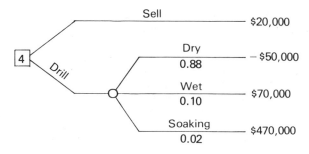

Diagram 12.6

Assuming that Flint reaches decision point 4
  EMV (Sell) = $20,000
  EMV (Drill) = $0.88 \times -\$50{,}000 + 0.1 \times \$70{,}000 + 0.02 \times \$470{,}000$
              = −\$27,600

Clearly, then if decision point 4 is reached, then Flint should sell.

Now that we have evaluated decision points 2, 3 and 4 we can evaluate decision point 1. Before we do this, however, it might be useful to summarise what we have concluded. If Flint does not employ the geologists services, then his best course of action is to drill (EMV = $56,000). If he does employ the geologist's services, then his action will depend upon the geologists report. If the report is good, then Flint should drill (EMV = $79,600) but if the report is bad he should sell (EMV = $20,000)

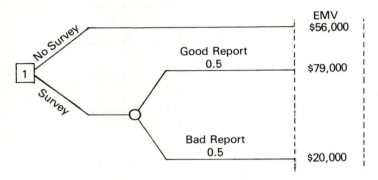

Diagram 12.7

Evaluating decision point 1 we have

EMV (no survey) = $56,000
EMV (survey)   = 79,600 × 0.5 + 20,000 × 0.5 = $49,800

Finally, then, we conclude that Flint should not engage the services of the geologists and drill for oil.

Before leaving this problem, you should notice that there is a chance (a 70% chance, in fact) that Flint faces a loss of $20,000 if he drills. Now if this loss is too great for Flint to shoulder, then he should sell to the property developer i.e. apply the maximin criterion.

Exercises to Chapter 12

12.1 An investment trust manager wishes to buy a portfolio of shares and he has sufficient funds to buy either portfolio A, portfolio B or portfolio C. The potential gains from the portfolios will depend upon the level of economic activity in the future, and the following estimates have been made (all figures in £000).

		State of Nature		
		Expansion	Stability	Contraction
Portfolio	A	100	50	−50
	B	50	100	−25
	C	−50	0	180

Which portfolio should be selected if the manager applies
a) the maximax criterion?
b) the maximin criterion?
c) the minimax criterion?

12.2 Suppose that the investment manager makes the following probability estimates for the states of nature in question 1

Expansion	Stability	Contraction
0.1	0.4	0.5

Which portfolio should the manager buy if he uses the EMV criterion?

12.3 If perfect information is available to the investment manager, how much should he pay for it?

12.4 An economic forecaster is prepared to predict the level of economic activity for the investment manager in question 1, but will charge a fee of £25,000. If the forecaster's predictions have been correct on 90% of occasions in the past, is the prediction worth buying?

12.5 A newsagent finds the demand for a certain weekly magazine varies between 10 and 16 copies. He pays the publisher 15p per copy and sells them for 25p per copy, but copies unsold at the end of the week have no value as they cannot be returned. Construct the payoff matrix facing the newsagent.

12.6 For the newsagent in question 5, how many copies should be held in stock according to
a) the maximax criterion?
b) the maximin criterion?
c) the minimax criterion?

12.7 Suppose the demand for the magazine has the following probabilities

Demand per week	10	11	12	13	14	15	16
Probability	0.05	0.10	0.15	0.25	0.20	0.15	0.1

How many magazines should the newsagent stock according to the EMV criterion?

12.8 Which criterion do you consider the more appropriate to the newsagent in question 5.

12.9 Mr. Wealthy has £50,000 invested at 10% per annum compound. He has been given the chance to buy the patent rights for an automatic rifle, and the patent rights will last for 10 years. All rifles made under this patent are sold to the Ministry of Defence, and the contract is worth £10,000 per year. Now it has been hinted that the government is contemplating an increase in defence spending, and should this happen the value of the contract would rise to £40,000 per year. Given that there is a 20% chance of increasing defence spending, and that the patent right would cost £50,000, should Mr. Wealthy buy the patent rights, or should he leave his money invested (NB ignore the income earning capacity of the income from royalties)
Hint: £50,000 invested at 10% per annum compound would grow to 50,000 $\times (1.1)^{10}$ = £129,687 in 10 year's time.

12.10 What would have to be the probability of an increase in defence spending to make it just worth while for Mr. Wealthy to buy the patent rights?

12.11 Suppose an official of the Ministry of Defence is prepared to reveal the government's intention on defence expenditure to Mr. Wealthy for a fee of £30,000. Is this information worth buying?

12.12 International Conglometerates Ltd. is considering which of two firms it should purchase. It could buy Allied Dog Foods for £3m. and this could be expected to yield £0.75m per year. However Parliament is considering legislation to restrict the number of dogs. A decision is not expected for two years, but if the number of dogs is restricted, then the annual receipts would fall to £0.1m. Alternatively, Guided Systems Ltd could be bought for £1m, and this would yield a cash flow of £0.5m per year. Within two years, the government must decide whether to replace the present early warning system, and Guided Systems would then have to decide whether or not to expand at a cost of £2m and tender for the replacement system. If the tender is successful, guided Systems cash flow would increase to £0.8m per year, but if the tender is unsuccessful (or if no tender is made) then the cash flow would fall to £0.1m per year. Any decision made now must last for the next ten years.

Draw a tree diagram of the alternative courses of action facing International Conglomerates. Insert the appropriate payoffs and hence decide the best course of action under the maximax and maximin criteria.

12.13 For question 12, find the best course of action according to the EMV criterion given the following probabilities
P (Government restricts number of dogs) = 0.4
P (Government replaces early warning system) = 0.7
P (Guided System's tender is successful) = 0.9.

12.14 If an official at the Home Office is prepared to release information on the government's intention towards restricting the number of dogs, how much is this information worth to International Conglomerates?

12.15 A manufacturer has spent £20,000 on developing a product, and must now decide whether to manufacture on a large or small scale. If demand for the product is high, then the expected profit during the product's life would be £700,000 for a high manufacturing level and £150,000 for a low manufacturing level. If the demand is low then the expected profit is £100,000 for a high manufacturing level jnd £150,000 for a low manufacturing level. The initial indication is a 40% chance of a high demand, but a market research survey could predict the demand with 85% accuracy. How much can the manufacturer afford to spend on market research?

Chapter Thirteen

Probability Distributions

In Chapter 11, we examined the basic concepts of probability. We can now use our knowledge to examine aspects of probability under certain rigidly defined conditions. The main difficulty that you have probably experienced so far in dealing with probability problems is that there is no one, assured method of dealing with them. However, when the conditions underlying a problem match the conditions we shall examine in this chapter, there will be a well-defined route to the solution. So although the concepts explained here are probably more involved than in Chapter 11, you can rest assured that the applications of the concepts will be much simpler. Our first task will be to examine what is meant by a probability distribution.

What is a Probability Distribution?

Suppose in an experiment we define all the outcomes, and choose one of the outcomes as an attribute. We could then form a frequency distribution by finding the probability that the attribute does not occur, occurs once, occurs twice − and so on. The distribution we have formed is called a *probability distribution*. An example will clarify what we mean by this. Suppose we spin a coin, and choose as our attribute the number of heads occurring. The probability distribution would look like this:

Number of Heads	Probability (Frequency)
0	$\frac{1}{2}$
1	$\frac{1}{2}$
	1

Notice that the sum of the frequencies is one − a feature of all probability distributions. This is because the number of times the attribute can occur (in this case zero or one), form mutually exclusive, collectively exhaustive events.

In forming the above probability distribution, we have been using *a priori* concepts: we calculated the probabilities *before* the experiment was performed. However, it is certainly possible to form a probability distribution using empirical evidence, i.e. after the experiment is performed. Suppose, for example, we know that in a large batch of components some of them will be defective. We decide to draw a sample of six items, so the sample could be free from defectives, or contain 1, 2, 3, 4, 5 or 6 defectives. We wish to find the probability of each outcome. Well, we could draw, say, 100 samples and count the number of defectives in each. Our result may look like this:

Number of defectives	0	1	2	3	4	5	6
Frequency	75	15	7	2	1	0	0

To form the probability distribution we divide each of the frequencies by 100 (the total frequency). Se we have:

Number of defectives	0	1	2	3	4
Probability	0.75	0.15	0.07	0.02	0.01

In this chapter we will examine certain standard types of probability distributions and use them as a basis for calculating a priori probabilities. You may feel that we are devoting too much time to a priori probability and ignoring the other forms. However, if we *suspect* that the outcome of an experiment would conform to one of the standard types of probability distributions then we can make some very useful predictions. We can always check our assumptions by obtaining empirical probabilities and comparing them with a priori probabilities. Later, we will show you how to test how well an a priori distribution fits or corresponds to the empirical evidence.

The Binomial Distribution

The first probability distribution we will examine is the so-called Binomial distribution, and we shall examine how it arises by considering a very simple example. Let us suppose that three coins are tossed together and take as our attribute the number of heads occurring. Well, the probability that none of the coins is a head can be calculated very simply using the multiplication law,

$$P_{(0)} = \tfrac{1}{2} \times \tfrac{1}{2} \times \tfrac{1}{2} = \tfrac{1}{8}$$

Now let us calculate the probability that one of the coins shows a head. The problem here is that whereas there is only one event where no heads occur, there are a number of events where just one head occurs, namely HTT, THT, TTH (where H is a head occurring and T is a tail occurring). Fortunately, each of the events is equally likely ($\tfrac{1}{2} \times \tfrac{1}{2} \times \tfrac{1}{2}$), so

$$P_{(1)} = 3 \times \tfrac{1}{2} \times \tfrac{1}{2} \times \tfrac{1}{2} = \tfrac{3}{8}$$

Two heads can occur in three equally likely ways (HHT, HTH, THH) so

$$P_{(2)} = 3 \times \tfrac{1}{2} \times \tfrac{1}{2} \times \tfrac{1}{2} = \tfrac{3}{8}$$

but three heads can occur in one way only (HHH) so

$$P_{(3)} = \tfrac{1}{2} \times \tfrac{1}{2} \times \tfrac{1}{2} = \tfrac{1}{8}$$

We have now calculated the full probability distribution, and we could represent it like this

No. of Heads Occurring	Probability	or Probability
0	$\frac{1}{8}$	0.125
1	$\frac{3}{8}$	0.375
2	$\frac{3}{8}$	0.375
3	$\frac{1}{8}$	0.125
	1	1.000

A binomial distribution is concerned with two terms (hence its name) – the probability that the event we are considering occurs (we will call this p) and the probability that the event does not occur (we call this $1 - p = q$). We have just considered an example where $p = q$ (the probability of obtaining a head is identical to the probability of obtaining a tail), but we can use a very similar analysis for cases where p and q are unequal *as long as the experiment is performed three times.* Suppose we have a very large consignment of components, and we know from past experience that 10% of components are defective. We randomly select three components from the consignment – what is the probability distribution of the number of defectives in the sample? We let p be the probability that an item in the sample is defective. Hence we have $q = 0.9$, $p = 0.1$.

$$P_{(0)} = q \times q \times q = q^3 = (0.9)^3 \qquad\qquad = 0.729$$
$$P_{(1)} = 3 \times q \times q \times p = 3q^2p = 3 \times 0.9^2 \times 0.1 \qquad = 0.243$$
$$P_{(2)} = 3 \times q \times p \times p = 3qp^2 = 3 \times 0.9 \times 0.1^2 \qquad = 0.027$$
$$P_{(3)} = p \times p \times p = p^3 = 0.1^3 \qquad\qquad\qquad = 0.001$$
$$\qquad\qquad\qquad\qquad\qquad\qquad\qquad\qquad\qquad 1.000$$

Notice that we are using the results of the previous example. There is only one way for the sample to contain no defectives, and only one way for it to contain three defectives. There are three equally likely ways that the sample can contain one defective, and three equally likely ways that the sample can contain two defectives.

Examining the above examples, we can see that to calculate the individual probabilities in a binominal distribution, we must find the number of equally likely events comprising the outcome we require, and multiply this by the probability of any one of the events. Suppose we increased the sample size in the question above to five items, and wished to find the probability that three of the items were defective. Now we know that this outcome can occur in a number of equally likely ways, and it is easy to calculate the probability of any one of these ways. For example

$$q \times q \times p \times p \times p = q^2p^3 = (0.9)^2(0.1)^3 = 0.00081$$

is the probability that the first two items are non-defective and the last three are defective. We now need to know the number of equally likely ways of obtaining this outcome. We could write out all the events (as we did previously) but this would be very tedious — there are 10 equally likely events in this case. Also, we could never be sure that we have included all of the outcomes! How do you think we could go on writing out all the outcomes of drawing, say, two defectives if the sample contained 100 items?

Pascal's Triangle

A very useful device for finding all the equally likely events in a binomial distribution is *Pascal's Triangle,* and an example of it is constructed below.

```
                    1   1                             n = no. of items
                    1   2   1                         2    in the
                    1   3   3   1                     3    sample
                    1   4   6   4   1                 4
                    1   5  10  10   5   1             5
                    1   6  15  20  15   6   1         6
Number of defectives 0   1   2   3   4   5   6
```

Reading off from the triangle, we see that if we draw a sample of 5 items ($n = 5$) there is one way that a sample can contain no defectives, there are 5 equally likely ways that the sample can have one defective, 10 ways for it to have two defectives — and so on. Using this information we can calculate the complete binomial distribution.

No. of Defectives	No. of Equally Likely Events	Probability of One of these Events	Required Probability	
0	1	$(0.9)^5$	$(0.9)^5$	$= 0.59049$
1	5	$(0.9)^4(0.1)$	$5(0.9)^4(0.1)$	$= 0.32805$
2	10	$(0.9)^3(0.1)^2$	$10(0.9)^3(0.1)^2$	$= 0.0729$
3	10	$(0.9)^2(0.1)^3$	$10(0.9)^2(0.1)^3$	$= 0.0081$
4	5	$(0.9)(0.1)^4$	$5(0.9)(0.1)^4$	$= 0.00045$
5	1	$(0.1)^5$	$(0.1)^5$	$= 0.00001$
				1.00000

So you can see that Pascal's Triangle is a useful way of finding the number of equally likely events. But if you are going to use Pascal's Triangle, it will be necessary for you to be able to construct it. The first thing you should notice is that the first and last number in any row is one. The rest of the numbers are obtained by adding the numbers in the previous row together in pairs. So the numbers for row six would be

```
      1    1+5   5+10  10+10  10+5   5+1    1
    = 1     6     15     20    15     6     1
```

and the numbers in row seven would be

```
      1    1+6   6+15  15+20  20+15  15+6   6+1    1
    = 1     7     21     35    35     21     7     1
```

The numbers that we are obtaining in the rows of Pascal's Triangle are called the *binomial coefficients,* and certainly it is easy to obtain them. But suppose we are drawing a sample of 20 items. If we don't have a copy of Pascal's Triangle available, it will be necessary to construct its first 20 rows, and this is going to be quite a task! It is possible, of course, but let us examine an alternative to Pascal's Triangle which we think that, in the long run, you will find easier.

Combinations

Most gamblers are aware that if you want the bookmaker to choose a stated number of results (say three) from a larger number of selections (say six) he will do so. In fact, this is one of the most popular forms of betting both on the football pools and on the racetrack. So popular is it that many bookmakers issue tables telling us how much it will cost to bet on any number of results chosen from a larger number. Such a table is a copy of Pascal's Triangle, but instead of reading off 'the number of defectives' we read off the 'number of correct results'. So gamblers can now look at the table and see that if they want to select six football teams and win the bet if any three of them draw, then we must make 20 bets. Bookmakers call these tables 'permutation tables', and instruct us to write our bets in the form 'Perm any 3 results from 6 selections = 20 bets'.

Now strictly, the bookmaker is wrong – when we are betting on three results from six we are dealing not with a permutation but with a combination, and we will now show you how to calculate how many combinations of 'r' results there are if you make 'n' selections. We can write this problem symbolically like this:

$$^nC_r$$

and to calculate the number of combinations we apply the formula

$$^nC_r = \frac{n(n-1)(n-2)\ldots[(n-r)+1]}{r(r-1)(r-2)\ldots \times 1}$$

This formula looks very fearsome, but in fact it is very easy to apply. If we examine Pascal's Triangle we see that if we select 7 football teams to win then there are 35 equally likely ways of just 3 of them winning. Now let us check this by using the formula for combinations. Here we have $n = 7$, $r = 3$, so $(n - r) + 1 = (7 - 3) + 1 = 5$, and

$$^7C_3 = \frac{7 \times 6 \times 5}{3 \times 2 \times 1} = 35$$

We are now in a position to derive a general expression for a binomial distribution. Suppose that within a population a proportion p has a certain attribute (call it 'defective') – it follows that $1 - p = q$ is the proportion of non-defectives. Now suppose we draw a sample of n items.

The probability that the sample is free from defective items $= P_{(0)} = q^n$.

The probability that the sample has one defective item is $P_{(1)} = {}^nC_1 q^{n-1} p$ (if one item is defective then $n-1$ must be non-defective)

The probability that the sample has two defective items is $P_{(2)} = {}^nC_2 q^{n-2} p^2$

We could continue in this way until we reach the probability that all the items in the sample are defective, which must be $P_{(n)} = p^n$. Now we know that all the probabilities must add up to one, so

$$q^n + {}^nC_1 q^{n-1} p + {}^nC_2 q^{n-2} p^2 + {}^nC_3 q^{n-3} p^3 + \ldots + p^n = 1$$

We also know that $q + p = 1$, so $(q+p)^n$ must equal one. Now we can write

$$(q+p)^n = q^n + {}^nC_1 q^{n-1} p + {}^nC_2 q^{n-2} p^2 + {}^nC_3 q^{n-3} p^3 + \ldots + p^n$$

and call the right-hand side of this expression the *expansion of the binomial* $(q+p)^n$. The individual terms of the right-hand side give the probabilities of 0, 1, 2 ... n defectives.

The great advantage of using the general expression for the binomial distribution is that we often do not wish to calculate the entire distribution, but just part of it. If we knew that a bag of seeds had a 95% germination rate, and we planted 10 of these seeds, we could quite easily calculate the probability that (say) 2 of the seeds fail to germinate. We have

$$q = 0.95, \quad p = 0.05, \quad n = 10, \quad r = 2, \quad (n-r)+1 = 9$$

$$P_{(2)} = \frac{10 \times 9}{1 \times 2}(0.95)^8 (0.05)^2 = 0.0746$$

We would like to point out that when we state that it is quite easy to calculate probabilities using the general expression for the binomial distribution, we really mean that it is easy to decide *what* to do. We are only too willing to admit that the actual arithmetic operations are most tedious.

The Mean and Standard Deviation of a Binomial Distribution

Let us begin by calculating the probabilities of the binomial distribution $(\frac{3}{4} + \frac{1}{4})^5$ (for example, sampling in groups of 5 from a population with 25% defectives). We shall use the general expression rather than Pascal's Triangle.

$$(\tfrac{3}{4} + \tfrac{1}{4})^5 = (\tfrac{3}{4})^5 + {}^5C_1(\tfrac{3}{4})^4(\tfrac{1}{4}) + {}^5C_2(\tfrac{3}{4})^3(\tfrac{1}{4})^2 + {}^5C_3 (\tfrac{3}{4})^2(\tfrac{1}{4})^3$$
$$+ {}^5C_4(\tfrac{3}{4})(\tfrac{1}{4})^4 + (\tfrac{1}{4})^5$$

$$= (\tfrac{3}{4})^5 + 5(\tfrac{3}{4})^4(\tfrac{1}{4}) + \frac{5 \cdot 4}{2 \cdot 1}(\tfrac{3}{4})^3(\tfrac{1}{4})^2 + \frac{5 \cdot 4 \cdot 3}{3 \cdot 2 \cdot 1}(\tfrac{3}{4})^2(\tfrac{1}{4})^3$$

$$+ \frac{5 \cdot 4 \cdot 3 \cdot 2}{4 \cdot 3 \cdot 2 \cdot 1}(\tfrac{3}{4})(\tfrac{1}{4})^4 + (\tfrac{1}{4}^5)$$

$$= 0.2373 + 0.3955 + 0.2637 + 0.0879 + 0.0146 + 0.001$$

We can now calculate the mean and standard deviation of the number of defectives in samples of 5 items.

Number of Defectives (x)	Probability (f)	fx	fx^2
0	0.2373	0	0
1	0.3955	0.3955	0.3955
2	0.2637	0.5274	1.0548
3	0.0879	0.2637	0.7911
4	0.0146	0.0584	0.2336
5	0.0010	0.0050	0.0250
	1.0000	1.2500	2.5000

$$\bar{x} = \frac{\Sigma fx}{\Sigma f} = \frac{1.25}{1} = 1.25 \text{ defectives per sample}$$

$$\sigma = \sqrt{\frac{\Sigma fx^2}{\Sigma f} - \left(\frac{\Sigma fx}{\Sigma f}\right)^2} = \sqrt{\frac{2.5}{1} - \left(\frac{1.25}{1}\right)^2}$$

$$= 0.968 \text{ defectives per sample}$$

In fact, there is no need to go through this procedure to obtain the mean and standard deviation. For any binomial distribution we know that

$$\text{mean} = np$$
$$\text{standard deviation} = \sqrt{npq}$$

We have just been considering the distribution $(\frac{3}{4} + \frac{1}{4})^5$, in which $p = \frac{1}{4}$, $q = \frac{3}{4}$ and $n = 5$. The mean is $np = 5 \times \frac{1}{4} = 1.25$, and the standard deviation is $\sqrt{5 \times \frac{1}{4} \times \frac{3}{4}} = 0.968$.

The Normal Approximation to the Binomial Distribution

Earlier, we considered the normal distribution, and used it to find the proportion in a distribution that was greater than or less than a given value. Now in fact, the normal distribution is another type of probability distribution. Earlier, we asked what *proportion* of electric lamps failed before (say) 1000 hours — but in fact we were also finding the probability that a lamp failed before 1000 hours. Can you see why? Surely, the proportion of lamps failing before 1000 hours can be taken as a measure of the probability that a particular lamp fails before this period. So when we are considering a normal distribution, *we can consider the words proportion and probability to be interchangeable.* The binomial distribution and the normal distribution, then, are two types of probability distributions. However, there is a subtle difference between them — the normal distribution is continuous in so far as the measurements considered can have any value (they are a continuous variable). The binomial distribution is discrete (we are concerned with integers only).

The greatest problem in using the binomial distribution is that it involves awkward arithmetic. Now if the sample size is very large, we can use the normal distribution as an approximation to the binomial distribution, and this will save a considerable amount of arithmetic. Suppose we draw a sample of 1000 packets of soap powder from a batch in which 20% of items are underweight, and we require to know the probability of obtaining more than 220 underweight packets in our sample. The appropriate binomial distribution is $(0.8+0.2)^{1000}$, and the probability we require is $P_{(221)} + P_{(222)} + P_{(223)} + \ldots + P_{(1000)}$. Obviously, this is going to be quite some task! Now the appropriate normal distribution to use is the one with the same mean and standard deviation as the binomial distribution. In this case, we have

$$\bar{x} = np = 1000 \times 0.2 = 200$$

$$\sigma = \sqrt{npq} = \sqrt{1000 \times 0.2 \times 0.8} = 12.65$$

In the diagram below we have drawn part of the histogram of the probability distribution $(0.8+0.2)^{1000}$ — it is *not* drawn to scale. If we want to find the probability of obtaining more than 220 packets, we must add the areas of the rectangles enclosing 221, 222, 223, ... 1000.

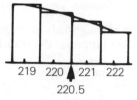

Diagram 13.1

The line joining the mid-points at the top of each rectangle represents the normal distribution that we are using as an approximation. If we wish to use this normal distribution as an approximation, then the diagram clearly shows that we require the proportion to the right of 220.5. This 0.5 adjustment is always used when we use the normal distribution as an approximation to a discrete distribution. So the situation looks like this:

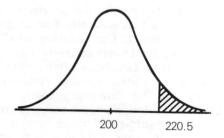

Diagram 13.2

The z score for 220.5 is

$$\frac{220.5 - 200}{12.65} = 1.62$$

and consulting the normal distribution tables we see that the probability we require is $100 - 94.74 = 5.26\%$.

Before we leave the binomial distribution, one very important point must be emphasised (though it is hoped that many of you will have realised it already). A binomial distribution assumes that the events are independent, that is, when we are drawing the sample the outcome of the first item drawn in no way affects the outcome of the second item drawn. This is only true if we are drawing from a large population.

The Poisson Distribution

Suppose we draw samples of two items from a population with 50% of items defective, and samples of 100 from a population with 1% of items defective. In both cases the mean is the same (one defective per sample) — but we could not expect to get the same probability distribution. In the first case, the number of defectives per sample can range between zero and two, but in the second case it can range between zero and 100. But if we draw large samples from a population containing a small proportion of defectives, then a very remarkable thing happens. The probability distributions for such samples tend to become the same *as long as the mean is constant*. To illustrate this feature we consider three cases: drawing samples of 100 from a population containing 1% defectives, samples of 1000 from a 0.1% defective population and samples of 10,000 from a 0.01% defective population. In each case then the mean is the same (one defective per sample).

$$(0.99 + 0.01)^{100} = 0.366 + 0.370 + 0.185 + 0.061 + 0.014 + 0.003 + 0.001 + \ldots$$
$$(0.999 + 0.001)^{1000} = 0.368 + 0.368 + 0.184 + 0.061 + 0.015 + 0.003 + 0.001 + \ldots$$
$$(0.9999 + 0.0001)^{10000} = 0.368 + 0.368 + 0.184 + 0.061 + 0.015 + 0.003 + 0.001 + \ldots$$

It would seem reasonable to suppose that we could discover the probability distribution for such samples as long as we know the mean number of defectives per sample. A third probability distribution — the *Poisson distribution* will enable us to do this.

Before we examine the Poisson distribution and discover how it works, let us examine another condition which calls for its use. So far, we have been considering a sample of known and determinable size, and counting the number of times that an event occurred. We could also count the number of times the event did not occur. However, there are many cases where we cannot count the number of times the event did not occur. Suppose, for example, we wished to investigate the incidence of industrial accidents in a particular trade. We could count the number of accidents within (say) one year and use this an an estimate of the mean number of accidents. But we cannot count the number of times the accident did not occur! For problems such as this, we would have to use the Poisson distribution, as we have no way of evaluating p, q or n.

The Poisson distribution looks like this:

$$e^{-x}\left[1 + x + \frac{x^2}{2!} + \frac{x^3}{3!} + \frac{x^4}{4!} + \ldots\right]$$

Let us look at this distribution and examine certain features of it that you might not have met before. Firstly, notice the numbers 2!, 3!, 4! (pronounce them '2 factorial, 3 factorial', etc.). This is just a convenient way of writing 'multiply the number n by $(n-1)$, then by $(n-2)$ and so on until finally we multiply it by one'. So

$$2! = 2 \times 1$$
$$3! = 3 \times 2 \times 1$$
$$4! = 4 \times 3 \times 2 \times 1$$
$$5! = 5 \times 4 \times 3 \times 2 \times 1$$

and so on. Secondly, there is the number e. This is a well-known constant and has a value 2.7183 (to four decimal places). In fact, most books of mathematical tables will have a table giving values of e^{-x}. Thirdly, the expression contains x, which is the mean of the distribution. We obtain probabilities from the expression like this:

$$P_{(0)} = e^{-x} \times 1$$
$$P_{(1)} = e^{-x} \times x$$

$$P_{(2)} = e^{-x} \times \frac{x^2}{2!}$$

$$P_{(3)} = e^{-x} \times \frac{x^3}{3!}$$

and so on. Now as e is a constant, the only variable in the expression is the mean x. So provided we know the mean of a distribution we should be able to calculate the probabilities. Let's try it and see.

In a particular industry, there are on average two fatal accidents per year. We want to find the probability that (a) the industry is free from fatal accidents and (b) the industry has three fatal accidents in a year.

In this case, we have $x = 2$, and the appropriate Poisson distribution is

$$e^{-2}\left[1 + 2 + \frac{2^2}{2!} + \frac{2^3}{3!} + \ldots\right]$$

$$P_{(0)} = e^{-2} \times 1$$

$$P_{(3)} = e^{-2} \times \frac{2^3}{3!}$$

Consulting a book of mathematical tables, we see that $e^{-2} = 0.1353$. However, in some examinations you are asked to calculate e^{-x} from first principles. To do this, we make use of the fact that

$$\log e^{-x} = \log 1 - x \log e$$

Using logarithm tables, $\log e = 0.4343$.

so
$$\log e^{-2} = 0 - (2 \times 0.4343)$$
$$= \bar{1}.1314$$
and
$$e^{-2} = 0.1353$$

Returning to our example.
$$P_{(0)} = 0.1353$$

so the probability that the industry is free from fatal accidents is 0.1353, or 13.53%

$$P_{(3)} = 0.1353 \times \frac{2^3}{3!}$$
$$= 0.1804$$

The probability that the industry has three fatal accidents in a year is 18.04%.

Conditions Necessary for using the Poisson Distribution

We can calculate probabilities using the Poisson distribution provided that we know the arithmetic mean of the distribution, and provided that p is small and n is large. A further condition is that the mean must remain constant. However, perhaps the most important condition is that the occurrence of the event must be purely at random. A good illustration of this is the flow of traffic along a highway. If we choose an isolated point on a highway and count the number of vehicles passing that point, then we would probably find a random flow of vehicles. However, if the point we choose is near a set of traffic lights, then we will not find a random flow — the flow will be 'bunched', i.e. the flow will be heavy when the lights show green. So we should be able to tell whether the occurrence of the event is random or not by using the Poisson distribution. In fact, this distribution was used for an important piece of statistical investigation during the Second World War. In 1944, London was subjected to bombardment by German V1 rockets (called doodlebugs). This was a pilotless vehicle packed with high explosive and equipped with sufficient fuel to carry it to London. When the fuel ran out, the vehicle would fall out of the sky and explode on impact. The problem was to decide whether the V1 rockets were guided to particular targets with a great degree of precision, or whether they fell on London at random. In other words, were the V1 rockets falling in clusters to a greater degree than could be ascribed to chance? R.D. Clark divided an area of 144 square kilometres into 576 equal squares, and counted the number of rockets falling in each square.

No. of bombs per square	0	1	2	3	4	5	Total
No. of squares	229	211	93	35	7	1	576

Now if we divide the frequencies by 576, then we can find the empirical probability distribution of the number of bombs per square

No. of bombs per square	0	1	2	3	4	5
Probability	0.3976	0.3663	0.1615	0.0608	0.0121	0.0017

If the bombs were falling at random, then we should be able to predict this probability distribution using the a priori Poisson distribution. To do this, we need to know the mean number of bombs per square.

x	f	fx	fx^2
0	0.3976	0.0000	0.0000
1	0.3663	0.3663	0.3663
2	0.1615	0.3230	0.6460
3	0.0608	0.1824	0.5472
4	0.0121	0.0484	0.1936
5	0.0017	0.0085	0.0425
	1.0000	0.9286	1.7956

For the moment, ignore the column headed fx^2. The mean is

$$\frac{\Sigma fx}{\Sigma f} = \frac{0.9286}{1} = 0.93 \text{ bombs per square}$$

and the Poisson distribution we require is

$$e^{-0.93}\left[1 + 0.93 + \frac{(0.93)^2}{2!} + \frac{(0.93)^3}{3!} + \frac{(0.93)^4}{4!} + \frac{(0.93)^5}{5!}\right]$$

Using the tables at the end of this book we find that

$e^{-0.93} = 0.3946$ so

$P_{(0)} = 0.3946$

$P_{(1)} = 0.3946 \times 0.93 = 0.3670$

$P_{(2)} = 0.3946 \times \frac{(0.93)^2}{2!}$ or using the previous term

$P_{(2)} = 0.3670 \times \frac{0.93}{2} = 0.1706$

$P_{(3)} = 0.3946 \times \frac{(0.93)^3}{3!}$

$= 0.1706 \times \frac{0.93}{3} = 0.0529$

$P_{(4)} = 0.3946 \times \frac{(0.93)^4}{4!}$

$$= 0.0529 \times \frac{0.93}{4} = 0.0123$$

$$P_{(5)} = 0.3946 \times \frac{(0.93)^5}{5!}$$

$$= 0.0123 \times \frac{0.93}{5} = 0.0023$$

Let us now write the empirical and Poisson probabilities adjacent to each other so that we can compare them.

No. of bombs per square	0	1	2	3	4	5
Empirical probabilities	0.3967	0.3663	0.1615	0.0608	0.0121	0.0017
Poisson probabilities	0.3946	0.3670	0.1706	0.0529	0.0123	0.0023

We feel sure that you will agree that there is a fantastically good agreement between the empirical and Poisson probabilities, and we must conclude that the V1 rockets were falling at random over the area.

Before we leave the Poisson distribution, we should note that its mean equals its variance (do you remember that the variance is the square of the standard deviation?). Now as the empirical and Poisson probabilities show such close agreement, we would expect the mean of the empirical probabilities to be approximately equal to the variance. We have already obtained the mean — it is 0.9286. The variance is

$$\frac{\Sigma f x^2}{\Sigma f} - \left(\frac{\Sigma f x}{\Sigma f}\right)^2$$

$$= \frac{1.7956}{1} - \left(\frac{0.9286}{1}\right)^2$$

$$= 0.9333$$

Again, we find a very close agreement between the mean and variance.

Of the two discrete probability distributions we have examined in this chapter, the Poisson distribution is the most widely used. It has been found to describe 'accidents", traffic flows, and the arrivals into queueing situations, very well indeed. Without doubt, both distributions are very important. However, the probability distribution that is the most important in statistical analysis is the one we examined in an earlier chapter — the normal distribution.

Exercises to Chapter 13
13.1 A chain is made of 10 links, and the strength of each link is equally likely to have any value between 50 and 60 lbs. What is the probability that the chain will break under a load of (a) 51 lbs, (b) 54 lb?

13.2 If half the electorate are supposed to vote Tory, and 100 investigators each ask 10 people their voting intentions, how many will report 3 or less voting Tory in their sample?

13.3 (a) A large company's records show that they have an average of 6% of their employees off work on any one day. They employ 6 van drivers. You may assume that the probability of absence from work of a van driver on any one day is the same as that for any other employee.
   (i) What is the probability that all their van drivers will be at work on a given day?
   (ii) What is the probability that at least 5 or their van drivers will be at work on a given day?

13.4 (a) The probability of a missile hitting a warship is $\frac{3}{5}$, and 3 hits are needed to put the warship out of action. If a salvo of six missiles are fired, show that the probability of putting the ship out of action is
$$\frac{19 \times 3^3}{5^4}$$
and evaluate to 4 decimal places.
(b) Razor blades are sold in packets of five. The distribution below shows the number of faulty blades in 100 packets.

No. of faulty blades	0	1	2	3	4	5
No. of packets	84	10	3	2	1	0

Calculate the mean number of faulty blades per packet. Assuming the distribution is binomial, estimate the probability that a blade taken at random from a packet will be faulty. (Hint: mean = np.) I.C.A.

13.5 It is expected that 10% of the production from a continuous process will be defective and scrapped. Determine the probability that in a sample of ten units chosen at random:
(a) exactly two will be defective; and
(b) at the most two will be defective; using in each case both:
   (i) the binomial distribution; and
   (ii) the Poisson approximation to the binomial distribution.
   (It is given that the value of e = 2,718.) C.I.P.F.A.

13.6 A buying department is considering an acceptance sampling scheme for incoming lots of a manufactured item that can be classified as either good or defective. The plan calls for a random sample of 50 items from each lot. If there is one or less defectives in the sample the lot is accepted, otherwise it is rejected.
(a) Find (i) from the binomial distribution, the probability of rejecting a lot that is 1% defective, and (ii) accepting a lot that is 10% defective.
(b) Find, using the Poisson distribution, the Poisson approximations to the above probabilities.
Comment on your results. I.C.A.

13.7 Ten per cent of males suffer from a certain disease. Use the normal approximation to the binomial distribution to find the probability that more than 60 men in a randomly selected group of 500 will suffer from the disease.

13.8 (a) The mean inside diameter of a sample of 400 washers produced by a machine is 8.92 millimetres and the standard deviation is 0.12 millimetres. Washers with inside diameters within the range of 8.74 to 9.10 millimetres are acceptable. Calculate the percentage of defective washers produced by the machine, assuming the diameters are normally distributed.
(b) Five per cent of the units produced in a manufacturing process turn out to be defective. Find the probability that in a sample of ten units chosen at random exactly two will be defective using:
(i) the binomial distribution,
(ii) the Poisson approximation to the binomial distribution. I.C.M.A.

13.9 An insurance salesman sells policies to 5 men, all of identical age and in good health. The probability that a man of this age will be alive in 30 years is $\frac{2}{3}$. Show that the probability of at least 32 men surviving the 30 years is:

$$\frac{2^3 \times 8}{3^4}$$

The probability that an individual suffers a bad reaction from an infection of a given serum is 0.001. Use the Poisson distribution to determine the probability that out of 2000 individuals, more than 2 will suffer a bad reaction. I.C.A.

13.10 A manufacturer uses high-speed weighing and packing machines with an accuracy adequate to ensure that only one bag in twenty is likely to be underweight when the machines are set to weigh a given weight. A customer receives a consignment of 60 bags. What are the changes that four or more will be underweight? (It is given that $e = 2.718$.) I.C.M.A.

13.11 At a certain time of day, the number of telephone calls coming in to a particular switchboard follows a Poisson distribution with a mean of 2 calls per minute. Calculate the probability of more than 4 calls in a minute.

13.12 The number of failures per week, of a certain type of machine, has been found to follow a Poisson distribution with mean 0.5. What is the probability that a given machine has three or more failures in a given week?

A firm owns five of these machines. What is the distribution of the total number of failures per week?

13.13 The demand for a component is 2 per month, and form a Poisson distribution. Stock is made up at the beginning of each month. What should be the stock level at the beginnning of each month so that the probability of a stockout is less than 5%? ($e^{-2} = 0.1353$.) I.C.A.

13.14 In a particular industry, 1 worker in 500 is fatally injured per year. What is the probability that a firm in the industry has not more than one fatal accident in a given year, given that the firm has a labour force of 300 men?

13.15 Suppose samples of 200 are drawn from a population with 0.5% of items defective. Use the Poisson Distribution to find the probability distribution for the number of defectives per sample. What percentage of samples contain a greater number of defectives than the mean plus $3 \times$ standard deviation?

13.16 | Number of defectives per sample | 0 | 1 | 2 | 3 | 4 | 5
| --- | --- | --- | --- | --- | --- | --- |
| Number of samples | 20 | 30 | 25 | 15 | 5 | 5 |

Fit a Poisson distribution to this data.

13.17 For mean values of 10 or more the Poisson distribution can be approximated by the normal distribution with the same mean and variance. The number of red blood cells per unit volume for a particular animal has the Poisson distribution with mean 15. Find the probability that the number of red cells in a given unit volume exceeds 20.

13.18 A complex television component has 1000 joints soldered by a machine which is known to produce, on average, one defective joint in forty. The components are examined, and faulty soldering corrected by hand. If components requiring more than 35 corrections are discarded, what proportion of the components will be thrown away? I.C.M.A.

13.19 In a certain company after long years of experience the following information is known about "debtor" accounts. Accounts have a mean value of £2,500 and a standard deviation of £300. With regard to collection on time, 5% of the accounts can be classified as "doubtful". Stating carefully any assumptions you might make, answer the following questions.
  i) If an account is randomly chosen, what is the probability that its value will be between £2600 and £2850?
  ii) If five accounts are randomly chose, what is the probability that there would be exactly two doubtful accounts?
  iii) One account is randomly chosen and it is learnt that its value is less than a certain amount with a 0.12 probability. What is that specific amount?
  B.A.(Business Studies)

13.20 Among the products of a certain lumber company are beadings of mean length 80cms and standard deviation 2cms. The production process is estimated to produce 10% "misshaped" beadings. It is specified also that the average number of knots per 100cms of beading is 1.5. Stating carefully any assumptions you might make, answer the following questions
  i) If one beading is chosen, what is the chance that its length is between 83cm and 85 cm?
  ii) If a sample of five beadings are selected, what is the probability that there are three misshapes?
  iii) What is the probability that one beading of mean length will contain two or less knots? B.A.(Business Studies).

Chapter Fourteen

Sample Design

You will probably be aware that most statistical information is obtained, not by examining the whole population, but by obtaining a sample and arguing that the characteristics of the sample are the same as those of the population as a whole. The sample result, obviously, will differ from that we obtain by taking a complete census, and later chapter will be concerned with the reliability of the sample results. Our objective here is not to discuss the results of a sample survey, but to explain to you how to select a sample which is truly representative of the population. You would not, for example, expect to obtain a reliable estimate of the heights of men in the United Kingdom if you chose your sample only from men in the Guards regiments; you could not possibly estimate the number of miles people travel by taxi each week if your sample consisted only of taxi-drivers.

Such statements as this are, we think, obvious, but they only tell us how not to select a sample; they give no information about how we should select one. So let us proceed step by step. The basic principle underlying any selection of a sample is that it should be a *random* selection — that is, that every single unit within the population should have an equal chance (or the same probability) of being chosen as a member of the sample. This is, of course, very difficult to achieve. Consider the simple case of drawing a raffle ticket to select the winners of five prizes. Why do you think that after each ticket is drawn the drum containing them is turned? It is, as you will know, to mix up the tickets once again. But would it make any difference if the drum were not turned? In our opinion, it would make a great deal of difference. The person drawing the ticket can easily introduce *bias* into the selection. Consider carefully how you yourself would select the winning tickets. Our observation shows that few people will choose tickets from the top of the drum. They tend rather to plunge into the pile of tickets and select one from near the bottom. Can you see that this sort of behaviour means that tickets near the top of the pile do not have an equal chance of selection with those nearer the bottom? The drum is turned after each ticket is selected in order to offset this type of bias. We cannot change human behaviour, but we can change the position of the tickets so as to give each one an equal chance of selection.

To return to sampling proper, in selecting a sample to determine the average number of miles travelled by taxi each week, the person who never takes a taxi must have just as much chance of being selected as have those who use taxis regularly — but no greater chance. In order to ensure this, statisticians have devised various methods of selection which, as far as possible, eliminate bias. If we are able to list the population in some sort of order and number them, we could determine the members of our sample by drawing numbers at random, drawing sufficient numbers to include, say, 5% of the total population. Better still, we could use a table of *random numbers* to determine who should be included in our sample. This table is so constructed that if you select any point at which to start and move through the table consistently in one direction, horizontally, vertically or diagonally, the digits you read off are randomly chosen and every number has the same probability of being selected. Suppose that the total population consisted of 1000 units. These could be numbered from 0 to 999 and you could select those to be included in the sample by reading off digits from the table of random numbers in groups of three. If, for example, the first three digits were 294, we would select the unit numbered 294 as a part of our sample. Similarly if the first three digits were 004, we would select the unit numbered 4.

If, on the other hand, we were examining industrial output for faulty units, we could use a form of pseudo-random sampling, by examining and checking the unit coming off the production line at predetermined time intervals. Alternatively, we could examine, say, five units, chosen at random from every carton of output. One way or another, however, we must ensure that our selection of the sample is random.

The Sampling Frame

Before we can draw a truly random sample we must be able to define the total population. The Electoral Register contains the names of all those entitled to vote in the United Kingdom; enrolment forms will list the names of all students in your college; business files will contain all the orders placed in a given financial year. Such lists, files or card indexes form what is known as a *sampling frame*. They will give basic details about every member of the population you are concerned with. Now, before drawing a sample, you must carefully examine the sampling frame to ensure that it is adequate for your purpose. If it is deficient in that a number of units are not included, those units have no chance of being selected and your sample cannot be truly random. Let us illustrate such deficiencies by looking at a very real problem.

It is the habit in Britain for public opinion polls to try to predict the results of government elections by interviewing a relatively small number of voters. The sampling frame for such a survey is the Electoral Roll, a list of all those eligible to vote who lived in the area last November. The list is produced in March each year, so when it is published it is already four months out of date. By the following February then, just before a new list is published, it is sixteen months out of date. Since about 0.5% of the population move into and out of the district each month (according to a Social Survey report) this would mean that about 8% of those who are on the Roll no longer live in the area, while about 8% of those living in the area are not on the Roll. Add to this the fact that about 4% of those who should be on the register are omitted because they do not return the necessary forms, and we find that the Electoral Roll is only about 88% accurate — not a good situation for a sampling frame. It is for this reason that investigators interview only those who actually go to the polling booth, or say that they intend to do so.

Systematic Sampling

In practice true random sampling is not possible unless there is a good sampling frame and the population is fairly small. So, for most practical work, investigators resort to methods of selection which are *quasi-random*, or not truly random. One of the most popular is to choose as a member of the sample every nth item on a list, the first sample unit being selected by some random method. If, for example, we wanted a sample consisting of 2½% of the population we would firstly select at random the first unit, say the 29th invoice in a file, and then every 40th invoice after that — the 69th, 109th and so on. Can you see why this type of sampling is not truly random? While the first item is chosen by random methods, every other item is then preselected. Nevertheless, this method of selection (called systematic sampling) approximates to random sampling sufficiently well to justify its widespread use. Once again we need a reasonably good sampling frame, but, in our opinion, having accepted that the sample is not truly random, it need not be 100% accurate. However, if we are to use systematic sampling, we must ensure that there are no 'cyclical' patterns in the sampling frame that match our choice of items. If, for example, we are taking a survey on a housing estate to assess the opinion of the residents about the 'open plan' layout, and every tenth house is a corner house, then a 10% systematic sample either always, or never, includes a corner house. Such a pattern as this could substantially affect the results we get.

Now, systematic sampling is useful so long as the population is homogeneous. It is suitable, for example, if accountants are selecting a sample of invoices to check for errors; or if the police are taking a sample of motorists passing a certain point to check that they have a valid driving licence. But you will appreciate that in much statistical work the population is heterogeneous. This is especially true if we are investigating opinions, because in many cases a person's opinions are formed by, or result from, the social class to which he belongs. If we are asking the simple question, 'Do you think that people earning over £10,000 a year pay too much in taxes?' the answer we get will often depend on the income of the person being questioned. Inevitably the rich will answer 'yes', while the poor are more likely to say 'no'.

Stratified Sampling

When the response we get to our questions is likely to depend in this way on the social group to which the person being questioned belongs, we will get far better results if we adopt *stratified sampling*. If we are investigating the social evils of traffic on the roads, it is quite possible that systematic sampling would result in a sample which contained only those who owned cars. The opinions of those who do not own cars would not be represented, and to that extent the sample results will be biassed. Stratified sampling has been designed to ensure that all important views are represented in the sample. In this type of sample each social group is represented in the sample in proportion to the size of that group in the population as a whole. Let us take a simple example to show how stratified samples are constructed. Suppose we are investigating opinions on education of those who are still at school or college. We may decide that these opinions will depend on whether the student is at university, state school, or private school. We know that 20% of students are at university, 10% at private school and 70% at a state school. If we are selecting a sample of 2000 students, our first step is to say that 20% of the sample, or 400 students, must be university students, 200 (10%) from private schools, and 1400 (70%) from state schools. We may now additionally decide that men and women have markedly different opinions, so we will now divide our three groups (or strata) into two sub-groups. If, for example, we know that 15% of university students are women, we would include 60 women among our sample of 400 university students. Each of these sub-groups can be further subdivided, perhaps in relation to age or to parents social class. It does not matter how far groups are subdivided, provided that we are stratifying according to a characteristic which is relevant to the survey. If, for example, we are investigating television viewing habits there is little point in stratifying the sample according to political affiliations. It is probably completely irrelevant. On the other hand, our viewing habits might well be affected by the size of our family, or whether we live in a rural or urban area and we would stratify according to these criteria.

Once we have decided on the number of people in each stratum of our sample, the persons to be included must, of course, now be chosen by some random method.

Now, one peculiarity can arise with this type of sampling. If our strata are many, it may be that an important sub-group of the population is entitled to be represented by only one or two members. It may be that we feel that we cannot obtain an adequate representation of the opinions of this group if we question one person only, and, if so, we should include more representatives of the group, say four or five. While this may seem reasonable, it means, of course, that members of this group have a far higher probability of being chosen as a part of the sample than have people in other strata. This type of sampling is known as sampling with a *variable sampling fraction*, and in assessing our final results we must make allowance for the difference in the representation of such strata. In many surveys, when samples are drawn from a markedly heterogeneous population the sampling fraction is variable rather than uniform. The principle finds its most important application in the evaluation of stock value by sampling methods. We can afford to take a very small sample of items costing only a few pence, since, if our sample is inaccurate it will make little difference to the total value of stock. But if the items of stock cost a great deal, even a small error may significantly affect our total stock valuation. Thus, in determining total stock value from a sample valuation of stock, accountants almost invariably ensure that high-value items have a greater probability of being selected than do low-value items.

Once again, with stratified sampling, we need a sampling frame which will give us a great deal of information about the social structure. Fortunately nowadays governments undertake so many social surveys and publish so many statistics that there is little difficulty in obtaining the necessary information. Hence stratified sampling is becoming increasingly popular whenever the objective is to assess people's attitudes or opinions.

Multi-Stage Sampling

One of the problems of simple random sampling is that if the sample selected is widely scattered over the country the interviewer may spend more time in travelling than he does in actually interviewing. Often, then, the cost of taking such a sample can be prohibitive. This is especially true in countries in which urbanisation has not yet developed. You can imagine the problems involved if you were to take a sample of people scattered all over central Canada, Malaysia or West Africa. Interviewers might have to spend weeks in travelling.

It is to overcome this problem that multi-stage sampling has been developed. Rather than spending time in travelling to interview 2000 people scattered all over Malaysia, is it not more convenient to interview say 100 people in each of twenty selected areas? Provided that we can ensure that the sample ultimately selected is representative, a great deal of time and expense can be saved.

In many ways the selection of a multi-stage sample is similar to the selection of a stratified sample, but we select primary groups and sub-groups geographically rather than on the basis of social characteristics. Typically, the first stage is to break down the area under survey into a number of standard regions. These may be areas such as the English county, the Canadian province or any other easily defined administrative area. The sample is then divided among these regions according to their population. The second stage is to select at random a small number of districts, say towns and villages, within the primary region. Once again it is necessary to allocate to each of these districts a number of interviews proportional to its population. Almost certainly at this stage, an element of stratification must appear if the sample is to be representative. Let us suppose that we are selecting six towns at random within each region. If ⅓ of the population live in cities of over 200,000 people, a further ⅓ in towns with a population of from 20,000 to 200,000 and the remainder in small towns and villages of under 20,000 inhabitants, it would be desirable to select two areas of each type.

Finally, having selected our towns and villages within each region, the sample in each town or village is chosen by some random method. Again, it may be felt to be desirable to choose a stratified sample within each town. How far this stratification is carried depends, of course, on the purpose of the survey and the homogeneity of the population. If we are examining housing conditions, and the type of housing ranges from one-roomed flats to 25-roomed mansions, some degree of stratification will be necessary. But if all the housing in the area is of an essentially similar type, as it may be in a small village, stratification is probably not necessary.

Every type of sampling discussed so far depends on the existence of a sampling frame. If no such frame exists, random sampling is not possible, and to undertake a census to derive a sampling frame is a very expensive process indeed. Unfortunately, in most underdeveloped countries, satisfactory sampling frames do not yet exist. Sometimes too, the very cost of conducting a survey may make it necessary to use a different type of sampling. To overcome these problems statisticians have designed two further sampling methods — cluster sampling and quota sampling.

Cluster Sampling

Cluster sampling was devised in the United States of America to try to overcome the problems of cost and the lack of a satisfactory sampling frame. Instead of selecting a random sample scattered over a wide area, it is a few geographical areas that are selected at random and every single household in each area is interviewed. Obviously the areas chosen must be relatively small. It is not possible to interview every household in cities like London and Singapore. The typical area selected might be three or four streets in a town, or perhaps an apartment block. The great advantage of this type of sampling is the saving in time and cost. Many interviews can take place within a short space of time with a minimum of travelling. Moreover it does not require any knowledge of the population before the survey is undertaken; that is, no sampling frame is necessary.

On the other hand, there is a basic problem with this type of sample. Whilst the population is heterogeneous, those who live in a small area or the same block of flats will tend to be homogeneous. They will tend to have the same opinions, the same characteristics, the same life style. Naturally one cannot be certain of this, but it seems to be highly probable, and since the statistician is selecting only a few such areas there is a great danger that the sample will be biassed. One way of trying to offset this tendency to bias is to increase the number of clusters in the hope that you will then include in your sample every important strata, but every increase in the number of clusters raises the cost of the survey. There comes a time when you will have to weigh the risk of selecting a biassed sample against the cost of avoiding the danger of bias. It is, however, a useful type of sample when you have no sampling frame, or when the cost of taking a random sample is too high.

Quota Sampling

There can be few people who have not been stopped in the street by an interviewer holding a questionnaire, or who at least know someone who has been so stopped. And the immediate reaction seems to be, 'Why did they choose me?'

The essence of this type of sample is that it is not preselected, but is chosen by the interviewer on the spot. He, or more usually she, is given a certain number (a quota) of questionnaires which it is her job to have completed during the course of the week, or month. She has a free choice of whom she asks to answer the questions. This choice, naturally, is subject to some general restrictions. One can imagine a young male interviewer stopping only girls of about his own age whom he thought to be pretty. Free choice does not extend as far as this. The interviewer is told that the completed quota of questionnaires must include a certain number of males and of females; a certain number from specified age groups such as, under 18, 18 to 40, 40 to 65. Controls such as this area easy to implement, but often the quota has to be further subsidivided by social class or occupation. This can make quota sampling very difficult as there is no real definition of what constitutes, say, the upper middle class, and certainly you can seldom tell a person's social class by looking at him. Such a requirement necessitates the interviewer being given detailed definitions and descriptions of what the survey body means by each term it uses. Subject, however, to controls such as this, the sample is chosen by the interviewer from people who pass in the street.

Now, obviously, such a technique of sampling can be open to a great deal of abuse and bias. When it was first introduced, many cases were reported of intervewers sitting at home filling in their own questionnaires without ever interviewing anyone. Of course this was soon remedied by the questionnaire asking for names and addresses, and the survey body making a spot check to ensure that such people had actually been interviewed. Deliberate fraud of this nature is very rare now. Much more important is the bias to which the quota survey may be subject. If the interviewer does not start work until, say, ten o'clock, the majority of people she can stop must of necessity be the housewife or the unemployed; if she starts at, say, nine o'clock, the vast majority will be clerical or managerial workers; to start even earlier means that she will probably meet largely manual workers. Thus, the interviewer must be prepared to spread her work throughout the day. It is little use trying to have all the questionnaires completed before the morning coffee-break. A further source of bias is that the people you meet in the street are either going somewhere or doing something; they resent being interrupted and their answers may be hurried and slapdash, given without serious thought. These problems necessitate further controls and advice being given to the interviewer. She may be told at what times to conduct the interviews or even where to conduct them. We would not wish to imply that interviewers are rascals or rogues — far from it — but we must point out that the success of such a survey depends very much on their following their instructions to the letter.

Since the quota sample has these weaknesses, why is it becoming increasingly common? Well, for one thing, it is cheap. It eliminates repeated calls to interview a person who may not be in at home on the first two or three occasions. In fact, it is estimated that each interview in a quota sample costs only about half as much as each interview in a random sample. Then, too, we have in every survey a number of people who do not respond. In a quota sample it does not matter. They are ignored and we pass straight on to the next person with no loss of time and no cost. In other types of sample, a substitute has to be found for each person who does not respond.

It might, in fact, be argued that this itself introduces bias into the sample. It may be that the man who is prepared to fill in questionnaires is some kind of extrovert, and we may be getting the views of only one type of individual.

Nevertheless, given the existence of the controls and checks we have mentioned, all the evidence points to the fact that in skilled hands quota sampling gives reasonably satisfactory results. Statisticians generally accept that it is not a substitute for random sampling statistically, but it is so quick and so cheap a method of carrying out surveys that it is not likely to be replaced for a very long time.

Finally, we must stress that the effectiveness of any sample survey, whatever the method of sampling used, depends to a very large extent on the questions which are asked. If they are ambiguous, so will be our answers; if the interviewer puts the questions with a bias, our results will be biassed. No survey result can be better than the questionnaire on which it is based and this question of questionnaire design is something we must now take up in the next chapter.

Chapter Fifteen

Planning a Sample Survey

During every general election newspapers are full of public opinion polls, all of which claim to tell us which way the electors are going to cast their votes. So dogmatic are some of the statements they make that one sometimes wonders if it is really necessary for us to go to the polling booth at all! Then on voting day, most of those forecasters who were wrong rapidly disappear, leaving the field to those few who have been reasonably successful. Yet those who have failed need not feel ashamed. No statistician will ever dare hope that his sample results will ever be one hundred per cent accurate, and hours of time are devoted to every sample survey merely to reduce the error to the irreducible minimum. Before we can explain how we do try to avoid errors and bias in our results, we must first look at the possible ways in which the sampling error may arise.

You will have appreciated already that one of the most important factors in any sample survey is the questionnaire — the form which poses the questions which give us the information we want. It is a highly flexible instrument. It can be filled in by the interviewer after a series of oral questions; it can be completed by the person being interviewed in the privacy of his own home; it can be completed on the doorstep, or by telephone, or it can be sent by mail. But in spite of this tremendous flexibility, it lends itself readily to bias and error.

Even before an interview takes place, error may have crept into the survey design. In constructing our questionnaire we may have failed to ask questions which ultimately turn out to be relevant. The field of sampling is littered by experienced researchers who cry, 'If only I had asked that question!' Once we pose our questions sources of error become far more numerous. Questions we ask may be misleading or not easily understood; they may mean different things to different people. A question on wages earned last week will be understood by some people to mean gross wages; by others to mean net wages; some will include overtime and bonus payments; others will merely state the basic wage; workers paid monthly may merely divide the monthly wage by four, ignoring the fact that in a month apart from February, there are more than four weeks, and that he really ought to divide his annual salary by 52. Then too there may arise a great deal of bias from the fact that the interviewee does not like to admit to things which he feels may put him in a bad light. If we were to ask, 'How often do you take a bath?', how many people do you think would have the courage to say 'Seldom', or 'Never'? We have a suspicion that the reaction would be something like this. 'Most people bath at least once a week or more. If I admit to less than this I am classifying myself as a dirty person. The investigator will never know anyway, so I will say once or twice a week'. Here, the response is adjusted to present a particular image which the respondent feels is necessary for his own self-esteem.

In the very nature of things, many questions posed in a survey must be personal or intimate questions. There is a very strong rumour that women approaching the age of 40 will remain 39 years old for an indefinite period of time; a man who is a junior clerk in a professional office may well give his occupation as an accountant or lawyer; an unmarried mother may be reluctant to admit that she has children; a man who has not told his wife how much he earns may not be willing to state his true salary. Every one of us has something to hide — something we may not be willing to admit publicly.

Thus, however careful we are, it is not likely that our questionnaire will give us a completely true picture of our sample, and it is always as well to regard the results of a sample survey as, at best, an approximation to reality. Nevertheless, an understanding of the possible sources of error is the first step in reducing the error that may arise in a survey.

Stages in the Sample Survey

Although surveys are designed for an infinite variety of purposes, the basic pattern is the same in every case. The quality of the data we derive depends particularly on three of the stages we will outline. The first is the selection of the sample, since, if this is not representative of the population, no reliable conclusions can be drawn. Secondly, as we have seen, the drafting of the questions to be asked demands very careful thought. Thirdly there is the all-important task of interviewing, since more errors will creep in at this stage than at any other. Most of our discussion in this chapter will centre around these three points. But first:

1. Define your objectives

It should be merely common sense to ask right at the beginning what we are trying to find out and how the survey can help us. Yet, all too often, this trite and obvious point is ignored, and as a result the questionnaire deviates from its objectives, asking questions which are irrelevant to the survey, and failing to ask questions which are essential. It is often useful initially to make a study of any already published statistics to see if it will simplify the survey (or perhaps make it unnecessary). Above all a study of secondary data will help you to clarify your mind as to the type of information which will or will not, be useful to you. Beyond this, the best advice we can give you is to bear in mind constantly *why* you are undertaking the survey. Ask yourself constantly in what way any given piece of information is going to help you to meet your survey objectives. You may still fail to get all the information which, with hindsight, is necessary; but you will be sure that what you do get is relevant.

2. Decide how the information is to be obtained

Here you have a choice of two main methods — the postal enquiry, and the survey using personal interviews. The great advantage of the *postal enquiry* is the relatively low cost. It can be sent to a very large number of people relatively cheaply, and so the sample size can be increased considerably without overspending. Additionally there is no risk of bias or mistakes which may occur if the interviewer influences the respondent to give particular answers. But when you get down to it, these apparent advantages are often fictitious. The great weakness of the postal enquiry is that very few people bother to complete and return the questionnaire. On average a response rate of 20% of questionnaires returned without reminders is considered satisfactory, while 40% is exceptionally good. Now, missing data due to non-response can be a major source of error. Unfortunately we cannot ignore those who do not reply and assume that they have the same characteristics as those who do reply. It is very likely that a group who do not bother to return the questionnaire will differ in some important way from those who do take the time and trouble to fill it in. Almost inevitably as postal survey will necessitate sending several reminders, and even then a very high response rate is seldom achieved. Thus, in the final analysis, the cost per completed questionnaire can be quite high.

Often, the response rate can be increased substantially by writing a letter to members of the sample explaining the importance of the survey, what it is hoped may result from it, and requesting their cooperation. Sometimes the promise of a copy of the results might do the trick, especially in business surveys, while, in the case of very important enquiries, it may be necessary to offer a small gift, or even a nominal cash payment. But whatever we do, we are not likely to get a response rate as high as we would like. W. A. Hendricks, in his book *The Mathematical Theory of Sampling,* suggests that it is legitimate to make adjustments to the results of a postal survey to allow for missing data. He quotes a survey undertaken to discover the average number of fruit trees on a particular type of farm. The original questionnaire was mailed, and followed up by two reminders to those who had not yet replied. The response at each stage was noted, and the average number of trees per farm was calculated at each stage. The results were as follows.

No. of Mailings	Response (%)	Ave. No. of Trees per Farm
1	13	456
1 + 2	26	408
1 + 2 + 3	39.4	385

It is obvious that those who replied late were those who had the fewer number of trees, and it could be argued that those who had not replied at all would have fewer trees still on their farms. One way of dealing with this data would be to construct a fully logarithmic graph, showing the relationship between the response rate and the average number of trees per farm. We could then project the trend shown and estimate the average number of trees for a 100% response rate. If we convert all our figures into logarithms, we get the following table.

	Logarithm of
Response Rate	Ave. No. of Trees
1.1139	2.6590
1.4150	2.6107
1.5995	2.5855

These figures can now be graphed and the trend shown projected.

On our diagram the log of the estimated number of fruit trees for a response rate of 100% (log 2.0000) is 2.525, which converts to an estimated average number of fruit trees per farm of 335. In fact the true value was known to be 329. Our estimate is not completely accurate (in fact it is almost 2% in error), but it is still very, very much better than our sample result with a response rate of only 39.4%. The assumption we made here was, in fact, that those who did not respond were similar to those who replied late, and in many surveys replies are divided into groups according to the date of reply in an effort to discover the main characteristics of non-respondents.

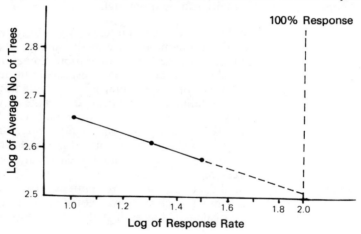

There is little doubt, however, that non-response is a major source of weakness of the postal survey, and the best strategy of all is to arouse and maintain interest in the survey so as to secure as high an initial response as possible. The experience of the British Social Survey points to the fact that response rates of 80% *can* be achieved, if the respondent's co-operation is secured, if the idea can be put over that the survey is of importance to him personally, and if the questionnaire is short and simple. But in spite of their limited success, the hard fact of the matter is that a self-selecting group who decide to return the questionnaire is not representative of the population as a whole. The only way in which non-response can be completely overcome is to make the return of the forms compulsory, and even then there is no guarantee that they are filled in accurately. Since only the government can exercise compulsion, they are the main users of this type of survey.

Perhaps the most widely used method of conducting the sample survey is the *personal interview*. Here a third party, the interviewer, acts as a communications link between the investigator and the members of the sample. He, or she, is responsible for contacting the respondent, for asking questions and for recording the responses. So important is the interviewer that it can well be argued that the representativeness of the sample depends on their ability to persuade their subject to co-operate in filling in the questionnaire, and the accuracy of the sample depends on their skill in extracting and recording information.

Unfortunately, it is all to easy for the interviewer to introduce bias. His tone of voice, even facial expression, and any tendency for him to indicate his own views may all lead to the interviewee expressing opinions which are those of the interviewer and not his own. There are times when, however simply a question is worded, it will be misunderstood and will have to be 'interpreted' by the interviewer. In this interpretation the precise meaning that was intended may well be lost. Then too, faced with a hesitant or tongue-tied member of the public, even the skilled interviewer may prompt, or even put answers into her mouth. Shortly before writing this, one of your authors was stopped, in a quota sample, and asked if he had recently seen a television advertisement for a particular brand of after-shave lotion. Since he was preoccupied thinking about this chapter he hesitated to collect his thoughts; to his horror, before he could speak, the young interviewer said, 'Perhaps you haven't seen it,' and underlined that phrase on the questionnaire. Can you see what went wrong? Busy people, stopped on their way to work, need a few minutes to collect their thoughts before questions are put; the interviewer suggested the response instead of waiting; and worst of all, she then accepted her own suggestion. In fact, your author not only *has* seen the advertisement in question, but actually uses that brand of after-shave lotion. We hate to think of the bias that will appear in that particular survey and can only hope that it was a practice survey carried out by students of statistics.

In spite of these problems, the personal interview does have a large number of advantages over the postal questionnaire. Non-response is minimised. In fact a response rate as high as 80% is not at all unusual. Inevitably, there will be some non-response as a few people may be out every single time the interviewer calls; they may be on holiday, or have left the district, or merely refuse to answer. We would distrust any interviewer who claimed to have a 100% response, but would not dispute a claim of 90% from a trained investigator. Then too, more questions (and more complicated questions) can be asked in an interview than can be asked in a postal questionnaire; questions can be asked of people who could not complete a written questionnaire — the very young, the very old, or the illiterates; often mere visual observation will enable the interviewer to check whether the answers are likely to be correct. Finally, in a personal interview, the order in which the questions are put is under control, whereas in a postal survey there is no control over the order in which they are answered. As we will see later, this can be a crucial factor in eliminating bias.

Nevertheless, some surprising things occur. Just after the Second Word War, in a survey to ascertain how many people would claim campaign medals, it was found that the age of the female interviewer affected the response of the ex-servicemen. The younger, and perhaps prettier, the interviewer was, the greater was the percentage of her interviewees who denied that they would claim their medals.

Obviously training can do a great deal to eliminate interviewer bias. But, however well your team may be trained, mistakes will still occur. It is not the obvious mistakes that matter. When we edit the schedules we will soon pick up the fact that John William Smith is a man, and not a woman as marked on his form. It is the small mistakes which we cannot detect which are important. It is very easy to misunderstand an answer, especially if it cannot be answered by a simple 'yes' or 'no'. It is just as easy to classify an answer wrongly, underlining *Never* for example, instead of *Infrequently*. The only way in which mistakes of this nature can be detected is if a sub-sample of an interviewer's respondents are interviewed a second time by the most highly skilled interviewers available. In fact, many bodies do this to check the potential accuracy of the completed questionnaires as a matter of course. It is, however, expensive, and in the long run the best solution may well be to give each interviewer as detailed instructions as are possible, and as much training and experience as can be afforded. Some investigators argue that tests have shown that on a normal sample survey, the trained interviewer is no more accurate than the amateur; but there is little doubt that in the modern 'in-depth' surveys the trained interviewer can elicit responses that the amateur can never hope to elicit.

3. The preparation of the questionnaire

We have stressed repeatedly that a badly designed questionnaire may ruin an otherwise well-planned survey. Essentially, questions and answers are the means of communication between the investigator and his subject. If you think about it, in this process there are five possible sources of error for every question. Firstly the question itself may not be phrased to enquire about what the investigator wishes to know; secondly the question may not be heard correctly; thirdly what is heard may be misunderstood; fourthly the reply may be misheard by the interviewer; and fifthly the reply may be misunderstood by the interviewer. Is it not obvious that what we need are short, unambiguous questions which allow of a limited number of unambiguous replies? Unfortunately, this is not always possible. If we desire questions which allow the simple answer 'yes' or 'no', there can be no misunderstanding of the answer, even if the question itself is ambiguous. But there is always the need for the third alternative answer, 'Don't know', and such questions are difficult to devise. Suppose we asked the simple question, 'Do you think that garages should give trading stamps and free gifts with sales of petrol?' Let us consider just how 'simple' this question is. Apart from the obvious replies, 'yes', 'no', and 'I don't know', some people may think that it depends on the number of trading stamps given with each gallon or on the nature of the free gift. Others may wish to say 'yes' to gifts and 'no' to trading stamps, or vice versa. Are we to assume in our answer that if the garages cease to give gifts or stamps that the price of petrol will be reduced, or will it remain the same? So you see, the question is not so simple, and considerable skill is called for in constructing questions to which the answer is a categorical 'yes' or 'no'. Usually this kind of question is confined to purely factual matters such as, 'Do you own a car?'

The second type of question is what we have come to know as a multiple-choice question. Here the respondent is asked a question such as, 'Why did you take your holiday in Spain last year?', and the question is followed by a number of possible reasons, one of which is to be underlined. The problem with this type of question is that the list of reasons must be limited, and of necessity we must include a category 'other reasons', leaving the respondent to reply as he wishes. Once again we have opened the door to ambiguity. A further problem is the implication in such a question that only one of the reasons given is important. Two or three, or perhaps all, the reasons may be equally important, or perhaps none are of importance. Here the order of the alternatives can be of importance. Experiments in varying their order show that there is a marked tendency to choose alternatives at the top and bottom of the list, ignoring those in the middle.

The final type of question is open-ended. leaving the reply entirely to the respondent. 'Why do you drink this particular brand of beer?' This is a very useful kind of question to obtain qualitative information, but it is the most open of all to ambiguity. If it is summarised by the interviewer what we are recording is what the interviewer thought the respondent meant. If the reply is quoted in full it may be rambling, incoherent and open to many interpretations, and this is just as dangerous.

In order to help us get over this problem of communicating with our sample, there are a number of rules we must follow when we are constructing the questions to ask. The problem is to devise questions which everyone understands, which mean the same thing to everyone, and which ask what they are meant to ask.

(i) Firstly, keep the questions short. Not only is a short question easier to understand, but also there is nothing more confusing to a person being interviewed than to forget the beginning of a question before the end is reached. The ideal type of question from this point of view is the extremely simple, 'What is your age?' As a general rule it is better to break down a more complicated question into two or three shorter ones. Consider the following question which you could be asked if you were filling in a car accident report form for an insurance company. 'If you were driving your own car give the reference number of your insurance policy; if not, state the owner's name, whether his insurance cover is comprehensive or third party only, and give the name of his insurance company.'

Such an example is of course grossly exaggerated. No one but an amateur (or perhaps a civil servant) would produce such a question, and in practice it would appear in this form:
1.   a.   Were you driving your own car?   Yes/No
    b.   What is the number of your insurance policy?   7K/..........

2. If your answer to question 1a is NO
    a. Give the name of the owner   Mr/Mrs/Miss
    b. What type of insurance cover does he/she have?
                                        Comprehensive/Third Party
    c. What is the name of his/her insurance company?   ....

You will see that each of these short questions concentrates on one thing at once and is not likely to confuse the person filling in the form.

(ii) Use simple language. In any sample there is likely to be a very wide variety of ability in the use of langauge, and we must make sure that everyone understands what we mean. It is not a question of underrating the intelligence of the members of the sample but of trying to ensure that the interviewer does not have to explain the question in his own words which may introduce some bias into the answer. In any case we feel it is better to say, 'Are you telling a lie?', rather than, 'Are you perpetrating a terminological inexactitude?' Simplicity has been the keynote of every great speech in history, from Julius Caesar's 'I came, I saw, I conquered', to Winston Churchill's great wartime speeches. They were great because everyone understood them. Follow the same policy in drafting your own questions.

(iii) Avoid questions which lead to a particular answer. It is far better to ask, 'What do you consider to be the best brand of toothpaste?', than to ask 'Do you consider White Shine to be the best brand of toothpaste?' The mere mention of a name is often enough to cause a particular response. Equally we must try to avoid emotive words. Socialist and Tory arouse more passion than do Conservative and Labour. At the beginning of the last war when a survey of people's attitudes in America was being taken, it was found that almost twice as many people claimed to be anti-Nazi than claimed to be anti-German. Try at all costs to avoid implying in your question that you would like a particular answer.

(iv) Make sure that the respondent has the information. It is of little use asking people in Central Africa what type of central heating they prefer. Even if the question is a simple one, the informant may not know the answer. A housewife may never have been told her husband's salary; or she may have forgotten when she last bought biscuits. Often she has to be led up to such things step by step. Firstly she should be asked if she has ever bought Brand X biscuits. If she replies 'yes', she can be asked if she has bought them in the last year, then the last month, then in the last fortnight and so on until she remembers clearly. The approach may seem cumbersome, but experience has taught us that we should not rely too much on people's memory.

(v) Consider if a respondent would be willing to tell the truth. He may know the answer, but be unwilling to reveal it, since certain answers may appear (to him) to lower his prestige in the eyes of the interviewer. Television viewers may claim to prefer the more serious programmes to the more frivolous ones; readers may claim to read classical novels when in fact they read cowboy stories. We can, of course, always insert test questions to check if these answers are truthful. We could, for example, ask them to summarise the content of two serious television programmes they have seen in the last week, or give the author and title of a couple of serious novels they have read. In 1960, when one of your authors was helping to undertake a survey, one housewife gave her age as 41, and then when asked for her date of birth, gave it as 1911, which, of course, made her age 49. Such a situation calls for a great deal of tact on the part of the interviewer. More interesting is a further American survey[1] in which the following questions were asked:

	Question	% Replying Yes
1.	Have you ever heard of the word 'afrohetin'?	8
2.	Do you recall voting last December in the special election for your state representatives?	33
3.	Have you ever heard of the famous writer John Voolson?	16
4.	Have you heard of the *Midwestern Life Magazine*?	25
5.	Have you heard of the Taft-Johnson-Pepper Bill on veterans' housing?	53

The interesting thing about this survey is that none of the things mentioned in fact ever existed, yet so many people said they had heard of them. Those who said 'yes' probably did so because they felt that if they had taken more interest in current affairs they *would* have heard of them.

(vi) Consider the order of the questions. Here, two things are important. The person being interviewed must not feel that the questionnaire is going to be too difficult to complete, or pry too much into his private life. The early questions should be easy and general, and only as confidence builds up should searching questions be introduced. Equally, wait until the interviewer has been able to build up some sort of friendship with the respondent before you introduce, slowly and carefully, questions on his private life. And if your questions are really personal, such as, 'Have you ever been unfaithful to your wife?', do stress that the survey is absolutely confidential. A question such as this would, of course, have been preceded by more general questions as to whether the respondent believed that fidelity in marriage was of great importance, or usual.

(vii) Finally, try to ensure that the answers given are capable of being interpreted in one way only. A very common category of answer is 'Don't know'. This, however, could mean, 'I do not understand the question', or 'I do not have the information', or 'I have the information, but cannot decide between alternatives', or 'I have the information, but am not going to tell you'.

It is important that interviewers should indicate which of these alternatives is meant, rather than lumping together such diverse information under one general heading.

---

1. Quoted in Elliot and Christopher, *Research Methods in Marketing*.

4. Select the sample

Since we have already dealt with many different types of sampling in the last chapter, we will merely stress here that the sample must be chosen to be of such a size and composition that it will yield the best possible results consistent with the expenditure you are permitted.

5. Undertake a pilot survey

However careful you have been so far, what you have done still remains to be tested in the field. It may be that the questionnaire we have designed will be unsatisfactory in practice. To test what we have done so far, we usually undertake a small pilot survey. How many people we interview does not matter so long as they are chosen at random; the results themselves are not as important as the lessons we learn. We may find, for example, that a question we thought to be simple and self-explanatory is misunderstood by most people, and will have to be altered. It may be that one section of the questionnaire results in replies so vague that it becomes obvious that we are asking questions about which few of our respondents have any knowledge. It would be better to leave it out altogether.

At this stage it is our work we are testing, and experience of the pilot survey will yield valuable information about the difficulties we are likely to meet in the field, and will save us from making serious mistakes in the much more costly survey proper.

6. Brief the interviewers

Before the fieldwork proper starts, all interviewers should hold a final meeting. They will already have a copy of their instructions, and great care should be taken to ensure that they understand them. The lessons learned, and the difficulties met with in the pilot survey should be carefully examined and a course of action determined to meet circumstances that may arise. We will never be able to foresee all the difficulties of every single interviewer, but we must ensure that every single one of them has every scrap of information which is available and knows exactly what he is about to do.

Now, finally, we do not pretend that if you follow all the above advice and rules you will produce the perfect sample survey. There will still be mistakes, but we hope that you will have avoided the worst. Nor will we claim that your sample results will be one hundred per cent accurate. Such a sample has yet to be devised. But you will have avoided some of the major sources of error.

Error there will be, however. If you have done your job well, it will be small. Later chapters will show you how to estimate the magnitude of the error that may remain: those that follow will show you how to assess the significance of your sample results and take decisions on them.

Exercises to Chapters 14 and 15

15.1 Statistical data may be collected by sample or census inquiry. Describe both methods and explain why sample inquiry is more frequently used than census inquiry.   A.C.A.

15.2 Describe three of the more important methods of sampling used in conjunction with statistical surveys. Suggest methods which may be employed in statistical surveys to validate the sample information obtained.
I.C.S.A.

15.3 Explain what is meant by (a) stratified, (b) quota sampling, illustrating by a brief example in each case how samples are selected. Comment on the main advantage(s) and disadvantage(s) of each method.
A.C.A.

15.4 Distinguish between random sampling and quota sampling. Indicate the advantages and disadvantages of each method, and state how non-response would be dealt with in the use of each procedure.
I.C.S.A.

15.5 Describe carefully how four of the following types of sample are taken: simple random, stratified random, multi-stage random, systematic, quota.

15.6 Identify the relative advantages and disadvantages of different techniques of sampling such as random, systematic, stratified, multi-stage and quota. Select the technique which you think would be most appropriate for a motor-car manufacturing company seeking information from car owners to enable the company to design 'the most popular family saloon car'. Support your selection with a reasoned argument.

15.7 (a) Why are samples used?
(b) Describe in detail two methods of selecting samples.
(c) What kind of errors arise in sampling?

15.8 Explain what is meant by (a) random sampling, (b) quota sampling. What are the principal arguments in favour of using some form of random sampling rather than a quota sample?

15.9 What is the object of sampling? Discuss the problems involved in selecting from a large population samples which typify these populations.

15.10 What is the object of 'sampling'? Explain briefly the principles on which sampling is based and discuss the problems involved in obtaining from large human populations samples which characterise them adequately.

15.11 Describe the following types of samples:
(a) random;
(b) stratified;
(c) systematic;
(d) multi-stage.
State the conditions under which each would be used and the advantages to be gained.

15.12 (a) What is meant by the term "sampling frame"? Give two examples of a sampling frame from which a human population could be sampled.
(b) Define each of the following and describe the advantages of each type of sample:
(i) stratified;
(ii) multi-stage;
(iii) quota.

15.13 (a) An inquiry is to be carried out among a population which has the following characteristics:

Proportion of bus travellers to travellers by other means   4:8
Proportion travelling less than five miles to that travelling five miles and over:
  (i) bus travellers   7:3
  (ii) travellers by other means   2:3

If each interviewer is to question a quota of sixty people, how many of each class of person should be selected per quota?

(b) What chief advantage has the quota method of sample selection when compared with other methods? What are its disadvantages?

(c) What advantages are derived from the use of statistical samples?

15.14 Discuss the sampling methods which you consider should be used to obtain the following information:
(a) Public opinion on the Common Market;
(b) Weights of workers (male, female and juvenile) employed in a very large factory;
(c) Heights of Canadian born persons resident in the U.K.;
(d) Health details relating to office workers employed by local authorities in the U.K.

15.15 (i) In what ways does quota sampling differ from random sampling?
(ii) What are the considerations involved in deciding whether the information which is usually derived from the population census should be obtained by a random sample of all households?

15.16 (a) Postal questionnaires and interviews are two methods of collecting data. List the advantages of each method.
(b) One of the main defects of surveys by interview is the problem of interviewer bias. What is interviewer bias and how can the problem be minimised?
(c) What are the main points to be considered in the design of a postal questionnaire?   I.C.M.A.

15.17 (a) What rules should be observed in constructing a questionnaire for collecting statistical information?
(b) What are the advantages and disadvantages of using questionnaires for collecting data?

15.18 Describe the main methods which are used in the collection of statistical data, and comment on the problems associated with each method.

15.19 Discuss the problems involved in using questionnaires for the collection of statistical data.

15.20 Give an account of the principles to be considered when drawing up questionnaires.

15.21 List, and then describe carefully with examples, general rules which should be followed when designing a questionnaire for a statistical survey.

A local authority is considering whether or not to construct a sports centre and intends to send to a sample of residents a short postal questionnaire in order to ascertain their views. Design a suitable questionnaire.

15.22 (a) What is meant by a *sample*? Describe briefly the following:
   (i) random sampling;
   (ii) multi-stage sampling;
   (iii) stratified sampling.
   (b) Describe briefly the ways in which bias may arise in response to each of the following questions when directed by an interviewer at members of the class of people indicated.
   (i) Do you sympathise with Communism? (general public)
   (ii) Do you use 'X' (a washing-up liquid)? (housewives)
   (iii) How many people are there in your family? (general public)

15.23 Describe carefully the way in which you would select a sample for an investigation into:
   (a) the journeys made by car and lorry drivers through a town;
   (b) the brands of pipe tobacco smoked by men in a particular residential suburb.

15.24 As the Personnel Officer of a factory employing 10,000 workers at three locations of approximately equal labour strength, you wish to carry out an inquiry among all the employees to find their views on a rearrangement of holidays. It has already been agreed that the present fourteen days per annum with six additional days at bank holidays should be increased to a total of twenty-seven days per annum. The inquiry is to find out how the workers prefer the new total to be distributed through the year.
   (a) Discuss the suitability of a postal questionnaire for this purpose, comparing this method with other possible methods. State your final choice of method and give your reasons.
   (b) Draft in general outline a questionnaire to be used with the method upon which you have decided.

15.25 (i) What are the advantages and disadvantages of employing random sampling methods instead of non-random sampling methods when carrying our a nationwide sample survey?
   (ii) Describe *in detail* how you would obtain a random sample of adults from a town of approximately 50,000 inhabitants. What are the main factors affecting the sample size? How would you deal with the problem of non-responders?

15.26 A large local authority with approximately 100,000 residents proposes to conduct a survey to obtain some idea of the expected use of a proposed new sports centre. Explain briefly:
   (a) the advantages of employing random sampling methods instead of non-random sampling methods;
   (b) the principal factors affecting sample size;
   (c) how you would deal with the problem of non-respondents;
   (d) how you would stratify the population if it was decided to employ a stratified random sample.

15.27 (a) State the main points which should be observed when designing a questionnaire.
(b) Draft a questionnaire (of about 8 questions), to be completed by a sample of motorists using a large firm's car service department, which will provide information regarding the views of those motorists on the organisation of the service department. (Technical questions are *not* required.)

15.28 You have been asked to conduct a survey on wages within a large company. Explain in detail how you would go about the designing, carrying out and reporting on the results of your survey.

15.29 Distinguish clearly between stratified random sampling, multi-stage random sampling, quota sampling and the deliberate selection of a sample. Explain the circumstances in which each would be advantageous. Discuss the problems involved in the use of sampling frames.

15.30 What do you understand by the term *sampling frame*? Give an example, in each case, of a sampling frame suitable for sampling from
(i) adults in the United Kingdom,
(ii) students in technical colleges.
You have been asked to advise on the design of a random sample of about 10,000 addresses in Great Britain. Describe how you would select your sample if you are to employ a multi-stage stratified sample. Pay particular attention to your sampling units and to your stratification.

15.31 (a) Outline the principles of questionnaire design.
(b) Draw up a questionnaire of about eight questions either
(i) to discover from part-time students in a technical college information relating to the course they are pursuing and its relevance to their future job prospects, or
(ii) to test public opinion on the proposal to set up a Welsh parliament.

15.32 (a) Give *two* occasions when the selection of a pure random sample would not be feasible. What types of sampling would overcome the difficulties?
(b) A manufacturer of electrical appliances wishes to carry out a survey of households in a given smokeless fuel zone to assess the sales potential for an electric central heating system. Answer the following questions, giving reasons for each answer.
(i) What would determine the size of the sample chosen?
(ii) What information may be used to provide a sampling frame?
(iii) Should the manufacturer use postal questionnaires or carry out door-to-door interviews?
(iv) Should questions be 'open-ended' or require a pre-printed answer to be ticked?

15.33 Either
(a) Explain what is meant by a pilot survey and justify its place in the planning of a social survey, or
(b) Explain what is meant by cluster sampling, illustrating your answer with an example.

15.34 Outline the steps involved in planning a statistical survey. Discuss three or them in detail.

15.35 (a) Describe how *bias* occurs in a statistical investigation. Which sampling technique would be used to obtain an unbiassed sample of:
- (i) ages of West Indian-born males resident in the United Kingdom;
- (ii) weights of male, female and juvenile employees in a large factory;
- (iii) health details of councillors in English district councils?

(b) Explain briefly the following:
- (i) sampling frame;
- (ii) quota sampling;
- (iii) systematic sampling.

Chapter Sixteen

Sampling Theory

You will have realised by now that the art of sampling is central to the study of statistics, and that we sample to learn something about the population from which the sample was drawn. As the sample we draw is likely to form only a fraction of the population, it is evident that any conclusion we draw about the population is subject to error, and, as we have seen, we can minimise the scale of such errors by using sound sampling methods. However, despite all the care we may take in designing a sample, the chance of making errors still exists, and we shall now turn our attention to attempts to quantify the scale of such errors. Before we start, it must be stated that our analysis of sampling errors will be limited in three ways. Firstly, we will confine our attention to large samples, as small samples create problems of their own which we shall examine later. This naturally begs the question as to what constitutes a large sample, and we will state now (without any attempt at justification) that *a large sample is a sample with more than 30 items.* The second limitation that we will place on our analysis of sampling is that we will confine our attention to simple random samples. Thirdly, we shall (for the present) either confine our attentions to infinitely large populations or, what amounts to the same thing, if we sample from finite populations then we will replace each item before drawing the next.

It is a fundamental truth about statistical theory that if we have prior knowledge of the population, we can make predictions about the behaviour of samples drawn from that population. Now you may object that this is of little use: we want to draw conclusions about the population from a sample. However, before we can assess the validity of such conclusions, it is necessary to understand thoroughly the theory of sample behaviour.

The Sampling Distribution of Sample Means

Let us start by assuming that we have a population with a known mean μ and a known standard deviation σ. From this population we will draw many samples of n items, and for each sample we will find the mean $\bar{x}$. Clearly, each time we draw a sample, we are not going to obtain the same value for the sample mean, so it would be sensible to arrange our sample means into a frequency distribution. We call this frequency distribution a *sampling distribution of sample means,* and we shall now attempt to make some predictions about this sampling distribution.

Suppose you were asked to guess the value of one of these sample means, what would your guess be? As we know that the population mean is μ, it would surely be sensible to guess that the sample mean would also be μ. For example, if we know that the average heights of all male adults in a certain town is 175 centimetres, then it would be sensible to guess that a sample of 100 men from this town would also have an average height of 175 centimetres. Now of course, in practice any *individual* sample mean cannot be expected to be exactly the same as μ: some sample means will be more than μ and some less than μ. So perhaps it would be more sensible to generalise, with a statement about all sample means like this:

the average value of $\bar{x}$ (the sample mean) is μ (the population mean)

We can expect, then, that the sample means will be clustered about the population mean, and it would be sensible to measure the degree of spread with a measure of dispersion. The measure we will use is the standard deviation, but you should note that to distinguish the fact that we are considering a sampling distribution rather than a sample or a population, in such cases as we call the standard deviation the *standard error* (s.e.). Now the size of the standard error will depend on the size of the sample drawn, and will lie within limits that we can define precisely. Let us consider the largest sample we can possibly draw: this would occur if we selected every item in the population. The sample mean would be exactly equal to the population mean, and we would obviously obtain the same value for the sample mean in every sample we selected. So the sample means would not show any spread, and the standard error of the mean would be zero. Now suppose we selected the smallest samples possible i.e. samples of one item only. The sample means would show exactly the same spread as the individual items in the population and the standard error of the mean would be σ – the population standard deviation. We can conclude, then, that the standard error of the mean must be at least zero, and at most σ. Moreover, as the sample size is increased, then the standard error decreases. In fact it can be shown that

The standard error of the sample mean $= \dfrac{\sigma}{\sqrt{n}}$

We do not intend to prove this formally, but you can satisfy yourself that it is true by undertaking the following experiment.

Experiment 1

Let the population consist of the digits 1, 2 and 3, so

the population mean $= \dfrac{1+2+3}{3} = 2$, and

the population standard deviation $= \sqrt{\dfrac{(1-2)^2+(2-2)^2+(3-2)^2}{3}} = \sqrt{\dfrac{2}{3}}$

If we drew samples of two items from this population then we would predict that

the average value of the sample means $= \mu = 2$, and

the standard error of the sample means $= \dfrac{\sigma}{\sqrt{n}} = \sqrt{\dfrac{\frac{2}{3}}{2}} = \dfrac{\sqrt{2}}{\sqrt{3}} \times \dfrac{1}{\sqrt{2}} = \dfrac{1}{\sqrt{3}}$

a) List all the samples of two items that can be drawn (there are nine of them) and find the mean of each sample. (Remember you are sampling with replacement).
b) Find the average value of the sample means and verify that it is 2.
c) Find the standard deviation of the sample means (i.e. the standard error of the mean) and verify that it is $\dfrac{1}{\sqrt{3}}$

The Central Limit Theorem

Let us summarise what we have learnt so far. If we draw samples of n items from a population with a known mean μ and a known standard deviation σ, then we can make predictions about the sampling distribution of the sample mean. We can predict that the average value of the sample mean is μ, and that the standard error of the sample mean is $\dfrac{\sigma}{\sqrt{n}}$. Now if the population itself is Normally Distributed, you would expect the sample means to be Normally Distributed, but what is rather curious is that even if the population is not Normally Distributed the sample means still turn out to be Normally Distributed *as long as the sample is reasonably large*. This strange feature of the sample mean is called the *Central Limit Theorem* and explains why we spend so much time considering the Normal Distribution. Again, the algebraic proof of the Central Limit theorem is rather complicated, but a simple experiment will demonstrate its truth. Suppose we take as our population the ten digits 0 to 9, and select single items from this population. Each digit has an equal probability of being selected, the distribution is by no means Normal – it is a rectangular distribution as shown in diagram 16.01.

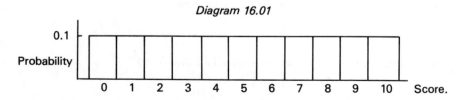

Diagram 16.01

Now let us suppose that samples of two digits are drawn (the first digit is replaced before the second is drawn), then 100 different samples are possible. For each sample we could add together the two digits drawn. The smallest sample total would be 0 + 0 = 0, and the largest 9 + 9 = 18. All the sample totals must fall within the range 0 to 18, so the sample means must be in the range 0 to 9. We must now find the number of ways of obtaining each sample total. A sample total of zero can be obtained in one way only – if both the digits drawn are zero. A sample total of one can be obtained in two ways (0 + 1 or 1 + 0). A sample total of 2 can be obtained in 3 ways (0 + 2, 2 + 0 or 1 + 1). Continuing in this way we can obtain the sampling distribution of sample means.

Sample Total	Sample Mean	No. of Ways	Probability
0	0	1	0.01
1	0.5	2	0.02
2	1.0	3	0.03
3	1.5	4	0.04
4	2.0	5	0.05
5	2.5	6	0.06
6	3.0	7	0.07
7	3.5	8	0.08
8	4.0	9	0.09
9	4.5	10	0.10
10	5.0	9	0.09
11	5.5	8	0.08
12	6.0	7	0.07
13	6.5	6	0.06
14	7.0	5	0.05
15	7.5	4	0.04
16	8.0	3	0.03
17	8.5	2	0.02
18	9.0	1	0.01

The sampling distribution of the sample means has been drawn in diagram 16.02. We can conclude that if samples of two items are drawn from a rectangular distribution, then the distribution of the sample means will be triangular.

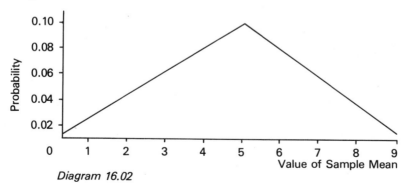

Diagram 16.02

If we increase the sample size to three, then the distribution of sample means would look like this:

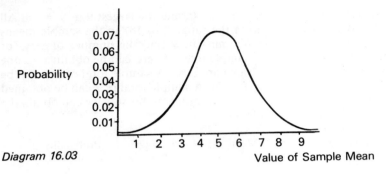
Diagram 16.03

We see, then, that as the sample size increases from one to two to three, then the distribution of sample means changes from rectangular to triangular to bell-shaped. In fact, it is not difficult to imagine that the greater the sample size, the closer will the distribution of sample means move to a Normal Distribution. So as long as we confine our attention to large samples then, we can use Normal Distribution tables to make predictions about sample means.

Example 1

What proportion of samples of 100 items drawn from a population, mean 100 and standard deviation 20, would have a mean greater than 105? We would predict that the average value of the sample mean would be 100, and the standard error of the sample mean would be $\frac{20}{\sqrt{100}} = 2$. The diagram below illustrates the proportion we require.

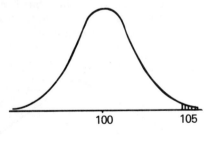
Diagram 16.04

The Z score for 105 is
$$\frac{105 - 100}{2} = 2.5,$$
and consulting the Normal Distribution tables we can expect only 0.621% of sample means to exceed 105.

Example 2

A confectioner produces cakes which he packs in cartons of 100 for the catering trade. He prints on the carton 'average weight per cake not less than 95.5 grams'. If cakes have a mean weight of 100 grams and a standard deviation of 20 grams, what proportion of batches contravenes the Trade Descriptions Act?

Every time the confectioner packs a carton, this is equivalent to selecting a sample of 100 cakes. We can consider, then, that the average weight of cakes in the cartons forms a sampling distribution of sample means, with a standard error $\frac{20}{\sqrt{100}} = 2$. As we wish to know the proportion of cartons with mean weight of cakes less than 95.5 grams, the problem can be illustrated like this:

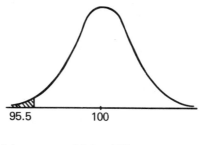

Diagram 16.05

The Z score for 95.5 is $\frac{95.5 - 100}{2} = -2.25$

and consulting the tables we find that only 1.222% of cartons would contravene the Act.

The Sampling Distribution of Sums and Differences

We shall now discuss sampling distributions formed by taking sums or differences. To understand just what this means, consider modern industry; a considerable part of industrial activity is based, not on making something but on assembling components that have been made in a number of different factories. This is especially true of the motor industry. Let us take a very simple case and imagine that we are joining end-to-end two straight metal bars produced in two workshops. The first workshop aims to produce bars with lengths of 20cm., but of course there will be variation in the actual lengths. We will assume that the lengths of bars are Normally Distributed with a mean of 20cm, and a standard deviation of 0.05cm. Bars produced in the second workshop we will again assume to be Normally Distributed, this time with a mean length of 10cm. and a standard deviation of 0.04cm.

When a workman picks up two bars for joining he is, in fact, sampling. He is drawing samples of two items; one from the first workshop and one from the second. Almost certainly, the length of the assembled bars will be important and this is the sampling statistic. It is obtained by taking the sum of the lengths of the two individual bars. Now as the length of the individual bars varies, then so will the lengths of the assembled bars. We could form a frequency distribution of the lengths of the assembled bars. As the bars have been obtained by *sampling,* and as their lengths have been obtained by *adding,* the frequency distribution of the lengths of assembled bars is called a *sampling distribution of sums.* What can we conclude about this sampling distribution? Well, as the lengths of the individual bars are Normally Distributed it would seem reasonable to assume that the lengths of the assembled bars will also be Normally Distributed. It is true that

> *if the populations are Normally Distributed so will be the distribution of sample sums formed from the populations*

Suppose you were asked to guess the mean length of the assembled bars — it is probable that you would argue something like this: as bars produced in the first workshop have a mean length of 20cm. and bars produced in the second workshop have a mean length of 10cm., then the mean length of the joined bars would be $20 + 10 = 30$cm.

> *The mean of the sample sums is the sum of the means of the populations from which they were formed i.e. mean of sample sums $\overline{x} = \mu_1 + \mu_2 + \ldots + \mu_n$.*

What can we conclude about the dispersion of the lengths of the assembled bars? Suppose we take $\overline{x} \pm 4\sigma$ as covering the effective limits of any Normal Distribution (consult the table and you will see that it covers all but 0.06%). The longest bar produced in the first workshop would be $20 + 4 \times 0.05 = 20.2$cm., and the shortest would be $20 - 4 \times 0.05 = 19.8$cm. This gives a range of lengths of $20.2 - 19.8 = 0.4$. Bars produced in the second workshop would have a maximum length of $10 + 4 \times 0.04 = 10.16$cm., and a minimum length of $10 - 4 \times 0.04 = 9.84$cm. This gives a range of $10.16 - 9.84 = 0.32$cm. When a workman picks up two rods for joining he could pick the two longest bars and produce an assembled bar of $20.2 + 10.16 = 30.36$cm. He could choose the two shortest bars and produce an assembled bar of $19.8 + 9.84 = 29.64$cm. So the range of assembled bars is $30.36 - 29.64 = 0.72$cm. *Notice that this is the sum of the ranges of the two populations* $(0.4 + 0.32 = 0.72)$. However, we do not usually measure dispersion with the range — we prefer to use the standard deviation (as we are dealing with a sample we should call this measure the standard error). Unfortunately, what is true of the range is not true of the standard error. We cannot add the two populations' standard deviations and conclude that the standard error of the sample sums is $0.04 + 0.05 = 0.09$cm. But although we cannot add the standard deviations we can add the variances.

> *The variance of the sample sums is the sum of the variances of the populations from which they were formed*

This additive property explains why statisticians often prefer to use the variance to the standard deviation. We can now conclude that

$$\text{standard error of the sample sums} = \sqrt{\sigma^2_1 + \sigma^2_2 + \ldots + \sigma^2_n}.$$

The standard error of the lengths of the assembled bars would be

$$\sqrt{0.04^2 + 0.05^2} = 0.064$$

Experiment 2

This experiment is designed to verify the statements we have just made about the mean and the standard error of sample sums. Let the population A be the digits 1, 2, and 3, which we know from experiment 1 has a mean 2 and a standard deviation $\sqrt{\frac{2}{3}}$. Let the population B be the digits 5, 7, and 9 which has a mean of 7 and a standard deviation of

$$\sqrt{\frac{(5-7)^2+(7-7)^2+(9-7)^2}{3}} = \sqrt{\frac{8}{3}}$$

Now suppose we form all possible samples by drawing one digit from each population. If we find the sample sums then we would predict their average value to be $2 + 7 = 9$, and the standard error to be

$$\sqrt{\sigma^2_A + \sigma^2_B}$$
$$= \sqrt{\frac{2}{3} + \frac{8}{3}} = \sqrt{\frac{10}{3}}$$

1) List all the samples that can be formed by drawing one item from each population.
2) Find all the sample sums, verify that their average is 9 and their standard error is $\sqrt{\frac{10}{3}}$.

Example 3

A manufacturer produces two types of metal rods. Type A has a mean length of 3.5cm. and a standard deviation of 0.03cm. Type B has a mean length of 4.5cm. and a standard deviation of 0.04cm. A rod of each type is selected, joined together, and the joined rods are then placed in a slot 8.1cm. long. What proportion of joined rods will not fit the slot?

Mean Length of joined rods = $3.5 + 4.5 = 8$cm.
Standard error of the lengths of joined rods = $\sqrt{0.03^2 + 0.04^2} = 0.05$cm.

The shaded area in diagram 16.06 illustrates the proportion of assembles that will not fit. To find this proportion we must first find the Z-score for 8.1cm.

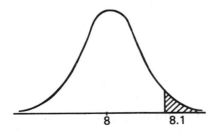

Diagram 16.06

$$Z = \frac{8.1-8}{0.05} = 2.$$

Consulting the tables we would expect 2.275% of assembled rods not to fit the slot.

We shall now turn our attention to considering the sampling distribution of sample differences, and we shall do this by developing example 3. Let us suppose that part of the assembly process of a certain machine involves placing a metal bar into a slot. The bars have lengths that are Normally Distributed with a mean of 8cm. and a standard deviation of 0.03cm. You should check for yourselves that all but 0.06% of bars will have lengths within the limits 7.88cm. to 8.12cm — a range of 0.24cm. Slots have lengths that are Normally Distributed with a mean 8.3cm. and a standard deviation of 0.04cm., so slots will have lengths within the limits 8.14cm. to 8.46cm. — a range of 0.32cm. Now there will be a gap between the slot and the bar — it is measured by subtracting the bar length from the slot length. The gaps then form a sampling distribution of sample differences. Now as you would imagine, the mean gap would be 8.3 − 8.0 = 0.3cm.

> *The mean of the sample differences is the difference between the means of the populations from which they were formed i.e. mean of sample differences* $\bar{x} = \mu_1 - \mu_2$

Turning now to the range of the gap, the largest gap will be obtained by taking the largest slot with the smallest bar, so the largest gap is 8.46 − 7.88 = 0.58cm. The smallest gap is obtained by taking the smallest slot with the largest bar, so the smallest gap is 8.14 − 8.12 = 0.02cm. So the range of gaps is 0.58 − 0.02 = 0.56cm. This is the sum of the ranges of the two populations (0.24 + 0.32 = 0.56). As variances are additive we can conclude that

$$\text{standard error of sample differences} = \sqrt{\sigma_1^2 + \sigma_2^2}.$$

Experiment 3

This experiment is designed to verify the statements we have made about the mean and standard error of sample differences. Find your results of experiment 2. If we form sample differences by subtracting the digit drawn from population A from the digit drawn from population B, then we would predict the mean difference to be 7 − 2 = 5, and the standard error of the difference to be

$$\sqrt{\frac{2}{3} + \frac{8}{3}} = \sqrt{\frac{10}{3}}$$

Form all the sample differences, verify that their mean is 5 and their standard error is $\sqrt{\frac{10}{3}}$.

Example 4

Full boxes of rice have weights that are Normally Distributed with a mean of 225 grams and a standard deviation 1.5 grams. The empty boxes have weights that are Normally Distributed with a mean 7 grams and a standard deviation 0.05 grams. Printed on the packet is the claim that the minimum net weight of the contents is 214 grams. What proportion of packets fail to meet this claim?

Mean weight of contents = 225 − 7 = 218 gms
Standard error of the weights of contents =
$\sqrt{1.5^2 + 0.05^2}$ = $\sqrt{2.2525}$ gms

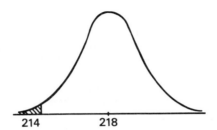

Diagram 16.07

The shaded area in diagram 16.07 represents the proportion we require. The Z score is

$$\frac{214 - 218}{\sqrt{2.2525}} = -2.665$$

The table tells us that a very low proportion — only 0.385% of packets are below the specified minimum weight.

The Sampling Distribution of Sample Proportions

There has been an increasing tendency in recent years to give the results of sample surveys in percentage form. A manufacturer may claim that not more than 3% of his output develops faults in the first year of use; public opinion surveys may report that a certain percentage of the population prefer filter tip cigarettes to plain cigarettes; or that 38.2% of the population can be expected to vote Labour at the next general election. But most statisticians prefer to use proportions rather than percentages; they would report that the proportion that can be expected to vote Labour is 0.382.

Let us suppose that we are considering a population, some members of which possess a certain attribute. For example, the population we are considering might be the female population of the United Kingdom, and the attribute might be blue-eyed blondes. Again, the population might be a very large batch of manufactured goods and the attribute might be that the good is defective. If we draw many samples from this population, calculate the proportion in each sample that has the attribute and record our results in a frequency distribution, then we will have formed a sampling distribution of sample proportions. Now if we know the proportion in the population which has the attribute (call this proportion π), then we can make the following predictions about the sampling distribution:

The mean proportion in the sampling distribution will be π — the same as the proportion in the population.

If the samples all contain n items, then the standard error of the proportions is $\sqrt{\dfrac{\pi(1-\pi)}{n}}$

Notice that, just like the standard error of sample mean, the larger the sample size the smaller will be the standard error.

Experiment 4

This experiment is designed to enable you to verify the statements we have just made about the sampling distribution of sample proportions. Suppose we have a population consisting of the items A, A, and B, where A means that the item is perfect and B means that the item is defective. So we have a proportion of defective items in the population of $\pi = \frac{1}{3}$. If we draw samples of two items from this population, then we would predict that

the mean proportion defective in the samples $= \pi = \frac{1}{3}$

the standard error of the proportion defective $=$

$$\sqrt{\dfrac{\pi(1-\pi)}{n}} = \sqrt{\dfrac{\frac{1}{3} \times \frac{2}{3}}{2}} = \sqrt{\dfrac{1}{9}} = \dfrac{1}{3}$$

a) List all the possible samples of two items and find the proportion defective in each
b) Verify that the mean proportion defective is $\frac{1}{3}$ and that the standard error of the proportions is $\frac{1}{3}$

Example 5

Suppose we know from the past experience that 5% of a certain output is defective. We intend to draw samples of 800 items and find the proportion defective in each. In this case, we have $\pi = 0.05$, so

The mean proportion defective in the samples $= \pi = 0.05$

The standard error of proportions $= \sqrt{\dfrac{0.05(1-0.05)}{800}} = 0.0077$

If we wish to know the proportion of samples that can be expected to contain more than 6% defective items then, as the sample is large, we can use the Normal Distribution

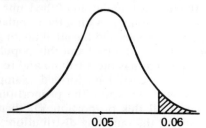

Diagram 16.09

$$Z = \frac{0.06 - 0.05}{0.0077} = 1.3$$

The table tells us that 9.68% of samples would have more than 6% defectives.

Confidence Limits

Let us return to the population we considered in example 1, i.e. the population with a mean of 100 and a standard deviation 20. If we draw samples of 100 from this population, then their means will be Normally Distributed with a standard error $\frac{20}{\sqrt{100}} = 2$. Let us now ask ourselves between what limits we would expect the central 95% of sample means to lie. Diagram 16.10 illustrates the problem. We are excluding the 2.5% of means at the extreme right of the distribution and the 2.5% at the extreme left, and we require the Z scores for the points a and b. Now b has a Z score that is exceeded only on 2.5% of occasions, so consulting our table we find that b must have a Z score of 1.96. In other words

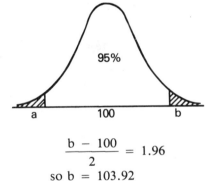

Diagram 16.10

$$\frac{b - 100}{2} = 1.96$$
$$\text{so } b = 103.92$$

As a and b are symmetrical about the mean, a must have a Z score of -1.96, so
$$\frac{a - 100}{2} = -1.96$$
$$\text{so } a = 96.08$$

Thus, 95% of sample means can be expected to fall within the range 96.08 to 103.92. *Such limits are known as confidence limits, and the one we have calculated is the 95% confidence limit.* Now let us generalise. If we draw samples of n items from a population with a mean μ and a standard error σ,

95% confidence limits for sample means $= \mu \pm 1.96 \dfrac{\sigma}{\sqrt{n}}$

Beware in your interpretation of 95% confidence limits. The concept does not mean that we are 95% sure that a single sample mean lies within these limits — all that we can say about a single sample mean is that either it will lie within the limits or it will not. *The 95% confidence limits mean that if we drew many samples, and find the mean for each, then we can expect 95% of sample means to lie within the stated limits.* Now of course, you may not be satisfied with 95% confidence limits, and you may want to be more sure than this. We could, for example use

99% confidence limits for sample means = $\mu \pm 2.576 \dfrac{\sigma}{\sqrt{n}}$

So for our example, we can expect 99% of sample means to lie within the range

$$100 \pm 2.576 \times \dfrac{20}{\sqrt{100}}$$

i.e. 94.848 to 105.152. So we notice that if we wish to increase the level of confidence, we must widen the range for the sample means.

So far, we have been taking the central range of sample means, but occasions might arise when we wish to assess how confident we are that sample means will not be greater than a certain value. In medicine, for example, it is often important that a dose of medicine does not contain more than a certain amount of a drug. No harmful effects occur if the patient takes less, but he must not take more. Suppose that in a distribution of sample means, we wish to know the value below which we would expect 95% of sample means to lie. Here, of course, we are still considering 95% of the possible sample means, but here we are excluding the 5% at

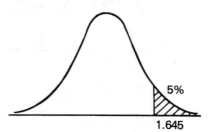

Diagram 16.11

the top rather than the 2½% at each end (see diag. 16.11). Reference to our table tells us that a Z score of 1.645 is the value below which 95% of sample means will lie. So an alternative 95% confidence limit can be expressed as

$$\mu + 1.645 \dfrac{\sigma}{\sqrt{n}}$$

Likewise, an alternative form for the 99% confidence limit is

$$\mu + 2.326 \dfrac{\sigma}{\sqrt{n}}$$

So we see that confidence limits can take two forms. On the one hand, we can state a central range for sample means – this is called a *two-sided confidence limit*. On the other hand, we can state a confidence limit that will be exceeded by a certain percentage of sample means – this is called a *one-sided confidence limit*. Perhaps an example will clarify the difference.

Example 6

Samples of 100 items are drawn from a population with a mean 1000 and a standard deviation 200. Set up 95% confidence limits for the sample means. Dealing firstly with the two sided confidence limit

$$\sigma \pm 1.96 \; \frac{\sigma}{\sqrt{n}}$$

$$1000 \pm 1.96 \times \frac{200}{\sqrt{100}}$$

$$= 1000 \pm 39.2$$

This is illustrated in Diagram 16.12, showing the range within which the central 95% of sample means will lie

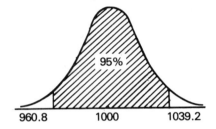

Diagram 16.12

The one-sided confidence limit could be expressed as an upper limit (see diagram 16.13, which shows the value above which only 5% of sample mean will lie)

$$\mu + 1.645 \times \frac{\sigma}{\sqrt{n}}$$

$$1000 + 1.645 \times \frac{200}{\sqrt{100}} = 1032.9$$

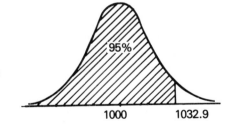

Diagram 16.13

Alternatively, the one-sided confidence limit could be expressed as a lower limit (see diagram 16.14, which shows the value below which only 5% of sample means will lie).

$$\mu - 1.645 \times \frac{\sigma}{\sqrt{n}}$$

$$1000 - 1.645 \times \frac{200}{\sqrt{100}} = 967.1$$

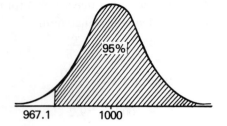

Diagram 16.14

The differences between one-sided and two-sided confidence limits are important — and we will meet them again in later chapters. However, it must be stated that *when a statistician talks of a confidence limit he usually means the two sided case.* Although we have examined confidence limits for sample means, they could be applied to sample proportions or sums and differences. For example, the 95% two-sided confidence limits for sample proportions is

$$\pi \pm 1.96 \times \sqrt{\frac{\pi(1-\pi)}{n}}$$

Summary

In this chapter, we have introduced concepts that are of fundamental importance in statistics, and it would be useful to summarise what we have learned. If we draw samples and derive some statistic (for example the sample mean or the sample proportion) and record our results in a frequency distribution, then we have formed a *sampling distribution*. The *standard error* measures the variation in our sample statistics, and we met three standard errors.

Standard error of the mean $= \dfrac{\sigma}{\sqrt{n}}$

Standard error of the proportion $= \sqrt{\dfrac{\pi(1-\pi)}{n}}$

Standard error of the sum (or difference) $= \sqrt{\sigma_1^2 + \sigma_2^2}$

A *confidence limit* is a prediction about sample results, and two types are possible. A two-sided confidence limit gives the central limits for sample results, so

95% two-sided confidence limits $= \pm 1.96 \times$ s.e.

99% two-sided confidence limtis $= \pm 2.576 \times$ s.e.

where s.e. is the standard error.

A one-sided confidence limit gives a limit that will be exceeded by a certain percentage of samples.

95% one sided confidence limit is either $+1.645 \times$ s.e. or $-1.645 \times$ s.e.

99% one-sided confidence limit is either $+2.326 \times$ s.e. or $-2.326 \times$ s.e.

In the next chapter, we will consider an important application of these concepts.

Exercises to Chapter 16

16.1 Write brief notes on each of the following:
   a) a sampling distribution
   b) standard error and sampling errors
   c) the Central Limit Theorem
   d) sampling sums and sampling differences.

16.2 Suppose samples of 100 items were drawn from a population with a mean 200 and a standard deviation of 30. What proportion of samples will have means
   a) greater than 202
   b) less than 199
   c) between 198.5 and 203.

16.3 Set up 90%, 95% and 99% two-sided confidence intervals for the sample means in example 1.

16.4 Set up 90%, 95% and 99% one-sided confidence intervals for the sample means in example 1.

16.5 A manufacturer produces metal rods which have a mean breaking strength of 60kg. and a standard deviation of 5kg. He wishes to pack the rods in bundles and guarantee that the average breaking strength of rods exceeds 59kg. How many rods should he pack in each bundle if he wishes no more than 5% of bundles to violate the guarantee?

16.6 A machine cuts rods to a specified length of 2cm. and it is known that the machine operates with a standard deviation of 0.02cm. If a sample of 100 rods has a mean length of 2.005cm., what would you conclude?

16.7 There are three distinct stages in the servicing of a particular machine

Stage	Mean time	S.D. of times
Cleaning	18 mins	1.0 mins
Greasing	12 mins	0.5 mins
Setting	10 mins	0.5 mins

Assuming that the time are independent and Normally Distributed, find the proportion of machines serviced in less than 37.5 mins.

16.8 Now you will have to think about this one! Part of the assembly of a machine involves placing a metal rod into a slot. Both the rods and the slots have lengths that are Normally Distributed with the following parameters

	Mean	Standard Deviation
Rods	16cms	0.03cms.
Slots	16.1cms	0.04cms.

Find the distribution of the clearance between the bars and the slots. If a clearance of under 0.2cms is satisfactory, find the proportion of assemblies that fail to meet this requirement.

16.9 Paddy Flynn has won a contract with an industrial estates corporation to erect a number of single-storey industrial units. He won the contract because he promised to complete each unit within 82 days of starting. Paddy's quantity surveyor produced the following estimates for each unit

	Mean time	Standard deviation
Foundation work	10 days	2 days
Brickwork	40 days	5 days
Roofing	20 days	3 days

The corporation has imposed a penalty of £100,000 for each unit that is completed after the 82 days promised. An insurance company will insure Paddy against exceeding the 82 days completion time for a premium of £5,000 per unit. Advise Paddy as to whether the insurance policy should be taken up.

16.10 In a certain country, men have a mean height of 5ft 6in, standard deviation 3in, and women have a mean height of 5ft 3in, standard deviation 2in. In a random sample of 100 married couples, the average height difference between husband and wife was 2in. What might you conclude from the result of this sample?

16.11 It is known that some 8% of components produced are defective. If components are packed in batches of 1000, what proportion of batches contain more than 8.5% defective items?

16.12 Set up 95% and 99% confidence intervals for the proportion defective in samples of 500 drawn from a population that contains 7% defective items.

16.13 The Ruritanian Ministry of Agriculture has set a standard 94% germination rate for grain seed, and only 2% of batches of seed must have germination rates less than the standard. If a certain crop of grain seed is known to have a 95% germination rate, how many seeds should be in each batch in order to meet the minimum requirement?

16.14 The Blotto Distillery has in the past maintained a 30% share of the market for gin in a certain country. The company wishes to monitor its market share by sampling 1000 gin buyers at weekly intervals. In a particular week, 290 buyers in the sample bought Blotto's gin — what would you conclude?

16.15 The Lightning Spark Plug Company has a contract to supply the Ministry of Defence with 100,000 spark plugs per week. If more than 1075 plugs are defective, then the company must pay a £1,000 penalty to the Ministry. On average, 1% of plugs produced by the company are defective. The company can employ an inspection scheme to ensure that the Ministry's specifications will be met, but this will cost £150 per week. Advise the company.

Chapter Seventeen

Quality Control

If you conducted a 'popularity poll' of all the jobs in a factory, it is an odds on bet that the inspector would be low down the list. The factory operatives dislike the inspector because they see him as the person who rejects part of their output, and so reduces bonus earnings. Management all too often (although incorrectly) see the inspector as a necessary evil − a cost that they would dearly like to dispense with! Some managers even call inspectors 'unproductive labour'. Fortunately, rational people recognise that inspectors perform a most important function. After all, quality *does* matter. Mass production has meant that goods are available to us as consumers at prices we can, on the whole, manage to afford. Unfortunately, mass production can also mean that a proportion of the goods produced will inevitably be of inferior quality. We have all had experience of annoying faults developing in the goods we buy. The inspector can, and should, be the guardian of quality.

Now we do not wish to give you the impression that inspectors should solely be seen as fulfilling an important social function − they also serve the firm well (despite the cynical attitude of some sectors of management). Quality also matters to the firm − for the firm that produces shoddy goods will eventually find its sales falling, and will find an unacceptable part of its output returned as rejects. The efficient inspector is a positive asset to the firm.

Having recognised the need for inspection, we must now decide on the form that the inspection will take. Some firms inspect finished goods in special inspection shops. The fault with this method is that as faults are found after the goods have been made, it is possible that the firm has produced large batches of defective items that have not been discovered until it is too late to do much about it. The purpose of inspection is not merely to identify defective items, but also to stop further defectives occurring. Inspection should occur as near to the point of production as possible, and inspection should be frequent. A fairly common, and commendable, system of inspection is the 'patrolling inspector'. The 'patrolling inspector' visits each machine as frequently as possible and inspects items produced by that machine, so if lapses from quality are occurring the trouble can be rectified before too many defectives have been produced. Also, the inspector who works to a sensible system will keep good records of how each machine is performing. This often enables him to spot the lapses from accepted standards *before* they occur.

In very few cases would there be 100% inspection — not only would this be costly, but in many cases the machines can produce faster than the inspectors can inspect! Quality control, then, in the main depends upon sampling theory. The actual inspection system used will depend on the nature of the possible faults that can occur.

Control System for Variate Sampling —

Industry abounds with examples of goods and components that must be within certain measurable dimensions (or tolerance limits) and we will examine a quality control system for such goods. We shall consider the case of a manufacturer producing metal rods, which must have diameters within the tolerance limits 1 cm to 1.01 cm, i.e. 1.005 ±0.005. When we dealt with the standard deviation we found it convenient to take an arbitrary origin, and remove the decimal point by taking a different unit of measurement. We shall resort to this method again. Taking 1 cm as an origin, and one ten thousandth of a cm as unit, the tolerance limit becomes 50 ± 50. How can we set up a suitable quality control system for this manufacturer?

Step One

Before we can set up any quality control system, we must have some idea of the mean and standard deviation of the diameters of the rods produced by the machinery. Suppose we measure a large number of rods, and obtain the following information

$$\text{Mean diameter} = \mu = 52$$
$$\text{Standard deviation of diameters} = \sigma = 13.76$$

Step Two

We now check that the machinery is capable of meeting the tolerances, because if it cannot, then there is little point in continuing. To do this, we will assume (quite reasonably) that the diameters of rods are Normally Distributed. If you consult the table at the end of this book you will learn that 99.8% of items in any normal distribution can be expected to lie within the range

$$\mu \pm 3.09\sigma$$

In this case, then, we can expect 99.8% of rods to have diameters within the range

$$\pm 3.09 \times 13.76 = \pm 43$$

Now to meet the tolerances, the diameters must be within the limits ± 50, so we can conclude that the machinery is capable of meeting the tolerances.

Step Three

We must now decide on just how the inspection will take place. Of course, there is nothing to stop the inspector randomly selecting a rod, measuring its diameter, and making some judgement as to the performance of the machinery. However, we have learnt enough about sampling to realise that it is most unwise to draw conclusions based on a sample of one item. It is far safer to draw a sample of a number of rods, and form judgements on the basis of the mean diameter. We will assume that the inspector will base his conclusions on the mean diameters of samples of five rods.

Step Four

In the previous chapter, we saw that if samples of n items were drawn from a population with a mean μ and a standard deviation σ, then the 95% confidence limit for sample means is

$$\mu \pm \frac{1.96\sigma}{\sqrt{n}}$$

Now strictly speaking, this will not apply to the system we are considering as the sample size we have chosen can hardly be considered large. However, it is the normal practice in quality control to assume that the Central Limit Theorem does apply to small samples. The justification for this is that many samples will be drawn, and these many small samples can be considered as equivalent to a large sample. In the case of rods, then, 95% of sample means will be within the range

$$52 \pm \frac{1.96 \times 13.76}{\sqrt{5}}$$
$$= 40 \text{ to } 64$$

In other words, we can expect 19 out of 20 sample means to be within the range 40 to 64. We call this range the *warning limits* for sample means.

In a similar fashion, we can deduce that 99.8% of sample means will lie in the range

$$\mu \pm \frac{3.09\sigma}{\sqrt{n}}$$

or in the case of our example

$$52 \pm \frac{3.09 \times 13.76}{\sqrt{5}}$$
$$= 33 \text{ to } 71$$

So we can expect 499 out of 500 sample means to fall within the range 33 to 71 and we call these limits the *action limits*.

We now have a basis for controlling the quality of output. As long as only one sample mean in 20 lies outside the warning limits, and one sample mean in 500 lies outide the action limits, we are producing a uniform quality and the process is said to be under control. We will consider later what action should be taken if these limits are violated.

Step Five

In step two we verified that the machinery was capable of meeting the tolerances. However, it is also necessary to ensure that the control system is capable of detecting any lapses from the tolerance limits. We might get this situation if the machine setting (i.e. the mean) was rather high. Suppose that the mean was 60. Now we already know that individual items have a range of ± 43, so if the mean was 60 then the individual diameters could be as high as 103, and the upper tolerance limit would be exceeded.

If we look carefully at diagram 17.01 the first thing we notice is that the upper tolerance limit minus the lower tolerance limit exceeds the range of measurement expected ($A > B$). According to the test conducted in step two, we can conclude that the machinery is capable of meeting the tolerances. However, the upper tolerance limit is below the upper range for individual items, so it is possible that the upper tolerance limit will be exceeded (i.e. individual lengths can penetrate into area B). If we were now to use the action limits as our quality control system, i.e. check that the means of samples do not fall outside these limits, then the samples could indicate that the production was under control and *lapses from the upper tolerance limit would go undetected*.

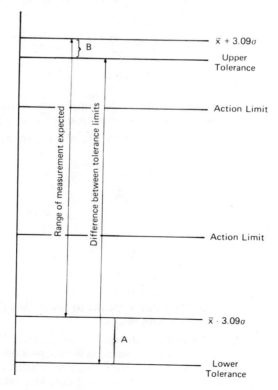

Diagram 17.01

Can you see that if the tolerances are outside one of the limits $\mu \pm 3.09\sigma$, then no undetected lapses from tolerances will occur? More formally, we can state that the tolerance limits must be at least

$$3.09\sigma - \frac{3.09\sigma}{\sqrt{n}}$$

outside the action limits if no undetected lapses from tolerance are to occur. Putting it another way, the upper action limit must be at least this amount below the upper tolerance limit, and the lower action limit at least this amount above the lower tolerance. This is called the *allowable width of the control lines* (A.W.C.L.). Writing T_U for the upper tolerance limit and T_L for the lower tolerance limit, we have

$$\text{A.W.C.L.} = T_L + \left[3.09\sigma - \frac{3.09\sigma}{\sqrt{n}} \right] \text{ to } T_U - \left[3.09\sigma - \frac{3.09\sigma}{\sqrt{n}} \right]$$

Substituting, we have

$$0 + \left[3.09 \times 13.76 - \frac{3.09 \times 13.76}{\sqrt{5}} \right]$$

to

$$100 - \left[3.09 \times 13.76 - \frac{3.09 \times 13.76}{\sqrt{5}} \right]$$

i.e. 24 to 76

As the action limits (33 to 71) lie well within this range, we conclude that no undetected lapses from tolerance will occur.

Step Six

We can now set up the *control chart for sample means*. In diagram 17.02 the action and warning limits have been inserted. We have assumed that the inspection system is under way; that twenty samples of five items have been drawn; and their means plotted on the control chart. All of the samples are well within the action limits. Notice that two sample means fall outside the warning limits — but we would only expect one to do so. Is this significant? Probably not — it is quite likely that all of the means of the next twenty samples will be within the warning limits.

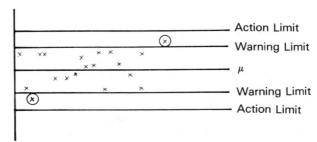

Diagram 17.02

Suppose we use the means chart as our basis for controlling quality. We draw samples of five items every (say) 15 minutes, and plot the means on the chart. Our results may look like this:

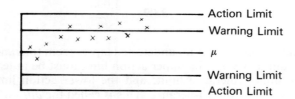

Diagram 17.03

Carefully compare the last two diagrams. In the first case, the sample means are clustered fairly evenly about the mean. In the second case, the means are clustering about the upper warning limit. What does this second case signify? It would appear that the mean diameter has increased — the setting of the machinery has drifted upwards. Would we be justified in halting production and lowering the setting of the machine? Now breaks in production are expensive, and we must be sure that drifting has occurred before a resetting is justified. Common sense tells us that in the first diagram there is insufficient evidence of a drift.

Suppose we now reset the machine and continue to draw samples. The first sample drawn has a mean less than the lower action limit. What shall we do? We expect only one sample in 500 to be outside these limits — yet the first sample drawn is outside the lower action limit. It is possible that the first sample drawn is that one sample in 500 — the rogue sample is just as likely to be the first as any other. However, exceeding the action limits by definition demands that we take some action. The appropriate action is to draw more samples immediately — we certainly would not be justified in waiting 15 minutes to draw the next sample. If their means are clustered about μ, then it seem likely that we have indeed met the odd 1 in 500 case. However, if the means are clustered about the lower warning limit, then trouble is indicated.

Estimating the Standard Deviation

In the previous example, we assumed that the standard deviation was carefully calculated. In practice, this is seldom done: instead the standard deviation is estimated from the sample range. To see how this is achieved, let us consider another example. Suppose again that we wish to set up a control system to check the diameters of metal rods. Twenty samples, each of five items were drawn, and their diameters measured and recorded in 10,000ths of a centimeter above 1 centimeter the results were as follows.

Sample No.	1	2	3	4	5	6	7	8	9	10
	59	24	58	58	54	67	55	22	68	53
	57	28	52	67	45	43	44	54	59	74
	44	66	32	46	79	33	45	62	49	39
	53	53	35	76	46	39	40	35	42	58
	82	43	11	46	69	53	74	69	44	67
Sample Total	295	214	188	293	293	235	258	242	252	291
Sample Mean	59	43	38	59	59	47	52	48	50	58
Sample Range	38	42	47	30	34	34	34	47	26	35

Sample No.	11	12	13	14	15	16	17	18	19	20
	52	66	42	46	63	71	43	34	47	53
	73	63	58	53	65	67	66	36	45	48
	48	30	44	40	70	55	52	41	78	64
	37	68	65	31	51	36	49	31	55	46
	62	49	50	33	58	52	40	47	77	53
Sample Total	272	276	259	203	307	281	250	189	302	264
Sample Mean	54	55	60	41	61	56	50	38	60	53
Sample Range	36	38	23	22	19	35	26	16	33	18

To estimate the mean diameter of all the rods, we calculate the mean of the sample means ($\bar{\bar{x}}$ — pronounced 'x double-bar').

$$\bar{\bar{x}} = \frac{59 + 43 + 38 + \ldots}{20} = \frac{1041}{20} = 52$$

You should realise that in the original units this represents 1.0052 correct to the nearest $\frac{1}{10,000}$ inch

The next stage is to calculate the range within each sample (in case you have forgotten, the range is the difference between the greatest and the smallest item. So the range of the first sample is $82 - 44 = 38$). Then we calculate the mean range $\bar{R}$

$$\bar{R} = \frac{38 + 42 + 47 + \ldots}{20} = \frac{633}{20} = 32$$

(or 0.0032 in original units).

We estimate the standard deviation by multiplying the mean range by an amount A_n. Now A_n is a statistical constant that varies according to the sample size. Unfortunately, the derivation of the distribution of the constant A_n is beyond the scope of this book, but tables are available, and an extract is printed below.

The distribution of A_n for samples of n items

Sample size n	A_n	Sample size n	A_n
2	0.8862	7	0.3698
3	0.5908	8	0.3512
4	0.4857	9	0.3367
5	0.4299	10	0.3249
6	0.3946		

We have drawn samples of 5 items, so the estimated standard deviation is

$$0.4299 \times 32 = 13.76$$

(or 0.001376 in original units)

We will now use $\bar{\bar{x}}$ as an estimate of the mean diameter of all metal rods, and 13.76 as an estimate of their standard deviation (we will examine the validity of these estimates in the next chapter). Armed with these estimates, we can now procede to set up a control system, starting at step two of the previous section.

Control System for Attributable Sampling

We shall now examine a quality control system for goods that are classified as either defective or non-defective. Much of what we have said for the first system also applies here — the Warning Limits and the Action Limits will be fixed at the same confidence intervals, and records will be kept in a similar fashion. The difference in this system is the lack of tolerance limits and so there will be no A.W.C.L.

Suppose electric light bulbs are packed in consignments of 500 and periodically a batch is exhaustively examined for defectives. So far, 50 batches have been examined, and have yielded the following number of defectives

10	8	7	6	6	6	8	8	6	6
8	6	6	7	5	8	8	7	8	8
6	7	9	8	8	6	6	9	6	7
6	7	3	7	5	7	6	7	9	8
6	8	7	6	6	6	8	8	6	6

We will use this information to set up a Quality Control system. Firstly, we put the data into the form of a frequency distribution, and calculate the mean and the standard deviation:

No of defectives per sample x	No. of Samples f	fx	fx^2
3	1	3	9
4	0	0	0
5	2	10	50
6	19	114	684
7	10	70	490
8	14	112	896
9	3	27	243
10	1	10	100
	50	346	2472

$$\bar{x} = \frac{346}{50} = 6.92$$

$$\sigma = \sqrt{\frac{2472}{50} - (6.92)^2} = 1.245$$

Now, in fact, a distribution of defectives is typically Binomial, but it is more usual to assume that a Normal Distribution is a good approximation. So, we have the Warning Limits at

$$6.92 \pm (1.96 \times 1.245) = 4.48 \text{ to } 9.36 \text{ defectives}$$

Only one sample in twenty can be expected to have defectives outside these limits. The Action Limits are

$$6.92 \pm (3.09 \times 1.245) = 3.07 \text{ to } 10.77$$

Only one sample in 500 can be expected to have defectives outside these limits.

On the chart the results of the first 50 samples have been recorded. As you can see, the process is well under control.

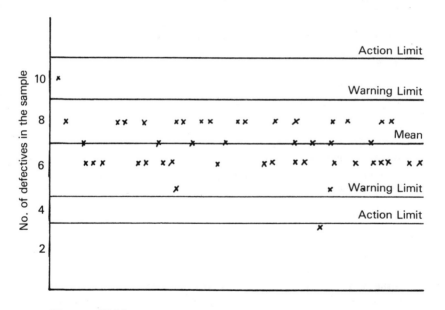

Diagram 17.04

You may be wondering why we have bothered with lower Warning and Action Limits: after all, if samples contain defectives less than those limits, then surely this is indicative of good performance. However, as we can expect one sample in 40 to fall below the lower Warning Limit, and one in a thousand to fall below the lower Action Limit, this acts as a good check on whether or not the samples drawn are truly random.

Exercises to Chapter 17

Use the distribution of A_n to estimate the standard deviation.

17.1 A machine produces and gaps spark plugs to tolerance limits of 25 thousandths of an inch ± 2 thousandths. Plugs outside this range would give erratic running of the engine. It is decided to set up a control chart for the machine. Records reveal that the average range for samples of 5 items is 1.05, and the mean of the sample means is 25.1. Set up a control chart, conducting a check that the tolerances are being met.

17.2 In step 2 we implied that

$$\text{tolerance limits} > 3.09\sigma,$$

and as $\sigma = \bar{R} A_n$ $\text{tolerance limits} > 3.09 A_n \bar{R}$

If we put $R = 3.09 A_n$, then

tolerence limits $> R \bar{R}$.

Evaluate R when $n = 5$, calculate the value of $R\bar{R}$ for exercise 1, and compare with the result in exercise 1.

17.3 The Warning Limits for sample means are:

$$\mu \pm \frac{1.96\sigma}{\sqrt{n}}$$

or $\mu \pm \dfrac{1.96 \bar{R} A_n}{\sqrt{n}}$

If $\dfrac{1.96 A_n}{\sqrt{n}} = W$, then

Warning Limits $= \bar{\bar{x}} \pm W \bar{R}$

Evaluate W when $n = 5$. Use the expression above to agree the Warning Limits obtained in exercise 1.

17.4 If the Action Limtis for sample means are

$\mu \pm A \bar{R}$

Write an expression for A and evaluate when $n = 5$. Agree the action limits obtained in exercise 1 using the constant A.

17.5 If the A.W.C.L. is $T_L + C \bar{R}$ to $T_U - C \bar{R}$

Write an expression for C_l and evaluate C when $n = 5$. Agree the A.W.C.L. found in exercise 1 using the constant c. You can now complete the following table and use it for the other exercises you will be asked to work

A_n	R	W	A	C
0.4299				

17.6 A machine is filling bags of fertiliser automatically. Samples of five were drawn, the grand mean weight being 25.02 lbs., and the mean range 0.06 lbs. Set up a quality control system. The firm receives an order on the understanding that bags will not contain less than 25 lbs. What conclusions would you draw?

17.7 Samples of 5 items are drawn at periodic intervals from the output of spark plugs. On a particular day, the first ten samples were:

Sample No.	1	2	3	4	5	6	7	8	9	10
	24.8	26.0	26.1	25.6	25.7	25.2	25.3	25.5	25.1	25.9
	25.6	25.4	25.0	25.5	25.8	26.1	25.9	25.6	25.8	25.7
	25.5	24.9	25.0	25.0	25.9	25.6	25.7	26.2	26.3	25.1
	25.8	24.8	25.4	26.0	25.5	25.4	26.0	25.1	25.2	25.8
	24.8	24.9	25.5	24.9	25.1	25.2	25.1	26.1	25.1	25.5

Plot these samples on a control chart. What action would you recommend?

17.8 A particular component is produced to a design dimension of 20 ± 12 (in some convenient units). A large number of samples of 5 items are drawn, giving $\bar{x} = 22.5$ and $\bar{R} = 8$. Calculate the Action Limits and the AWCL. What do you notice? What action would you recommend?

17.9 For the machine quoted in exercise 1, find the minimum tolerance to which it can operate.

17.10 Again for the machine in exercise 1, find the maximum shift that could occur in the mean before there was a danger of exceeding the tolerances.

17.11 A manufacturer knows that a supplier of components produces one defective item in 100. He instructs his quality control department to open each batch delivered and randomly select 100 items. If more than one defective item is found, the batch is returned to the supplier. The manager of the quality control department criticises this instruction — why?

17.12 A manufacturer makes metal washers, and a large number of samples of 5 items have been kept. The mean range of washers is 0.0002 cms. The firm receives an order for washers within the tolerance limits of 1 ± 0.0003 cms. Should the order be accepted?

17.13 Show that the Warning Limits for samples of 2 items are in approximately the same position as the Action Limits for samples of 5 items. (assume the same population is used for both types of samples).

17.14 The diameter of cable is specified as 0.75 in ± 0.025 in. A pilot production run is undertaken and 20 groups of 5 measurements were taken, all the individual results satisfying the specifications. The following results (in units of 0.001 in.) were obtained.

Sample	1	2	3	4	5	6	7	8	9	10
Mean	756	763	751	756	748	752	757	755	759	752
Range	30	12	13	16	24	17	21	15	24	21
Sample	11	12	13	14	15	16	17	18	19	20
Mean	762	746	753	756	761	753	758	754	748	764
Range	15	12	11	22	17	15	23	10	12	14

a) Set up a control chart for sample means
b) Does the pilot run indicate satisfactory performance of the plant?

17.15 A machine is programmed to produce ball-bearings having a mean diameter of 17.5 mm and a standard deviation of 0.06 mm.
  a) Set up a control chart for the means of samples of 9 items.
  b) Calculate the minimum tolerance limits that the machine can meet.
  c) The first 9 samples yielded the following results

Sample	1	2	3	4	5	6	7	8	9
Mean	17.47	17.50	17.49	17.52	17.54	17.57	17.51	17.52	17.50

  Plot the results on the control chart and comment.

17.16 A chemical plant shows the following production levels over a ten hour period.

Hour	1	2	3	4	5	6	7	8	9	10
Production	140	131	142	121	134	145	131	148	132	136

  Set up a suitable control system.

17.17 In order to set up a quality control system, 50 samples were exhaustively examined for the number of defectives. The results were:

No. of defectives	5	6	7	8	9	10	11	12	13
No. of samples	2	0	8	12	16	8	0	3	1

  Set up a quality control system for the number of defectives per sample.

17.18 a) In a sample of 1000 items produced, 42 were rejected. It was decided that control samples of 200 would be taken at frequent intervals. Find the Warning Limits and Action Limits.
  b) The first 15 control samples yielded the following results

Sample No.	No. of rejects
1	8
2	11
3	9
4	6
5	12
6	8
7	13
8	11
9	14
10	15
11	18
12	10
13	8
14	12
15	7

  Plot the results of these samples onto a control chart. What would you conclude?

17.19 From a particular productive process twenty samples, each of twenty items were selected. The following number of defectives were recorded: 3, 1, 0, 2, 1, 0, 0, 1, 1, 2, 0, 0, 4, 1, 0, 3, 0, 0, 1, 0. Set up a control chart for the number of defectives in a sample. *Note:* The mean number of defectives per sample is too small to be able to use a Normal Approximation, so the Poisson Distribution must be used. Find the appropriate Poisson Distribution, and use this to find the number of defectives that will be exceeded on only 5% of occasions (this gives the Warning Limit) and the number of defectives that will be exceeded on just 1% of occasios (this gives the Action Limit).

17.20 Motor vehicle bodies are produced assembled by a spot welding operation, and some joints will need rewelding. The table below gives the number of defective welds found on a vehicles which were selected at hourly intervals during a three day period. The number of defects can be assumed to form a Poisson Distribution.

1st Day	2nd Day	3rd Day
2	5	6
4	3	4
7	7	3
3	11	9
1	6	7
4	4	4
8	9	7
9	9	12

The results are those of one welder, and for each day the first four refer to morning work and the second four to afternoon work.

a) What is the mean number of defectives?
b) Find the Action and Warning Limits for the number of defective welds (Hint: as the mean is high, we can use a Normal Approximation. Remember that for a Poisson Distribution mean = variance)
c) Make any comments you consider relevant.

Chapter Eighteen

Estimation

In Chapter 16 we concerned ourselves with making predictions about sample behaviour and we did this by introducing the idea of sampling distributions and confidence limits. In order to form sampling distributions, we required knowledge of the appropriate population parameters. If we were variate sampling then we needed to know the population mean and variance, and if we were attribute sampling then we needed to know the population proportion. Now in this chapter, we will reverse the process: we will assume that the population parameters are unknown to us and that we want to predict them from sample evidence. We will use the technique known as *statistical estimation*. It goes without saying that we require any estimate made to be a 'good' one — an estimate that a statistician would call unbiassed. Before we start, then, we need to define precisely what is meant by an unbiassed estimate. *An unbiassed estimate is one that is just as likely to be too great as it is to be too small.* In other words, if we drew all possible samples size n and made an estimate of the population parameter from each, then the average value of our estimate would be equal to the true value of the population parameter. Obviously this is a sensible definition — the more data that we have then the more accurate would be our estimate.

Estimating a population mean

Let us suppose that we draw a sample of n items, and we use the sample to estimate the population mean. We can use the symbol $\wedge$ to mean "an estimate of", so the estimated population mean is denoted by $\hat{\mu}$. Think back to experiment 1 in chapter 16 — here we drew all possible samples from an infinite population, and found that the average value of the sample mean was the same as the population mean. The sample, mean, then, fits our criterion for an unbiassed estimate precisely, so we can state that

$$\hat{\mu} = \bar{x}$$

The single estimate of μ is called a *point estimate,* and we are going to be extremely lucky if our estimate is bang on target. It would seem reasonable to adopt the policy of Chapter 16 and state a confidence interval, or *interval estimate* of the population mean.

To set up this confidence interval, we must decide on how sure we want to be that our interval estimate does include the true value of the population mean and as we saw in Chapter 16, it is usual to apply a 95% level of confidence. To see how the method works, let us suppose we know that a population has a mean 100 and a standard deviation 20, though we keep this information to ourselves. We now ask a statistician to estimate the mean of this population by sampling. If the statistician decides to draw 100 items for his samples we can use confidence limits to predict that 95% of the means of such samples would be in the range.

$$100 \pm 1.96 \times \frac{20}{\sqrt{100}}$$

or (roughly) within the range 96 to 104.

The statistician now draws his samples, and sets up a 95% interval estimate from each sample. Suppose his first sample had a mean of 99.5, then his interval estimate for the population mean would be

$$\hat{\mu} = 99.5 \pm 19.6 \times \frac{20}{\sqrt{100}},$$

or between 95.5 and 103.5 — an interval estimate that does in fact bracket the true value of the population mean. If his second sample had a mean of 102.5, then his interval estimate would be

$$\hat{\mu} = 102.5 \pm 1.96 \times \frac{20}{\sqrt{100}}$$

or between 98.5 and 106.5 — which again brackets the true value of the population mean. Diagram 18.01 illustrates the situation, and a glance at the diagram will show that provided the sample mean lies between

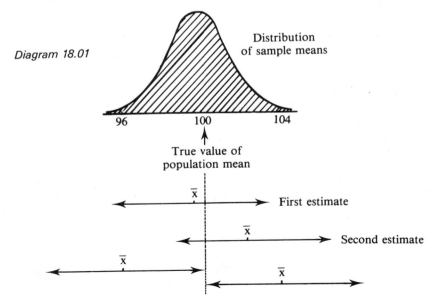

Diagram 18.01

96 and 104, then the interval estimate will bracket the true value of the population mean. Now sampling theory tells us that we can indeed expect 95% of sample means to lie within these limits, so we can expect 95% of interval estimates to bracket the true value of the mean. This is what is meant, then, when we talk of a 95% interval estimate for the population mean. If we draw many samples from a population, and set up interval estimates for the population mean, then 95% of our interval estimates will include the true value of the population mean.

Now if you think carefully about what we have learnt so far, you will realise that there are two snags with this analysis. Firstly, we have used the true value of the population standard deviation to set up interval estimates for the population mean — but in practice we will not know the true value of the population standard deviation, and we will have to use the sample standard deviation (s). Our interval estimate is, then

$$\hat{\mu} = \bar{x} \pm \frac{1.96s}{\sqrt{n}}$$

Now as s is only an estimate of σ, to retain 95% confidence, we really should widen our estimate a little. However, you can rest assured that as long as our sample is large (greater than 30), the amount by which we would have to widen our estimate is insignificantly small. We will examine this more closely later.

Secondly, we have assumed that the statistician has drawn many samples to estimate the population mean — but in practice he will draw just one. Once the estimate has been made, then the statistician is either right or wrong — there is no way he can be right on 95% of occasions. What, then, would an interval estimate mean in this case? The statistician is using a method that has a 95% chance of success in the long run — he knows that if he uses this method every time he is asked to estimate a population mean, then 95% of his interval estimates will bracket the true value of the population mean. He will be right on 95% of occasions.

Estimating a Population Proportion

Quite often, we will find it useful to use a sample to estimate the proportion of a population with a certain characteristic ($\hat{\pi}$). For example, we might wish to estimate the proportion in the population who intend to buy a certain product, or the proportion of defective items in a large batch. To do this, we can use a method essentially similar to estimating a population mean — i.e. we can set up a 95% interval estimate. Now the point estimate of the population proportion is the sample proportion (which we shall call p), so

$$\hat{\pi} = p$$

Moreover, Experiment 4 in Chapter 16 demonstrates that p satisfies the criterion for an unbiased estimate. The experiment demonstrated that if samples of n items are drawn from a population with a proportion 'defective', $\hat{\pi}$, and if the proportion 'defective' p is calculated for each sample, then the average value of p will be $\hat{\pi}$. To calculate an interval estimate, we will need the standard error of the proportion. Using the sample proportion, this is

$$\sqrt{\frac{p(1-p)}{n}}$$

So our 95% interval estimate for the population proportion is

$$\hat{\pi} = p \pm 1.96 \sqrt{\frac{p(1-p)}{n}}$$

Example 1

In oder to estimate its share of the market, a detergent manufacturer randomly selects 500 women and asks them whether they use the manufacturer's products. Two hundred and sixty report that they do so. Estimate the manufacturer's market share.

$$\hat{\pi} = p = \frac{260}{500} = 0.52$$

95% interval estimate for market share =

$$0.52 \pm 1.96 \sqrt{\frac{0.52(1-0.52)}{500}}$$

$$= 0.52 \pm 0.0437$$

or, using percentages

$$52\% \pm 4.37\%.$$

Determining the sample size

We have seen that we can give a 95% interval estimate for a population parameter by calculating

Sample statistic ± 1.96 × standard error,

and the quantity ± 1.96 × standard error is called the *sampling error*. Now clearly, the smaller is the size of the sampling error, the more precise will our interval estimate be. We shall now ask what determines the size of the sampling error, and see what we can do to reduce its size. If we multiply the standard error by a constant smaller than 1.96 then we will reduce the sampling error — but we will no longer have the interval estimate we require as our level of confidence would be less than 95%. If, then, we are to reduce the size of the sampling error we must turn our attention to the standard error.

Let us turn our attention first to estimating the population mean. In this case, the standard error is $\frac{s}{\sqrt{n}}$, and if we reduce the size of s or increase the size of n, then the standard error (and so the sampling error) would be reduced. Now although we cannot reduce the size of s (after all, the sample standard deviation is quite outside our control) we can increase the size of n. This will improve the precision of our estimate. Think about this for a minute and you will realise that it makes sense, because after all we should have more confidence in large sample results than in small sample results.

Suppose a sample of 100 items has a mean of 50 and a standard deviation of 20. Our interval estimate (with 95% confidence) for the population mean is

$$\mu = 50 \pm 1.96 \times \frac{20}{\sqrt{100}}$$

i.e. 50 ± 4 (approximately).

If this was not sufficiently precise, we could draw a larger sample and reduce the sampling error. If we wanted the sampling error to be 2 rather than 4, then

$$2 = 1.96 \times \frac{20}{\sqrt{n}}$$

so $\frac{2}{1.96} = \frac{20}{\sqrt{n}}$

and $n = \left(\frac{20 \times 1.96}{2}\right)^2$

$= 384.$

If we now draw another sample of 384 items, this should give us the degree of precision we require for our interval estimate. We can see then, that if we are prepared to state in advance the size of the sampling error that would satisfy us (call this *r*) and then undertake a pilot sample to determine the size of the sample standard deviation(*s*), then the sample size n that will give the required sampling error is

$$n = \left(\frac{1.96s}{r}\right)^2$$

with 95% confidence, or

$$n = \left(\frac{2.576s}{r}\right)^2$$

with 99% confidence.

We can use a similar analysis to find the sample size necessary to obtain a sampling error of $\pm$ r when estimating a population proportion. This is

$$n = \frac{(1.96)^2 p(1-p)}{r^2}$$

with 95% confidence or

$$n = \frac{(2.576)^2 p(1-p)}{r^2}$$

with 99% confidence.

Example 2

A sales manager has conducted a survey, and found that 5% of his competitors intend to switch to his product on their next purchase. He urges the board to sanction a capacity expansion. Before doing this, the board want to be 99% certain that the sales manager's estimate is correct to within ± ½%. What size sample should be examined to check the sales manager's claim?

$$n = \frac{(2.576)^2 \times 0.05(1 - 0.05)}{(0.005)^2} = 12{,}608$$

Obviously, such an exceptionally large sample would cost a lot of money!

Estimating a Population Variance

Let us turn our attention to estimating a population variance. From what we have learnt in this chapter you would expect that the unbiassed estimate of the population variance ($\hat{\sigma}^2$) is the sample variance s², but unfortunately this is not so. This can be demonstrated with a simple experiment.

Experiment 5

Find your results of experiment 1, where you drew samples of two items from the population 1, 2, 3.

$$\mu = \frac{1+2+3}{3} = 2$$

$$\sigma^2 = \frac{(1-2)^2 + (2-2)^2 + (3-2)^2}{3} = \tfrac{2}{3}$$

a) For each sample, find the variance s²
b) Show that the average value of the sample variance is $\tfrac{1}{3}$

So we see from this example that we must not use the sample variance as an estimate of the population variance — the sample variance is a *biassed* estimate as it *underestimates* the population variance. To obtain an unbiassed estimate of the population variance, we must increase the sample variance, and we do this by applying Bessel's correction: multiply the sample variance by $\frac{n}{n-1}$ i.e.

$$\hat{\sigma}^2 = s^2 \times \frac{n}{n-1}$$

or what amounts to the same thing

$$\hat{\sigma}^2 = \frac{\Sigma(x - \bar{x})^2}{n - 1}$$

Experiment 6

Suppose we have a population comprising the digits 5, 9

$$\hat{\mu}^2 = \frac{5+9}{2} = 7$$

$$\sigma^2 = \frac{(5-7)^2 + (9-7)^2}{2} = 4$$

a) Draw all possible samples of three items from this population (there are eight) and find the variance of each sample.
b) Show that the average value of the sample variance underestimates the population variance.
c) Multiply each sample variance by $\frac{n}{n-1}$ and show that the average value of the adjusted sample variance is 4 — the same as the population variance.

Let us examine again the expression for estimating the population variance

$$\hat{\sigma}^2 = s^2 \times \frac{n}{n-1}$$

Now the adjustment we must make to the sample variance to get an unbiassed estimate of the population variance will depend on the size of the sample we draw. The larger the value of n, then the smaller the adjustment we must make — this is clearly illustrated in the table below

n	$\frac{n}{n-1}$
2	2.000000
5	1.250000
10	1.111111
50	1.020408
100	1.010101
1000	1.001001

So we can see that when the sample size is two, we must increase the sample variance by 100% in order to obtain an unbiassed estimate of the population variance. If the sample size increases to 50 items, then the size of the adjustment falls to 2%, but if we increase the sample size to 1000 then the adjustment necessary is only 0.1%.

We have applied Bessel's correction to variate sampling, but it applies equally to attribute sampling. If the sample is large, then the variance of the sample proportion is a good approximation to the population proportion — but if the sample is small then Bessel's correction must be applied i.e.

$$\hat{\sigma}^2 = \frac{p(1-p)}{n} \times \frac{n}{n-1}$$

$$= \frac{p(1-p)}{n-1}$$

Now the importance of Bessel's correction is the implications it has when we either predict sample behaviour, or make estimates of the population mean or proportion. So far, we have used Z scores from the Normal Distribution when we set up confidence intervals or made interval estimates, but this is only justified as long as the population standard deviation is known or can be estimated with great precision i.e. if the sample is large. However, if the sample is small then the Normal Distribution will not apply: we will have no justification at all for using Z scores for the calculation of confidence intervals. Suppose we have a sample of n items with a mean $\bar{x}$ and a standard deviation s, then

$$\hat{\mu} = \bar{x}$$
$$\hat{\sigma}^2 = s^2 \times \frac{n}{n-1}$$
$$\hat{\sigma} = s \times \sqrt{\frac{n}{n-1}}$$

The estimated standard error of the mean will be

$$\frac{\hat{\sigma}}{\sqrt{n}} = s \times \sqrt{\frac{n}{n-1}} \times \frac{1}{\sqrt{n}}$$
$$= \frac{s}{\sqrt{n-1}}$$

Now if the sample is large, we can make a 95% interval estimate like this:

$$\hat{\mu} + \bar{x} \pm 1.96 \frac{s}{\sqrt{n-1}}$$

but suppose the sample is small — then the constant 1.96 cannot be used. You see, the problem is that by using small samples we have introduced further sampling errors into our estimates. In estimating the population mean we have also had to estimate the population variance — an estimate which itself is subject to error. The error we quote in our interval estimate, then, is too small and we should increase it; we must multiply the standard error by a constant *greater than* 1.96. Just how much greater than 1.96 should be our constant depends on the accuracy of our estimate of the population variance — which in turn depends on the sample size.

This problem of the size of the constant to use was investigated by W.S. Gosset in 1908, who published his results under the pseudonym "A Student" (his employers would not allow him to use his real name). He realised that Standard Normal Z scores would not apply to small samples, so he replaced them with a statistic he called t. Gosset discovered the sampling distribution of t, which takes into account the variations in s as well as the variations in $\bar{x}$. For each sample size, then, there is a different t distribution, so it is vitally important that we use the right one. Now Gosset did not tabulate the t distribution against the sample size — instead he tabulated it against the number of *degrees of freedom* in the sample. Don't worry about what this means just yet — for the moment just accept that the number of degrees of freedom is the denominator in Bessel's correction i.e. one less than the sample size.

Turn to the tables at the end of this book and you will find some values of the *t* distribution (reasonably complete tables as we gave for the Normal Distribution would fill an entire book). Tabulated in the first column is the number of degrees of freedom, and in the first row are the varying probability levels. To see how to use these tables, let us consider an example.

Example 2

A sample of 10 items has a mean 35 and a standard deviation 10. Estimate the mean of the population from which the sample was drawn.

Firstly, let us make the point estimate

$$\hat{\mu} = \bar{x} = 35$$

To obtain an interval estimate, we must estimate the standard error of the mean

$$\text{s.e. of sample mean} = \frac{s}{\sqrt{n-1}} = \frac{10}{\sqrt{10-1}} = 3.333$$

The interval estimate of the population mean is

$$\hat{\mu} = 35 \pm t \times 3.33$$

The appropriate value of t depends upon the degrees of freedom in the sample, and the level of confidence we require for the interval estimate. In this case, the sample has 9 degrees of freedom, so the t value we require will be given in the ninth row of the table.

Suppose we wished to set up a 95% interval estimate: the t value we require will be in the column headed Pr. = 0.025 *not* Pr. = 0.05 because, like the Normal Distribution Table the t table is concerned with the area in the right hand tail

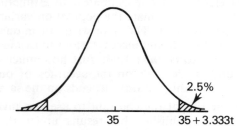

Diagram 18.01

So the appropriate t value for this interval estimate is 2.262, and the 95% interval estimate of the population mean is

$$\hat{\mu} = 35 \pm 2.262 \times 3.333$$
$$= 35 \pm 7.54$$

Likewise, the t value for a 99% interval estimate would be found in the column headed Pr. = 0.005, i.e.

$$\hat{\mu} = 35 \pm 3.250 \times 3.333$$
$$= 35 \pm 10.833$$

As the sample size increases, the t distribution becomes closer to the Normal Distribution. In fact, if the sample has 120 degrees of freedom (i.e. a sample of 121 items) then the 95% interval estimates for the population mean is

$$\hat{\mu} = \bar{x} \pm 1.98 se$$

whereas for a 'large' sample it would be

$$\hat{\mu} = \bar{x} \pm 1.96 se$$

— hardly any difference. In fact, most statisticians argue that if the sample has at least 30 items, it makes little difference whether the Normal Distribution or the t distribution is used. This is the reason why statisticians call samples greater than 30 'large' samples.

Degrees of Freedom

The time has now arrived to examine what is meant by degrees of freedom — a concept that creates problems for most students of statistics. The concept arises because sample results vary: if we use samples to estimate (say) a population mean, then we will not obtain the same estimate from each sample. Moreover, the smaller the sample size chosen, then the greater will be the variation in our estimates. Suppose we drew samples of 5 items from a certain population and found that the sample mean $\bar{x} = 20$. Of course, another, different, sample of 5 items could also yield the same estimate, and we will now examine the values that this sample could contain. if the sample is to contain 5 items and have a mean of 20, then the sample total (Σx) must be $5 \times 20 = 100$. In fact, we are perfectly 'free' to assign arbitrary values to any four of the five items in the sample. Suppose the first four items sampled are 10, 25, 30 and 18: then the last item must be $100 - (10 + 25 + 30 + 18) = 17$ — it couldn't possibly be any other value. So we see that if we 'fix' the sample mean at $\bar{x}$ in a sample of n items, we can arbitrarily assign values to $n-1$ items in the sample — we have $n-1$ degrees of freedom left in the sample after fixing the sample mean.

A similar analysis can be applied to the sample variance. Suppose a sample of 4 items has a variance 62.5 — again we have $4 - 1 = 3$ degrees of freedom in constructing a similar size sample with the same variance. Suppose three of the items sampled are 90, 95 and 105 (and the last is x), then the sample mean is

$$\bar{x} = \frac{90 + 95 + 105 + x}{4} = \frac{290 + x}{4}$$

$$s^2 = \frac{\Sigma(x - \bar{x})^2}{n}$$

so $\Sigma(x - \bar{x})^2 = ns^2$

$$\left(90 - \frac{290+x}{4}\right)^2 + \left(95 - \frac{290+x}{4}\right)^2 + \left(105 - \frac{290+x}{4}\right)^2 + \left(x - \frac{290+x}{4}\right)^2 = 4 \times 62.5$$

$$(70 - x)^2 + (90 - x)^2 + (130 - x)^2 + (3x - 290)^2 = 4000$$

$$12x^2 - 2320x + 110000 = 0$$

which solves to give $x = 110$ or $83\frac{1}{3}$

If we wish to fix both the mean and variance, then we will have less choice in arbitrarily assigning values to a sample — we will have $n - 2$ degrees of freedom. Suppose we want a sample of four items to have a mean 20 and a variance 10 — we can fix any two of the items in the sample. Suppose we fix the first two items at 16 and 18

$$\bar{x} = 20, \text{ so } \Sigma x = 4 \times 20 = 80$$

If we call the first item x, then the fourth must be

$$80 - (x + 16 + 18) = 46 - x$$

The formula for the sample variance is

$$s^2 = \frac{\Sigma x^2}{n} - \left(\frac{\Sigma x}{n}\right)^2$$

so $\frac{\Sigma x^2}{4} - (20)^2 = 10$

which gives $\Sigma x^2 = 1640$

so $16^2 + 18^2 + x^2 + (46 - x)^2 = 1640$
which reduces to

$$x^2 - 46x + 528 = 0$$

and solves to give either $x = 22$ or $x = 24$
So the sample must be 16, 18, 22, 24

Generalising, if we fix m statistics (for example the mean, variance etc.) in a sample of n items, then we will have $n - m$ degrees of freedom in fixing the values in the sample.

'Pooled' Estimates of the Population Variance

If we draw two samples from the same population we may find that the first sample of n_1 items has a variance s_1^2. From this sample we would estimate the population variance to be

$$\frac{s_1^2 n_1}{n_1 - 1}$$

Suppose that we draw a second sample of n_2 items with a variance s_2^2 the estimated population variance from this sample is

$$\frac{s_2^2 n_2}{n_2 - 1}$$

Now it would seem logical to argue that we can get a better estimate of the population variance by pooling the results of both samples than we could from either sample alone. We pool the variances like this:

$$\hat{\sigma}^2 = \frac{s_1^2 n_1 + s_2^2 n_2}{n_1 + n_2 - 2} \text{ and } \hat{\sigma} = \sqrt{\frac{s_1^2 n_1 + s_2^2 n_2}{n_1 + n_2 - 2}}$$

In this case, we have $n_1 + n_2 - 2$ degrees of freedom associated with our estimate.

If the samples were large then there would be not need to include the correction factor (-2) in the denominator as the difference in our results would be negligible. Suppose we draw a sample of 1000 items with a variance of 90, and a second sample of 1200 with a variance of 84, then our estimates of the population variance is:

$$\hat{\sigma}^2 = \frac{90 \times 1000 + 84 \times 1200}{1000 + 1200 - 2} = 86.806, \text{ which incorporates the adjustment}$$

or $\hat{\sigma}^2 = \dfrac{90 \times 1000 + 84 \times 1200}{1000 + 1200} = 86.72$, without the adjustment

The difference in our estimates is extremely small. But if the samples were small, say 10 and 16 items respectively

$$\hat{\sigma}^2 = \frac{90 \times 10 + 84 \times 16}{10 + 16 - 2} = 93.5, \text{ which incorporates the adjustment}$$

or $\hat{\sigma}^2 = \dfrac{90 \times 10 + 84 \times 16}{10 + 16} = 86.31$, without the adjustment

The difference is alarming and serves as a warning not to ignore the adjustment when we are dealing with small samples.

Experiment 7

This experiment is designed to verify that the unbiassed estimate of the population variance from two samples is given by

$$\hat{\sigma}^2 = \frac{n_1 s_1^2 + n_2 s_2^2}{n_1 + n_2 - 2}$$

Let the population be the digits 1,2. We could draw samples of 2 items from this population, and the values of $n_1 s_1^2$ would be 0, $\frac{1}{2}$, $\frac{1}{2}$, and 0. We could draw eight samples of three items from this population and the values of $n_2 s_2^2$ would be 0, $\frac{2}{3}$, $\frac{2}{3}$, $\frac{2}{3}$, $\frac{2}{3}$, $\frac{2}{3}$, $\frac{2}{3}$ and 0. Now suppose we estimate the population variance by pooling a sample of two items with a sample of three items, then four different estimates could be obtained

i) $\quad \dfrac{0+0}{3+2-2} = 0$

ii) $\quad \dfrac{0+\frac{2}{3}}{3+2-2} = \frac{2}{9}$

iii) $\quad \dfrac{\frac{1}{2}+\frac{2}{3}}{3+2-2} = \frac{7}{18}$

iv) $\quad \dfrac{\frac{1}{2}+0}{3+2-2} = \frac{1}{6}$

Find the number of ways it is possible to make each estimate; find the average value of the estimate and verify that it is equal to the population variance.

Just as we can pool two sample variances to estimate the population variance, we can also pool two sample means to estimate the population mean. If a sample of n_1 items has a mean $\bar{x}_1$, and a sample of n_2 has a mean $\bar{x}_2$, then we would estimate the population mean like this:

$$\hat{\mu} = \frac{n_1 \bar{x}_1 + n_2 \bar{x}_2}{n_2 + n_2}$$

Experiment 8

If we draw samples of two items from the population 1, 2, then the possible sample totals $n_1 \bar{x}_1$ = 2, 3, 3 and 4. If we draw samples of 3 items then the possible samples totals $n_2 \bar{x}_2$ = 3, 4, 4, 4, 5, 5, 5, and 6. Find all the possible estimates of the population mean by pooling samples of two items with samples of three items. Find the average value of the estimates and verify that it is equal to the population mean

Finite Population Adjustments

So far, we have been assuming that we sample from an infinite population, or (what amounts to the same thing) if we sample from a finite population then we sample with replacement. This is not a realistic assumption: although populations under consideration may be very large, they will still be finite, and it is difficult to imagine why a statistician would practice replacement sampling. We must now examine the effect of finite populations on sampling distributions. Well, the first thing to notice is that if the population is finite and we use the formulas developed so far, then we will be *overstating* the situation. For example, if we are variate sampling, drawing a sample of n items from a population of N items, then the expression $\frac{\sigma}{\sqrt{n}}$ would overstate the standard error of the mean. This can be demonstrated with a simple experiment.

Experiment 9

Suppose we have a population comprising the digits 4, 5 and 6.

$$\mu = \frac{4+5+6}{3} = 5$$

$$\sigma = \sqrt{\frac{(4-5)^2 + (5-5)^2 + (6-5)^2}{3}} = \sqrt{\frac{2}{3}}$$

If we draw samples of two from this population then

$$\text{standard error of sample mean} = \frac{\sigma}{\sqrt{n}} = \sqrt{\frac{2}{3}} \times \frac{1}{\sqrt{2}} = \frac{1}{\sqrt{3}}$$

1) Draw all possible samples of two items, but this time sample without replacement (there are six possible samples).
2) Find the sample mean and the standard error of the mean.

Now if you have performed this experiment correctly, then your calculated value for the standard error of the sample mean will be less than $\frac{1}{\sqrt{3}}$. Can you see why $\frac{\sigma}{\sqrt{n}}$ must overstate the standard error? If we practice replacement sampling then the smallest sample mean will be $\frac{4+4}{2} = 4$, and the largest sample mean will be $\frac{6+6}{2} = 6$. But if we practice sampling without replacement then the smallest sample mean will be $\frac{4+5}{2} = 4.5$, and the largest sample mean will be $\frac{5+6}{2} = 5.5$. So sampling without replacement reduces the range of sample means. The correct value for the standard error or the mean can be obtained using the formula

$$\frac{\sigma}{\sqrt{n}} \times \sqrt{\frac{N-n}{N-1}}$$

where N is the population size. So we would predict the standard error of the sample mean to be

$$\frac{1}{\sqrt{3}} \times \sqrt{\frac{3-2}{3-1}} = \frac{1}{\sqrt{3}} \times \sqrt{\frac{1}{2}} = \frac{1}{\sqrt{6}}$$

which should agree exactly with the standard error calculated from the samples in experiment 9.

The quantity $\sqrt{\frac{N-n}{N-1}}$ is called the finite population adjustment factor and it must also be applied to the standard error of proportions i.e.

$$\sqrt{\frac{\pi(1-\pi)}{n}} \times \sqrt{\frac{N-n}{N-1}}$$

Sometimes, it makes sense to ignore the finite population adjustment factor. Suppose that N is very large in comparison to n, then the adjustment factor will be very close to 1, and it will not be worthwhile using the adjustment factor. For example, if we draw samples of 100 from a population of 1000 then the size of the adjustment would be

$$\sqrt{\frac{1000-100}{1000-1}} = 0.949$$

but if we draw samples of 100 from a population of 100,000 then the adjustment would be

$$\sqrt{\frac{100,000-100}{100,000-1}} = 0.9995$$

We can use the finite population adjustment factor to show that it is the absolute size of the sample that is important and not its size relative to the size of the population. Let us suppose that a firm receives components in batches, and it wishes to verify that the proportion of defectives in the batches is remaining constant. The method used is to take a 5% sample, and use this to estimate the proportion in the batch. We know that the standard error is a measure of the accuracy of the estimate: the smaller the standard error then the better is the estimate. Suppose that a batch of 2000 items is received, then a sample of 100 would be examined to give a standard error

$$\sqrt{\frac{p(1-p)}{100}} \times \sqrt{\frac{2000-100}{2000-1}}$$

The next batch to arrive contains 10,000 items: according to our sampling scheme we would have to examine 500 items — a very large sample indeed. Is it really necessary to examine such a large sample? Suppose that the standard error obtained from the previous batch gave a satisfactory degree of accuracy — what size sample from a batch of 10,000 would give the same degree of accuracy? If this sample has n items, then the standard error is

$$\sqrt{\frac{p(1-p)}{n}} \times \sqrt{\frac{10{,}000-n}{10{,}000-1}}$$

and if this standard error is to be the same as previously then

$$\sqrt{\frac{p(1-p)}{n}} \times \sqrt{\frac{10{,}000-n}{10{,}000-1}} = \sqrt{\frac{p(1-p)}{100}} \times \sqrt{\frac{2000-100}{2000-1}}$$

Squaring both sides of this expression and cancelling

$$\frac{\cancel{p(1-p)}}{n} \times \frac{10{,}000-n}{9999} = \frac{\cancel{p(1-p)}}{100} \times \frac{1900}{1999}$$

$$\frac{10{,}000-n}{9999n} = \frac{19}{1999}$$

$$n \simeq 104.$$

In other words, a sample of 104 items from a population containing 10,000 items can be expected to yield the same degree of accuracy as a sample of 100 from a population of 2000. If the population size increases by 500% then we need only increase the sample size by 4%! Sampling schemes that rely on drawing a fixed percentage from the population are wasteful and inefficient — you should concentrate on the absolute size of the sample and not its size relative to the population.

Conclusion

In this chapter, we have examined the principles of statistical estimation. We have distinguished between biassed and unbiassed estimates, and have seen that interval estimates are more useful than point estimates. Finally, we have introduced the idea of finite population adjustments, and have seen that as long as the population size is large in comparison to the sample size, then such adjustments are not really necessary.

Exercises to Chapter 18
18.1 Write brief notes on each of the following:
   a) Point estimates and interval estimates
   b) A sampling error
   c) Biassed and unbiassed estimates
   d) Degrees of Freedom

18.2 A simple random sample of 100 sales invoices was taken from a very large population of sales invoices. The average value was found to be £18.50 with a standard deviation of £6.00. Obtain a 95% interval estimate for the true average value per sale.

18.3 A population of sales invoices is to be sampled so that the mean value per sale can be estimated. One hundred sales invoices are selected at random from this population and classified by value as shown in the following frequency distribution.

Value (£)	No. of sales invoices
0 – 50	10
50 – 100	9
100 – 150	15
150 – 200	22
200 – 250	13
250 – 300	7
300 – 350	9
350 – 400	5
400 – 450	6
450 – 500	4

Use the sample information to construct interval estimates for the population mean at the 95% and 99% levels of confidence.

18.4 An auditor randomly selects 100 invoices and checks for error in calculation. Errors in the firm's favour he marks with a plus and errors in the supplier's favour he marks with a minus. He tabulates his results as follows:

Error (£)	Frequency
−40 to −30	1
−30 to −20	2
−20 to −10	16
−10 to 0	49
0 to 10	23
10 to 20	6
20 to 30	2
30 to 40	1

Find the mean and standard deviation.

Assuming a normal distribution, estimate limits within which you would expect the mean error of all invoices to be. Assuming the firm handles 10,000 invoices per annum, estimate the most likely, most favourable and least favourable errors for one year.

18.5 A manufacturer wishes to estimate the mean weight of sacks of carbon black. He would be satisfied if his estimate was within ± 5lb of the true mean weight and be 99% sure of his estimate. An initial sample gives an S.D. of 15 lbs. What size sample yields the required estimate?

18.6 From a random sample taken over the period of a year a shopkeeper calculated his average daily sales to be £218 and the standard deviation of the sample to be £18. The number of days' sales in the sample was 60.
   (i) Estimate the mean daily sales for the year at the 99.8% level of confidence.

(ii) On the assumption that the shopkeeper would be unable to obtain any further stock for another twenty-four hours, estimate the value of stock he should hold in order to be 95% sure that his stock would last for twenty-four hours. State any other assumptions you would find necessary to make in order to calculate your estimate.

18.7 Ordit and Partners are carrying out a study of three independent paperback book retail firms that each have a large number of shops around the country. From samples taken from shops of each firm, the following information was determined:

Firm	Sample Size	Sample Average Selling Price (£)	Sample Standard Deviation Selling Price (£)
A	81	0.85	0.20
B	49	0.76	0.18
C	36	0.87	0.08

(a) Ordit wish to estimate the average selling price of paperbacks for each of the three firms. Use the data given to set up a confidence interval estimate for the true average paperback selling price of each of the three firms, using a confidence level of 90%. Comment on your results.
(b) If, for all three firms, Ordit wished to be within £0.05 of the true average price of paperbacks with 90% confidence, what size of sample should they have taken from the shops of each firm?
(c) In the light of your answer to (b) comment on the appropriateness of the data originally collected for the purpose as given in (a).

18.8 A local election is being held and there are two candidates, Smith and Jones, for one vacancy. A random sample of the electorate reveals that 55% of them will vote for Smith. How confident can you be that Smith will be elected if the sample size was 100? How large a sample would be needed for the 55% sample results for Smith to make you 99% confident of a win for him?

18.9 Your sales manager excitedly informs you that he has conducted a survey of competitors' customers, and he has discovered that 5% intend to switch to your product on the next purchase. He advises you to invest now to take advantage of this increase. The sales manager interviewed 100 people. What would you tell him? To be sure that it is worthwhile to expand capacity, you must be convinced within ½% of your sales increase. What size sample should be examined? What would you conclude?

18.10 In a random sample of 200 garages it was found that 79 sold car batteries at prices below that recommended by the manufacturer.
(i) What is a random sample? Describe briefly how one can be selected.
(ii) Estimate the proportion of all garages selling below the list price.
(iii) Calculate 95% and 99% confidence limits for this estimate and explain what these mean.
(iv) What size sample would have to be taken in order to be 95% certain that the population parameter could be estimated to within 2%?
(v) What size sample is necessary in order to be completely certain that the percentage could be estimated to within 2%. Explain your answer.

18.11   Nine bags of sugar were randomly selected and carefully weighed. The mean weight was 1.004 Kg and the standard deviation 0.0002 kg. Find the 95% interval estimate for the mean weight of all bags.

18.12   A sample of 10 items has a mean 20 and a standard deviation 0.5. A second sample of 15 items has a mean 19.5 and a standard deviation 0.4. Assuming that both samples came from the same population, estimate the population mean and standard deviation.

18.13   Show that the maximum value for the standard error of the proportion is

$$\sqrt{\frac{0.25}{n}}$$

N.B. you will need calculus to show this.

18.14   A sample of five items has values 12, 13, 17, 23 and x. Find x given that the sample mean is 18.

18.15   For the sample quoted in example 14, find x given that the sample variance is 21.2.

18.16   A sample of five items has values 15, 18, 24, x and y. Find x and y given that $\bar{x} = 20$ and $s^2 = 18$.

18.17   A civil engineering contractor buys a job lot of 10,000 pipes which have a nominal length of 10 metres each. A random sample of 100 pipes were chosen and their lengths were carefully measured. The mean length was 9.59 meters and the standard deviation 0.01 meters. Give a 95% interval estimate for the total length of piping in the job lot.

18.18   For the job lot of pipes quoted in exercise 17, find the sample size necessary to predict the total length of piping within ± 15 meters, with 95% confidence.

18.19   A bye-election has been declared at Anytown, which has 40,000 voters. A random sample of 100 voters reported that 53% can be expected to vote Democrat. Assuming that 70% of registered voters do actually vote, give a 95% interval estimate for the proportion voting Democrat at the election. What size sample should be examined to be 95% sure of a Democractic Victory?

18.20   Suppose we are sampling to estimate a population proportion. If the population contains 500 items then a sample of 50 gives a satisfactory degree of accuracy to the estimate. What size sample should be examined if the population contains 5000 items?

Chapter Nineteen

Hypothesis Testing

In chapter 12 we developed criteria that were intended to help us in the decision taking process. We assumed that all the information we needed to apply the criteria was available to us. But how often will this happen in the real world? It is likely that in many cases we will have to resort to sampling to give us some idea of the information we need. Now as you already know, we cannot be completely sure of sample results, so if we base our decisions on sample evidence, it is quite conceivable that we make the wrong choice. It is a major task of statistics to evaluate the probability of making an incorrect decision, and to keep this probability to a minimum.

Type I Errors

We will now return to a problem that faces many manufacturers: how to control the quality of the output. Let us suppose that the manufacturer produces woodscrews by a completely automatic process. One of the problems with the automatic process is that occasionally a screw misses the threading stage, so such screws should be rejected. Now there is little that can be done about this, and the manufacturer has installed machinery that would automatically remove such rejects. From past experience it is known that about 5% of output would be rejected by this machine. Now, although there is not much chance of bettering this rate, machine malfunction can cause the rate to worsen considerably. The problem experienced so far is that it has been difficult to detect just when the rate of rejects has increased because the quantity of output is so great. Obviously, the manufacturer wants to keep the rate as near to 5% as possible – after all, rejects cost money! What he decides to do is to install an automatic sampling device which randomly selects 1000 screws per hour, sorts them, and counts the number of rejects.

The purpose of drawing these samples is, of course, to detect when the reject rate has worsened, and do something about it. Taking remedial action involves stopping the process, and this will be expensive as output will be lost. So if remedial action is to be taken, the manufacturer wants to be pretty sure that the reject rate has worsened. What he wants is some *decision rule* to determine when to stop the process and make the necessary adjustments. Now as 5% of the output is rejected, the manufacturer can expect, on average, 50 rejects in each sample. Suppose he errs on the cautious side, and takes as his decision rule – 'stop the process if the number of defectives exceeds 60'.

Unfortunately, this rule does not exclude the possibility of taking the wrong decision; i.e. stopping the machine when the reject rate has not increased. Using our knowledge of the binomial distribution, we should be able to calculate the probability of taking the wrong decision. It will be the wrong decision if the number of defectives exceeds 60, but the reject rate is still 5%. The appropriate binomial distribution is $(0.95+0.05)^{1000}$ and the probability we require is $P_{(61)} + P_{(62)} + P_{(63)} + \ldots + P_{(1000)}$. Now obviously it would be advisable to use the normal approximation to the binomial distribution. Firstly, we require the standard deviation

$$\sigma = \sqrt{npq}$$

$$n = 1000, \; p = 0.05, \; q = 0.95$$

$$\sigma = \sqrt{1000 \times 0.05 \times 0.95}$$

$$= 6.89$$

The probability that the sample contains more than 60 defects is shown in the diagram by the shaded area (remember that we must make a 0.5

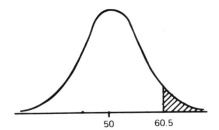

Diagram 19.01

adjustment when using a normal approximation to the binomial distribution). The Z score is

$$\frac{60.5 - 50}{6.89} = 1.52$$

Consulting the normal distribution tables, the probability we require is 6.43%. So if the manufacturer decides to stop the process if the number of defectives in the sample exceeds 60 on the assumption that the reject rate has risen above 5%, then the probability that he has made the wrong decision is 6.43%.

Is this, then, an acecptable risk? Well, this is really a decision that only the manufacturer can take. To decide whether this risk is too high, he must weigh up what is at stake. Can you see how the manufacturer can reduce the risk of taking the wrong decision? All that he need do is to alter his decision rule, taking a number of defectives greater than 60 as the borderline. If, for example, his decision rule was to stop the process if the number of rejects in the sample exceeds 65, then the Z score would be

$$\frac{65.5 - 50}{6.89} = 2.25$$

and the probability of making the wrong decision would be 1.22%.

The relationship between the probability of taking the wrong decision and the number of defectives in the decision rule is something we will examine in more detail later. What we shall now do is express the problem in statistical jargon. The manufacturer formulates the *hypothesis* that the overall number of rejects is 5%. He uses the following decision rule: if the number of rejects in a sample exceeds 60, reject the hypothesis and stop the process, otherwise accept the hypothesis. Notice that we have not altered the philosophy of his decision rule — all that we have done is to introduce the word 'hypothesis'. Now we have seen that there is a danger of making the wrong decision and needlessly stopping the process. Using statistical jargon, there is a danger of *rejecting the hypothesis when it should be accepted*. Statisticians refer to this risk as a *Type I error*. We have seen that the risk of a Type I error will vary according to the number of rejects quoted in the decision rule. We can quantify this risk by calculating the Z score for each number of rejects quoted in the decision rule.

No. of Rejects Quoted in the Decision Rule	Z Score	Probability of a Type I Error
52	0.36	0.3594
54	0.65	0.2578
56	0.94	0.1736
58	1.23	0.1093
60	1.52	0.0643
62	1.81	0.0351
64	2.10	0.0179
66	2.39	0.0084
68	2.69	0.0036
70	2.98	0.0014

It is useful to draw a graph of this data, and this is done in Diagram 19.02. The manufacturer can see at a glance the risk of incurring a Type I error as he varies the number of defectives in the decision rule. Of course, it is quite possible to state the degree of risk of a Type I error that would be acceptable, and read off from the graph the number of defectives in the decision rule that would yield this degree of risk. Suppose, for example, he would consider as acceptable a 5% chance of a Type I error, then consulting the graph we see that the decision rule should be 'stop the process if the number of defectives exceeds 61'. Using statistical jargon, we say that the graph is the *performance chart* for Type I errors in our decision rule. Notice that as we increase the number of defectives in our decision rule, the probability of a Type I error decreases.

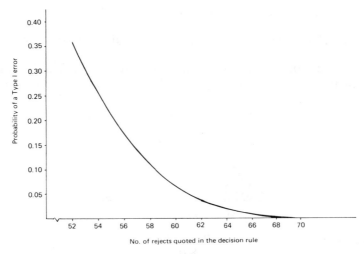

Performance chart: probability of a Type I error
Diagram 19.02

Type II Errors

Let us summarise what we have done so far. We have formulated the hypothesis that the number of rejects is 5%, and we use as a decision rule that if the number of defectives in a sample exceeds a certain number n, then we would reject the hypothesis and stop the process, otherwise we accept the hypothesis and allow the process to continue. We have seen that in operating this decision rule there is a danger of rejecting the hypothesis when it should be accepted — so we would needlessly stop the process. This is because it may be possible that the number of rejects in the sample exceeds n even though the overall number of defectives is still 5%. We call this a Type I error, and the performance chart shows how the risk of a Type I error varies with the size of n in our decision rule.

Now the Type I error is not the only risk the manufacturer runs when operating the decision rule. Originally, he operated the rule: Stop the process if the number of defectives exceeds 60, otherwise allow the process to continue. The philosophy underlying this rule is that if the number of defectives is 60 or less, then the overall number of defectives produced is 5%. Now it is quite possible for the overall rate of defectives to rise above 5%, but the sample he draws will contain 60 defectives or less. Should this be the case, then he will be allowing the process to continue when really he should stop the process and make the necessary adjustments. Returning to statistical jargon, there is a danger of *accepting a hypothesis that should be rejected*. Such risks are referred to as *Type II errors*, and we should be able to calculate the probability of such errors.

Suppose the overall rate of rejects produced rose to 7%. For samples of 1000, we have $n = 1000$, $p = 0.07$ and $q = 0.93$. So the appropriate binomial distribution is $(0.93 + 0.07)^{1000}$.

$$\text{mean} = np = 1000 \times 0.07 = 70$$

$$\text{standard deviation} = \sqrt{npq} = \sqrt{1000 \times 0.07 \times 0.93} = 8.07$$

We can now use the normal approximation to the binomial distribution to obtain the probability of obtaining 60 or less defectives in a sample of 1000 items.

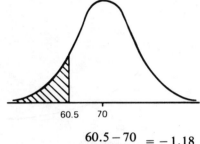

Diagram 19.03

The Z score is
$$\frac{60.5 - 70}{8.07} = -1.18$$

and consulting the normal tables we see that the probability of a Type II error is 0.119. In other words, there is an 11.9% chance that the decision rule will fail to detect a rise in the overall rate of defectives to 7%, i.e. there is an 11.9% chance that the manufacturer makes the wrong decision.

The probability of making a Type II error will depend on the actual rate of defectives produced by the process. So if we list all possible reject rates, we can calculate the appropriate probabilities of Type II errors.

Proportion of Rejects Produced by the Process	Mean (np)	S.d. $\sqrt{npq}$	Z Score	Probability of Type II Error
0.055	55	7.21	0.76	0.7764
0.060	60	7.51	0.07	0.5279
0.065	65	7.80	−0.58	0.2810
0.070	70	8.07	−1.18	0.1190
0.075	75	8.33	−1.74	0.0409
0.080	80	8.58	−2.27	0.0116
0.085	85	8.82	−2.78	0.0027
0.090	90	9.05	−3.26	0.0004
0.095	95	9.27	−3.72	0.0001
0.100	100	9.49	−4.16	0.0000

Diagram 19.04 shows this information graphically. Notice as the percentage of defectives produced by the process increases, the smaller is the probability of making a Type II error. Notice also that to calculate the probability of making a Type II error we must make an assumption about the value of the overall rate of defectives produced by the system. For Type I errors, this was not necessary, so it is easier to calculate the performance chart for Type I errors. Also, the performance chart for Type II errors is not in a very convenient form − it does not show the manufacturer how he can lessen the probability of a Type II error.

Earlier, we calculated that if the overall rate of defectives produced by the process was 7%, then the probability of a Type II error, given our decision rule, is 11.9%. Suppose we changed the decision rule to read: 'stop the process if a sample contains more than 55 defectives'. Now if the overall rate of defectives is 7% and a sample contains 55 defectives or less, then the process will not be stopped when it should be, and the probability of this happening can again be calculated.

$$Z = \frac{55.5 - 70}{8.07} = -1.80$$

and from the normal tables, the probability of taking the wrong decision is 3.59%. So if we reduce the number of defectives in our decision rule, this decreases the probability of a Type II error. Unfortunately, this action would automatically increase the probability of a Type I error!

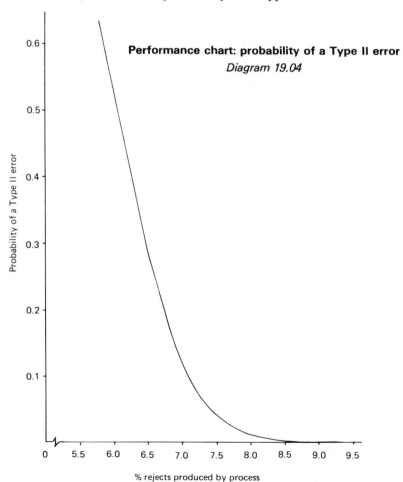

Performance chart: probability of a Type II error
Diagram 19.04

% rejects produced by process

Summarising then, a Type I error occurs when we reject a hypothesis that should be accepted, and a Type II error occurs when we accept a hypothesis that should be rejected. Many statisticians summarise the distinction in a table like this:

		Decision	
		Accept	Reject
Hypothesis is	True	Correct decision	Type I error
	False	Type II error	Correct decision

If we state our hypothesis and decision rule, we can immediately calculate the probability of a Type I error, but we cannot calculate the probability of a Type II error without introducing other assumptions. In other words, to calculate the probability of Type I errors, we keep the hypothesis constant and vary the decision rule; but to calculate the probability of Type II errors, we keep the decision rule constant and vary the hypothesis. We can reduce the probability of a Type I error by increasing the number of rejects in our decision rule, but unfortunately this would increase the probability of a Type II error. Now this is quite a problem, and in the next section we will try to reconcile this problem.

Significance Testing

In the last section we found no difficulty at all in calculating the probability of a Type I error. The reason for this was the form taken by our hypothesis: that the overall rate of rejects was 5%. So we could put $p = 0.05$ into the binomial distribution and calculate the probability of a Type I error with ease. Notice that in applying this hypothesis, we are assuming that the reject rate found in the past has continued; i.e. that the reject rate has not changed. We have done this despite the fact that the purpose of our investigation is to detect when the reject rate worsens. Because our hypothesis assumes that 'nothing has happened', it is called a *null hypothesis* and we use the symbol H_0 to stand for a null hypothesis. Now you may object that this seems rather strange — the purpose of sampling is to detect when the reject rate worsens, so we should be testing the hypothesis that the reject rate is greater than 5%. Now this is all very well, but it will mean that we are unable to calculate the probability of a Type I error without stating by *how much* the reject rate has worsened. In other words, stating that $p > 0.05$ means that we do not have a value for p to insert in the binomial distribution, and so we will be unable to calculate the probability of a Type I error. It is for this reason that we will ensure that from now on all the hypotheses we test will be null hypotheses.

The testing of null hypotheses has two implications that we must now examine. Firstly, the null hypothesis involves stating the complete opposite to what you are trying to detect. In the example we have been considering so far, we were trying to detect whether the reject rate had worsened, but the null hypothesis is that the reject rate is still 5%. Again, suppose we are trying to test for bias in a coin − the appropriate null hypothesis would be that the coin is unbiassed. It would follow from this that the probability of a head is 0.5, so we could then calculate the probability of a Type I error. We can now state a rule for formulating null hypotheses − if you are trying to detect change, then the appropriate null hypothesis is that no change has occurred.

The second implication of formulating a null hypothesis is that it is no help to us whatsoever, in calculating the probability of a Type II error. To do this we must state an *alternative hypothesis*. This is precisely what we did when calculating the probabilities of a Type II error in the last section. In fact, we stated a number of alternative hypotheses ($p = 5.5\%$, $p = 6\%$, $p = 6.5\%$, etc.). Now obviously there is a problem here: if we reject the null hypothesis it would seem reasonable to accept the alternative hypothesis: but which alternative hypothesis should we accept?

Let us now see if we can find a way round this problem. We incur Type II errors if we accept a hypothesis that should be rejected − so if we never accept a hypothesis we cannot make a Type II error! Let us restate the problem we have so far been considering in order to investigate this situation more closely. We know from past experience that the process is, on average, rejecting 5% of all screws produced. Now suppose we know that this reject rate can change either way; i.e. improve or get worse, and we wish to sample to detect such changes. The reason for this is that perhaps we can then account for fluctuations in reject rates − they may depend on the quality of steel used, or which shift is operating the machinery. In any case, we will adopt the null hypothesis that the reject rate is 5%, and operate the decision rule 'reject the null hypothesis if the number of defectives in the sample is greater than 60 or less than 40, otherwise *reserve judgement'*. Notice that we never accept the null hypothesis, and in many statistics textbooks you will come across the conclusion 'no reason to reject the null hypothesis' − which of course doesn't mean that it should be accepted! The philosophy behind this argument is that samples which contain between 40 and 60 defectives do not show a *significant* departure from the expected 50 defectives. In other words, the range 40-60 can be attributed to chance or sampling fluctuations. However, defectives outside this range cannot attributed to chance, the *sample result is significant* and we must reject the null hypothesis. Notice the use of the word 'significant' − many statisticians give the title *significance testing* to the methods we have been describing.

So you see it is possible to avoid making Type II errors altogether. However, you may complain that the reasoning behind this is very suspect. Initially, we were trying to detect when the overall reject rate rose so that we could do something about it. Our null hypothesis was that the overall reject rate was 5%, and our decision rule was 'stop the process if the number of defectives exceeds 60'; i.e. reject the null hypothesis. Now suppose that the number of rejects was 60 or less, and that this would induce us to conclude that there is no reason to reject the null hypothesis. Now think carefully about this! If the number of defectives in the sample exceeds 60, we stop the process; if the number does not exceed 60 we reserve judgement. But can we reserve judgement? If we try to do this, we will allow the process to continue — which is tantamount to accepting the null hypothesis! In such circumstances, then, we cannot avoid the risk of a Type II error!

Let us now see if we can devise an acceptable methodology for significance testing. Firstly, we will formulate a null hypothesis (H_0) in such a way that we can calculate the probability of a Type I error. Also, we will formulate an alternative hypothesis (H_1) so that rejection of H_0 automatically involves the acceptance of H_1. We will then specify the probability of a Type I error that we would be prepared to consider an acceptable risk. We call this probability the *level of significance,* and it is usual to specify probabilities of 5% and 1%. We then construct a decision rule based on the hypothesis and level of significance we have chosen. Whenever possible we word the alternative hypothesis in such a way that if we do not reject the null hypothesis we reserve judgement. If this is not possible, then the wording of the alternative hypothesis is such that the alternative to rejecting the null hypothesis is accepting it. In this latter case, we keep our fingers crossed that the probability of a Type II error is not too large. If Type II errors would be really serious, then we must examine the performance chart to see how likely they are.

One-sided and Two-sided Tests

We will now apply this methodology to the problem we have so far been considering. Firstly, we specify the hypotheses

H_0: $\pi = 0.05$, i.e. the overall reject rate is 5%
H_1: $\pi > 0.05$, i.e. the overall reject rate is greater than 5%
Level of significance = 0.05, i.e. the risk of a Type I error is 5%.

Consulting the performance chart for Type I errors, we see that for a 5% probability we must quote 61 defectives in our decision rule, so our decision rule is 'reject H_0 if the number of defectives in the sample exceeds 61, otherwise accept H_0'. We could represent this situation diagrammatically as in diagram 19.05.

Summarising this situation, then, we can say that on the basis of our null hypothesis we expect, on average, 50 defectives in samples of 1000. If the number of defective is 61 or less, then we can say that the deviation from 50 can be attributed to chance. But if the number of defectives exceeds 61 we cannot attribute this to chance; i.e. the results is significant, and we reject the null hypothesis concluding that the reject rate is greater than 5%. We realise that in using this decision rule we run a 5% risk of making a Type I error.

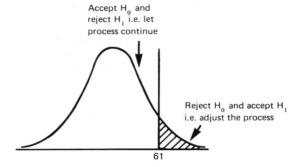

Diagram 19.05

In this example, we are interested in detecting shifts in the reject rate in one direction only. If the reject rate increases, then we must take some action. But if it decreases – so much the better. Tests to detect changes in one direction only are called *one-sided tests* or *one-tail tests,* and a glance at Diagram 19.05 will explain why this is so. However, there will be many cases when it will be more meaningful to detect changes in either direction.

To take an example, suppose we know that 40% of the electorate vote Socialist, and we wish to detect changes in the Socialist vote by sampling. It will be of equal interest to know whether the Socialist vote has increased or decreased. The appropriate hypotheses here would be

$$H_0: \pi = 0.4$$
$$H_1: \pi \neq 0.4$$

Notice that in this case the alternative hypothesis is stated in a 'not equal to' form. Without going into the mechanics of this problem (we will do this in the next chapter) suppose the decision rule was: 'in a sample of 1000, if the number voting Socialist is between 370 and 430 accept H_0 and reject H_1, otherwise accept H_1 and reject H_0'. Diagrammatically, the decision rule could be represented as in diagram 19.06.

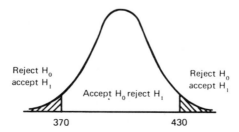

Diagram 19.06

Tests such as this are called *two-sided* or *two-tail tests,* again for obvious reasons. So if you wish to test for a change in an *unspecified direction* you should use a two-sided test, but if you wish to test for a change in a *specified direction* you would use a one-sided test. The type of test used will be reflected in the formation of the alternative hypothesis. Which type of test is appropriate will depend on the problem under consideration. There is no hard-and-fast rule to follow: as we shall see in the next chapter the choice is really a matter of common sense.

Chapter Twenty

Statistical Significance

In the last chapter, we indicated how statistics can help us come to the right decision. The methodology we used was as follows. We formulated two hypotheses: a null hypothesis and an alternative hypothesis, in such a way that rejecting one meant accepting the other. The wording we chose for the null hypothesis enabled us to calculate the probability of a Type I error. We then stated a decision rule, for example 'reject the null hypothesis and accept the alternative hypothesis if the number of defectives in the sample exceeds x'. If we rejected the null hypothesis, we then calculated the probability of a Type I error, and we called this probability the level of significance. This, then, is the methodology of the statistical technique called *significance testing*. In this chapter, we will move from methodology to the technique itself. In particular, we will be dealing with tests involving sample means and sample proportions. Before we do this, however, we must examine the choice of an appropriate level of significance.

We saw in the last chapter that when we altered our decision rule in such a way that we reject the null hypothesis only if the number of defectives is greater than previously stated, we then reduced the probability of a Type I error. We must now decide on the risk of a Type I error that we could consider acceptable. Now, of course, the acceptable level of risk is a personal choice, and it will vary with the problem we are considering. We would want, for example, a lower level of risk if we were testing the efficacy of a new drug than we would want if we were testing a marksman for bias in shooting. Despite this essentially subjective nature of the level of risk, statisticians do apply conventions. They consider a 5% probability of a Type I error as being acceptable, and construct their decision rules accordingly. If the hypothesis is rejected on the basis of this decision rule, we state that the sample result is *significant at the 5% level*. In other words, we do not think that the sample result can be attributed to chance, and our probability of making a Type I error is 5%. Also, it is usual to construct a second decision rule based on the probability of a Type I error being 1%, so if the null hypothesis is rejected on the basis of this rule, we state that the sample result is *significant at the 1% level*. So applying our decision rules to sample results, there are four possible conclusions we could reach:

(a) Accept the null hypothesis and reject the alternative hypothesis. In this case, we are saying that the sample results can be attributed to chance. However, as we saw in the last chapter, reaching this conclusion leaves us wide open to Type II errors, and the only way to avoid them is to conclude that

(b) there is insufficient reason to reject the null hypothesis – a conclusion that does not conclude anything and suggests the necessity for further sampling. However, if anything useful is to be achieved, then sooner or later we must risk the possibility of a Type II error and come to some definite conclusion.

(c) Reject the null hypothesis and accept the alternative hypothesis at the 5% level, but not at the 1% level. Here we are stating that the occurrence of the sample results cannot be attributed to chance, and the probability of a Type I error is less than 5% but greater than 1%.

(d) Reject the null hypothesis (and so accept the alternative hypothesis) at the 1% level. In this situation we are taking a less than 1% chance of incurring a Type I error.

Now let us look at the actual wording of the decision rule. We have implied that the decision rule will be stated in terms of the sample statistic, for example 'reject the null hypothesis if the proportion defective exceeds x'. How can we find the value of x? To do this, we must state the level of significance (which we will assume to be 5%), the sample size (call this n) and the null hypothesis about the population proportion (call this π). The value of x, then, is the proportion that would be exceeded in our samples on only 5% of occasions, and we can obtain this value from a one-sided confidence interval i.e.

$$x = \pi + 1.645 \sqrt{\frac{\pi(1-\pi)}{n}}$$

(If you cannot see that x is obtained from the one sided confidence interval, then you should read the chapter on Sampling Theory again). Alternatively, we may be testing for a change in the population proportion – a movement in either direction. Here, a two-sided decision rule would be appropriate – for example "accept the null hypothesis if the proportion defective in the sample is outside the range x_1 to x_2, otherwise reject it". Keeping the same level of significance, this range can be obtained from the two-sided confidence interval

$$\pi \pm 1.96 \times \sqrt{\frac{\pi(1-\pi)}{n}}$$

So we see that the decision rule is merely a confidence interval. In fact, if you think about it you will realise that a confidence interval is a prediction about the sample statistic on the assumption that the null hypothesis is true.

Now although there is absolutely nothing wrong with using decision rules to test hypotheses, statisticians do not, in fact, use them. Taking a one-sided test, a statistician would argue something like this: the null hypothesis would be rejected if

$$p > \pi + 1.645 \times \sqrt{\frac{\pi(1-\pi)}{n}}$$

where p is the sample proportion. Rearranging this condition, the null hypothesis would be rejected if

$$\frac{p - \pi}{\sqrt{\frac{\pi(1-\pi)}{n}}} > 1.645$$

You should recognise the left hand side of this condition: it is the Z score of the sample proportion, and because it is being used to test the null hypothesis we call the Z score a *test statistic*. The right hand side of the condition is the value for the Z score that would be exceeded only on 5% of occasions: if it is exceeded by the test statistic then we would reject the null hypothesis. So the right hand side of this expression is a kind of a *critical value* for our test statistic.

This, then, is how we shall undertake significance testing. We will set up a null hypothesis and an alternative hypothesis. If we assume that the null hypothesis is true, then we can calculate a test statistic (which basically is the standard score of our sample statistic), and compare the test statistic with its critical value. If the test statistic is greater than its critical value, then we would reject the null hypothesis and accept the alternative hypothesis. However, if the test statistic is less than its critical value, we would accept the null hypothesis and reject the alternative hypothesis.

Test of Population Means Using Large Samples

When we discussed sampling theory, we saw that if we drew large samples from a population with a known mean and standard deviation, then the sample means will be Normally Distributed. So, if we use large samples to test hypothesis about a population mean, then the test statistic would be the Z-score of the sample mean, i.e.

$$\text{test statistic } Z = \frac{|\bar{x} - \mu|}{\text{standard error}}$$

Notice that the test statistic is always taken as positive. We can use our knowledge of confidence intervals to state the critical values for Z. The critical value will depend upon the level of significance and on whether the test is one-sided or two sided.

	Critical Values for Z	
	5% level	1% level
one sided test	1.645	2.323
two sided test	1.96	2.576

Can you see why we always take the absolute value of the test stastic? If we did not, then we would have to state *two* critical values for each level of significance for a two-sided test — for example at the 5% level the critical values would be "greater than 1.96" and "less than −1.96". Now as, say, a test statistic of 2 and −2 are equally significant, it makes sense to take the absolute value of the Z score.

Example 1

A sugar refiner packs sugar into bags weighing, on average, 1 kilogram. Now the setting of the machine tends to "drift" i.e. the average weight of bags filled by the machine sometimes increases and sometimes decreases. if the mean weight of bags increases, then in effect he is actually giving away some sugar. If the mean weight decreases then he may find himself in trouble with the Weights and Measures Inspector. So it is really important that he controls the average weight of bags of sugar. He wishes to detect shifts in the mean weight of bags as quickly as possible, and reset the machine. In order to detect shifts in the mean weight, he will periodically select a sample of 100 bags, weigh them, and calculate the sample mean.

In effect, the refiner is performing a significance test, adopting as his null hypothesis that the population mean is 1, i.e.

$$H_0: \mu = 1.$$

Now shifts in μ in either direction are important, so the test is two-sided and the appropriate alternative hypothesis is

$$H_1: \mu \neq 1.$$

What other information do we need to know? Clearly, we need to know the variance of the weight of bags of sugar — suppose this is 0.01 kg. We can now calculate the standard error of the mean for samples of 100 bags

$$\frac{\sigma}{\sqrt{n}} = \frac{0.1}{\sqrt{100}} = 0.01 \text{ kg}$$

and if we drew a sample of 100 bags and use the sample mean $\bar{x}$ to test the null hypothesis, then the test statistic Z would be

$$\frac{|\bar{x} - \mu|}{\text{s.e.}} = \frac{|\bar{x} - 1|}{0.01}$$

as the test is two-sided, we would compare the test statistic with the critical values 1.96 and 2.576

Suppose the sample mean was 1.03 kg., then

$$\text{Test statistic } Z = \frac{|1.03 - 1|}{0.01} = 3,$$

and as this exceeds the critical value we would reject the null hypothesis and accept the alternative hypothesis at the 1% level of significance. We would conclude that the overall mean weight was no longer 1kg, and run a less than 1% chance of a Type I error.

Example 2

A certain highway in a city centre has been investigated in the past for noise pollution by vehicles. As a result of many measurements it has been found that between 16.30 and 17.30 on any weekday the average noise level is 130 decibels, with a standard deviation of 20 decibels. The residents are convinced that the noise level is getting worse, and having taken 50 readings, obtain an average of 134 decibels. Is their claim justified?

Clearly, this is a one-sided test, so we can formulate

$$H_0: \mu = 130$$
$$H_1: \mu > 130$$

$$\text{Test statistic } Z = \frac{|\bar{x} - \mu|}{\text{s.e.}} = \frac{|134 - 130|}{20/\sqrt{50}} = 1.41$$

As this is a one-sided test, the critical values for Z are 1.645 (at the 5% level) and 2.326 (at the 1% level). So we see that the difference between the population mean and the sample mean is not significant – it can be accounted for by sampling fluctuations. We would accept the null hypothesis and reject the alternative hypothesis, and we cannot conclude that noise levels have increased.

The two examples we have considered so far assume that we have a prior knowledge of the population – in particular they assume that the population mean and standard deviation are known. Had we not known the population standard deviation, we could have estimated it using s, the sample standard deviation. No adjustment to s^2 would have been necessary because in both cases we had large samples. Now of course, it will not always be the case that we know the population mean, and if we cannot compare a sample with a population then we must compare a sample with a sample. Tests of this nature come under the general heading of the *significance of the difference betwen the means of two samples*. Let us see how this test works by considering an example.

Example 3

The authors of this book agree on most matters, be they academic, social or cultural. They are both keen supporters of association football. However, on one thing they will never agree – which is the best team in the English Football League. One of the authors (the misguided one) is convinced that the finest team is Manchester United, but the other author (the one you must agree with) is, of course, a Liverpool F.C. supporter. To settle this argument once and for all, the authors agreed to apply statistical analysis to this problem. They will examine a sample of football matches, count up the number of attacking moves made per match, and so obtain a mean and standard deviation of the number of attacking moves made per match for both teams. Now we have a real problem here – is this test one-sided or two-sided? Suppose that $\bar{x}_1$ is the mean number of attacking moves made by Manchester United, and $\bar{x}_2$ is the mean number of attacking moves made by Liverpool. We can take $\bar{x}_1$ as the mean of a sample drawn from a population with a mean μ_1. (the long run mean number of attacking moves made by Manchester United). Likewise, we can take μ_2 as the long run mean number of attacking moves made by Liverpool. The appropriate null hypothesis is that there is no difference in the mean number of attacking moves, i.e.

$$H_0: \mu_1 = \mu_2 \text{ or }$$
$$H_0: \mu_1 - \mu_2 = 0.$$

Now the Manchester United supporter is convinced that his team is the better, so the alternative hypothesis that he wishes to test is

$$H_1: \mu_1 > \mu_2$$

i.e a one-sided test. Likewise, the Liverpool supporter would wish to test the alternative hypothesis

$$H_1: \mu_2 > \mu_1$$

Again a one sided test. But what of you, the unbiassed reader (at least, we will assume that you are unbiassed). You really don't know what to expect, so the appropriate alternative hypothesis from your viewpoint would be

$$H_1: \mu_1 \neq \mu_2$$

i.e. a two-sided test! For the purpose of this problem we will adopt the two-sided alternative hypothesis.

Let us suppose that sampling from the matches played we have

	Number of Attacking Moves		Number of Matches
	Mean	Standard Deviation	
Manchester United	$30 = \bar{x}_1$	$6 = s_1$	$50 = n_1$
Liverpool	$32 = \bar{x}_2$	$5 = s_2$	$60 = n_2$

Now the standard error of the mean for Manchester United is

$$\frac{s_1}{\sqrt{n_1}} = \frac{6}{\sqrt{50}} = 0.849$$

and the standard error of the mean for Liverpool is

$$\frac{s_2}{\sqrt{n_2}} = \frac{5}{\sqrt{60}} = 0.645$$

In making these calculation, we assume that the sample results for Manchester United come from a population with a standard deviation of 6, and the sample results for Liverpool come from a population with a standard deviation of 5. (This assumption is reasonable: the samples are large so the sample variance is a legitimate estimate of the population variance). We can now use the sample difference theorem to calculate the standard error of the difference between the means, i.e.

$$\text{Standard error of the difference} = \sqrt{\left(\frac{s_1}{\sqrt{n_1}}\right)^2 + \left(\frac{s_2}{\sqrt{n_2}}\right)^2}$$

$$= \sqrt{\frac{s_1^2}{n_1} + \frac{s_2^2}{n_2}}$$

$$= \sqrt{(0.849)^2 + (0.645)^2}$$

$$= 1.066$$

Now the actual difference between the means is $|\bar{x}_1 - \bar{x}_2| = 2$, but if the null hypothesis is correct, then the difference does not deviate significantly from zero. In other words, we can consider the difference of 2 as coming from a population of sample mean differences, with a mean zero and a standard error 1.066. We now have all the information necessary to undertake a significance test.

Test statistic $Z = \frac{2-0}{1.066} = 1.876$. As we are conducting a two-sided test, the critical values for Z are 1.96 (at the 5% level) and 2.576 (at the 1% level). So we must accept the null hypothesis: the samples do not provide evidence that there is a difference in the long run mean number of attacking moves made by the two football clubs.

We must state, though, that to use this method of significance testing, the samples must be *independent,* i.e. both samples are assumed to have come from the same population, and the selection of the first sample in no way affects the selection of the second. This method, then, could not be used for 'before and after' types of experiments, i.e. when the sample is considered under different conditions. Later we will show you how to deal with samples that are not independent (statisticians call them *paired samples*).

In this section, we have introduced two types of significance tests and it would be useful to introduce general expressions for the test statistic Z. In examples 1 and 2, we tested a hypothesis about a population mean μ using a sample mean $\bar{x}$. The test statistic in this case was

$$Z = \frac{|\bar{x} - \mu|\sqrt{n}}{\sigma}$$

In example 3 we tested for a difference in population means μ_1 and μ_2 using sample means $\bar{x}_1$ and $\bar{x}_2$, and sample standard deviations s_1 and s_2. The test statistic in this case was

$$Z = \frac{|\bar{x}_1 - \bar{x}_2| - 0}{\sqrt{\frac{s_1^2}{n_1} + \frac{s_2^2}{n_2}}}$$

Tests of Population Means using Small Samples

Let us now suppose that we use small samples to test hypotheses about population means. Now if we use the methods of the last section, we will run into trouble on two accounts. Firstly, our small sample will understimate the population variance, so our test statistic will be wrong. Secondly, the means of small samples are not Normally Distributed, so our critical values will be wrong. We have learnt that the means of small samples have a t-distribution, and the appropriate t-distribution will depend on the number of degrees of freedom in estimating the population variance. If we use large samples to test a hypothesis, then the critical values we use will depend upon the type of test (i.e. one-sided or two-sided). But if we use small samples, then the critical values will depend upon the degrees of freedom in the sample as well as the type of test. Fortunately, we already know how to use tables of the t-distribution (if you have forgotten, then refresh your memory by re-reading Estimation, Chapter 18.) Using small samples, then, both the test statistic and the critical values will be t-scores.

Example 4

The expected lifetime of electric light bulbs produced by a given process was 1500 hours. To test a new batch a sample of 10 was taken which showed a mean lifetime of 1410 hours. The standard deviation is 90 hours. Test the hypothesis that the mean lifetime of the electric light bulbs has not changed, using a level of significance of

(a) 0.05
(b) 0.01 (I.C.M.A.)

This question asks us to test that the mean has not changed, so we must employ a two-sided test.

$$H_0: \mu = 1500$$
$$H_1: \mu \neq 1500$$

In this example, we do not know the population standard deviation, so we must estimate it using the sample standard deviation. In an earlier chapter we saw that

$$\sigma^2 = \frac{ns^2}{n-1}$$

So the estimated population variance is

$$\frac{10 \times 90^2}{9} = 9000$$

and the estimated standard deviation is

$$\sqrt{9000} = 94.87$$

So the test statistic t for our sample mean is

$$t = \frac{|1410 - 1500|\sqrt{10}}{94.87} = 2.999. \,(=3)$$

Now for the critical values of t, we enter the t-table with $10 - 1 = 9$ degrees of freedom, remembering that this is a two-sided test. The critical values are 2.262 (at the 5% level) and 3.25 (at the 1% level). Our test statistic is significant at the 5% level but not at the 1% level. We reject H_0 and accept H_1, and conclude that there is some evidence to suggest that the mean lifetime of all bulbs has changed.

In fact, we can find the test statistic t directly without first estimating the population variance if we use the expression

$$\text{Test statistic } t = \frac{|\bar{x} - \mu|\sqrt{(n-1)}}{s}$$

In the last example, this would give

$$t = \frac{|1410 - 1500|\sqrt{10-1}}{90} = 3, \text{ as before.}$$

Example 5

After treatment with a standard fertiliser, the average yield per hectare is 4.2 tonnes of wheat. A super fertiliser is developed and administered to 10 hectares. The yields from the 10 hectares were 4.3, 6.0, 4.9, 6.1, 6.2, 5.4, 4.1, 4.2, 3.8 and 3.9 tonnes. Does this fertiliser give a higher yield?

Firstly, we should notice that in this case a one-sided test is appropriate, so we have

$$H_0: \mu = 4.2$$
$$H_1: \mu > 4.2$$

We now need the sample mean and variance

x	x^2
4.3	18.49
6.0	36.00
4.9	24.01
6.1	37.21
6.2	38.44
5.4	29.16
4.1	16.81
4.2	17.64
3.8	14.44
3.9	15.21
48.9	247.41

$$\bar{x} = \frac{\Sigma fx}{\Sigma f} = \frac{48.9}{10} = 4.89$$

$$s^2 = \frac{\Sigma x^2}{n} - \left(\frac{\Sigma x}{n}\right)^2$$

$$= \frac{247.41}{10} - (4.89)^2$$

$$= 0.8289$$

$$s = 0.9104.$$

$$\text{Text statistic } t = \frac{|4.89 - 4.2|\sqrt{10-1}}{0.9104} = 2.27$$

For a one-sided test with 9 degrees of freedom, the critical values of t are 1.833 (at the 5% level) and 2.821 (at the 1% level), so the sample mean is significant at the 5% level but not at the 1% level. We reject H_0 and accept H_1, concluding that there is evidence to suggest that the super fertilizer is effective.

When we dealt with large samples, we found it was possible to test the significance of the difference between the means of two populations. Using the *t*-distribution, we can also do this for small samples. In some cases, it is necessary to test the mean difference by calculating the difference between the items in the sample. For us to be able to do this, the samples must be 'paired', i.e. the same item must appear in both samples, though treated under different conditions. We would use this method for the 'before and after' type of statistics so favoured by advertising agencies (i.e. the samples are not independent).

Example 6

Before coaching, five candidates scored 38, 41, 52, 27 and 18% respectively in a statistics examination. After coaching, they scored 40, 45, 49, 30 and 24% respectively. Does this evidence suggest that coaching is effective?

Although this problem appears similar to Examples 3, we cannot use the same method as that method requires independent samples. In this case, we are looking at the same sample but under different conditions. What should we do?

Subtracting the scores, we obtain the change following coaching

$$x = 2, 4, -3, 3, 6. \quad \bar{x} = 2.4, s = 3.007.$$

Suppose we adopt as our null hypothesis that coaching is ineffective: then the mean difference $\bar{x}$ could well be zero.

$$H_0: \mu = 0$$

We are asked to test whether coaching is effective, i.e. increases the marks gained by students, so a one-sided test is appropriate and the alternative hypothesis is

$$H_1: \mu > 0.$$

$$\text{Test statistic } t = \frac{|2.4 - 0|\sqrt{5-1}}{3.007} = 1.596$$

For a one-sided test with 4 degrees of freedom, the critical values for t are 2.132 (at the 5% level) and 3.747 (at the 1% level) so. We must conclude that the test statistic is not significant, we accept the null hypothesis and reject the alternative hypothesis. There is insufficient evidence to suggest that coaching is effective.

In most cases, samples will be independent rather than paired, and if we want to test the significance of the difference between their means, we must use the method employed earlier in this chapter. Suppose we have two samples, the details of which are as follows:

	Sample Size	Mean	Variance
First sample	n_1	$\bar{x}_1$	s_1^2
Second sample	n_2	$\bar{x}_2$	s_2^2

We will assume that the first sample comes from a population with a mean μ_1, and the second sample comes from a population with a mean μ_2. We can now formulate the null hypothesis

$$H_0: \mu_1 = \mu_2 \text{ or}$$
$$H_0: \mu_1 - \mu_2 = 0.$$

We have a variance for each sample, and using the methods of the chapter on estimation we can pool the variances of the samples to estimate the variance of the population, i.e.

$$\hat{\sigma}^2 = \frac{n_1 s_1^2 + n_2 s_2^2}{n_1 + n_2 - 2}$$

Having estimated the variance of the population, we can now estimate the standard error of the difference between two independent means to be

$$\sqrt{\frac{\hat{\sigma}^2}{n_1} + \frac{\hat{\sigma}^2}{n_2}}$$

For samples, then

$$\text{Test statistic } t = \frac{|\bar{x}_1 - \bar{x}_2| - 0}{\sqrt{\frac{\hat{\sigma}^2}{n_1} + \frac{\hat{\sigma}^2}{n_2}}}$$

Before examining an example, we must now ask how many degrees of freedom are associated with this test. There are $(n_1 - 1)$ degrees of freedom associated with the first sample, and $(n_2 - 1)$ associated with the second sample. Now as we are pooling both samples, there are $n_1 + n_2 - 2$ degrees of freedom associated with estimating the population variance.

Example 7

Lumo Ltd. manufacture electric light bulbs, and claim that on average their lamps last longer than the lamps of their competitor Brighto Ltd. A random sample of ten lamps made by Lumo had the following lives (in hours) before failing.

$x_1 = 200 \quad 210 \quad 190 \quad 200 \quad 190 \quad 200 \quad 180 \quad 200 \quad 200 \quad 210$

and a random sample of eight lamps made by Brighto yielded

$x_2 = 190 \quad 200 \quad 210 \quad 190 \quad 180 \quad 190 \quad 200 \quad 190$

Does this evidence substantiate Lumo's claim?

Firstly, we use suitable short-cut methods to calculate the mean and variance for each sample.

x_1	$\frac{x_1 - 200}{10}$	$\left(\frac{x_1 - 200}{10}\right)^2$	x_2	$\frac{x_2 - 200}{10}$	$\left(\frac{x_2 - 200}{10}\right)^2$
200	0	0	190	-1	1
210	1	1	200	0	0
190	-1	1	210	1	1
200	0	0	190	-1	1
190	-1	1	180	-2	4
200	0	0	190	-1	1
180	-2	4	200	0	0
200	0	0	190	-1	1
200	0	0		-5	9
210	1	1			
	-2	8			

$$\bar{x}_1 = 200 + \frac{-2 \times 10}{10} = 198 \text{ hrs} \qquad \bar{x}_2 = 200 + \frac{-5 \times 10}{8} = 193.75 \text{ hrs}$$

$$s_1^2 = 100 \times \left[\frac{8}{10} - \left(\frac{-2}{10}\right)^2\right] \qquad s_2^2 = 100 \times \left[\frac{9}{8} - \left(\frac{-5}{8}\right)^2\right]$$

$$= 76 \text{ hrs} \qquad\qquad\qquad = 73.44 \text{ hrs}$$

We will test the null hypothesis that there is no difference between the mean lengths of lives of lamps, against the alternative hypothesis that Lumo's lamps last longer than Brighto's lamps, i.e.

$$H_0: \mu_1 = \mu_2 \text{ and } H_1: \mu_1 > \mu_2.$$

As we are in fact assuming that both samples could have been drawn from the same population, the estimated variance of this population is

$$\hat{\sigma}^2 = \frac{n_1 s_1^2 + n_2 s_2^2}{n_1 + n_2 - 2}$$

$$= \frac{10 \times 76 + 8 \times 73.44}{10 + 8 - 2}$$

$$= 84.22 \text{ hrs, with 16 degrees of freedom}$$

So the test statistic t is

$$t = \frac{|198 - 193.75| - 0}{\sqrt{\frac{84.22}{10} + \frac{84.22}{8}}} = 0.976$$

For a one-sided test with 16 degrees of freedom, the critical values of t are 1.746 (at the 5% level) and 2.583 (at the 1% level).

So we see that the difference between the means is not significant. We accept the null hypothesis: there is insufficient evidence to suggest that Lumo's lamps last longer.

Tests of Proportions

Tests involving sample proportions are extremely important in practice. In the last chapter, we saw the basis of such a test applied to quality control. Again, many market researchers express their results in terms of proportions, e.g. '40% of the population clean their teeth with Gritto'. It will be useful to design tests that will detect changes in proportions. As you can imagine, the binomial distribution forms the basis of such tests. Let us first examine how we can test the significance of the proportion of a small sample.

Example 8

Suppose it is claimed that in a very large batch of components, about 10% of items contain some form of defect. It is proposed to check whether this proportion has increased, and this will be done by drawing a sample of 20 components. Adopting our null hypothesis in the usual way we have

$$H_0: \pi = 0.1$$
$$H_1: \pi > 0.1$$

Assuming that H_0 is true, then we can calculate the probability distribution for samples of 20 components like this:

$$(0.9)^{20} + {}^{20}C_1(0.9)^{19}(0.1) + {}^{20}C_2(0.9)^{18}(0.1)^2 + {}^{20}C_3(0.9)^{17}(0.1)^3 + \ldots$$

Evaluating this expression, we have

No. of Defectives	Probability	Cumulative Probability
0	0.1215	0.1215
1	0.2700	0.3915
2	0.2850	0.6765
3	0.1900	0.8665
4	0.0897	0.9562
5	0.0319	0.9881
6	0.0089	0.9970
7	0.0020	0.9990
8	0.0004	0.9994
9	0.0001	0.9995

Examining this table, we see that the probability of obtaining more than 4 defectives is $1 - 0.9562 = 4.38\%$, and the probability of obtaining more than 6 defectives is $1 - 0.9970 = 0.3\%$. So if we obtain more than 4 defectives in a sample, we can reject the null hypothesis at the 4.38% level of significance, and if it contains more than 6 defectives we can reject at the 0.3% level of significance.

Now calculations involving the binomial distribution are very tedious. The way to avoid them is to draw a large sample, and we can then use the normal approximation. Suppose that in the last example a sample of 150 contained 20 defectives — would this indicate that the proportion defective had increased! Again, we have

$$H_1: \pi = 0.1$$
$$H_1: \pi > 0.1$$

And on the basis of our null hypothesis

mean $= n\pi = 0.1 \times 150 = 15$ defectives per sample

$$\sigma = \sqrt{n\pi(1-\pi)} = \sqrt{150 \times 0.1 \times 0.9} = 3.67$$

As the sample is large, our test statistic is the Z score

$$\frac{20-15}{3.67} = 1.36.$$

For a one-sided test, the critical values for Z are 1.645 (at the 5% level) and 2.326 (at the 1% level), so we would conclude that the proportion defective in the sample is not significant. We would accept the null hypothesis that the proportion defective is 0.1.

Example 9

Finally, let us examine how we could investigate the significance of the difference between two proportions in two samples. Suppose that in a sample of 300 people from Liverpool, 220 cleaned their teeth with Gritto, and in Manchester 240 out of 350 use Gritto. We wish to determine whether there is any difference between the proportion using Gritto. Clearly, this is a two-sided test, so we have

$$H_0: \pi_1 = \pi_2 \text{ (or } \pi_1 - \pi_2 = 0)$$
$$H_1: \pi_1 \neq \pi_2$$

The proportion using Gritto in Liverpool is $\frac{220}{300} = 0.733$, and the proportion using Gritto in Manchester is $\frac{240}{350} = 0.686$. If the null hypothesis is true, then we can regard both samples as coming from the same population. The best estimate we have of the proportion using Gritto in this population is

$$\hat{\pi} = \frac{220+240}{300+350} = \frac{460}{650} = 0.708$$

In an earlier chapter, we saw that the variance of the proportion of a sample of n items is

$$\frac{\hat{\pi}(1-\hat{\pi})}{n}$$

So the variance of the proportion of samples of 300 is

$$\frac{0.708 \times 0.292}{300}$$

and the variance of the proportion of samples of 350 is

$$\frac{0.708 \times 0.292}{350}$$

Using the theorem of sample differences, the variance of the differences between these proportions is

$$\frac{0.708 \times 0.292}{300} + \frac{0.708 \times 0.292}{350}$$

$$= 0.708 \times 0.292 \left(\frac{1}{300} + \frac{1}{350}\right)$$

$$= 0.00128$$

and the standard error of the difference is $\sqrt{0.00128} = 0.0358$. Now if the null hypothesis is true, then the difference between the proportions in the sample could well be zero. The test statistic Z is

$$\frac{|0.733 - 0.686| - 0}{0.0358} = 1.31$$

For a two-sided test, the critical values of Z are 1.96 (at the 5% level) and 2.576 (at the 1% level), so we would accept the null hypothesis that the proportion using Gritto in Liverpool and Manchester could well be the same.

As you can see, there are a wide variety of significance tests for testing means and proportions. Certainly, there are more than those mentioned in this chapter — there are tests, for example, for testing several means and proportions. However, the tests covered in this chapter will take you quite a long way in statistical analysis. Before we finish this chapter, let us stress again the importance of the sample size. You will have noticed that large samples impose fewer restrictions on testing, and their results will be much more reliable. It is for this reason that you should use large samples wherever possible.

Exercises to Chapters 19 and 20

20.1 Discuss briefly the following statements:
  (i) in hypothesis testing it is more important to minimise the Type I error than to worry about the Type II error;
  (ii) in hypothesis testing it is essential that the decision to use either a one-or a two-tailed test should be taken before the sample is drawn.
C.I.P.F.A.

20.2 From past experience it has been found that 5% of a certain output is rejected. To test the null hypothesis $H_0: \pi = 5\%$, a sample of 400 items are drawn. Form decision rules for testing H_0 at the 1% level if (a) the alternative hypothesis is one-sided and (b) the alternative hypothesis is two-sided.

20.3 Suppose that in 20.2 we adopt the one-sided alternative hypothesis, and the true reject rate rises to 7%. What is the probability of a Type II error?

20.4 Metal plugs are turned to a mean diameter of 5 cm, standard deviation 0.01 cm. We wish to test for change in the mean diameter at the 5% level of significance by drawing a sample of 100 plugs. Devise an appropriate decision rule. What would your decision rule have been if we wished to detect increases in the mean diameter?

20.5 A manufacturer produces cables with a mean breaking strength of 2000 lb and a standard deviation of 100 lb. By using a new technique the manufacturer claims that the breaking strength can be increased. To test this claim, a sample of 50 cables produced by the new technique is tested, and the mean breaking strength found to be 2050 lb. Can the manufacturer's claim be supported at a 0.01 level of significance?
I.C.M.A.

20.6 If the sample quoted in 20.5 had contained 9 cables, could the claim be justified?

20.7 What is the minimum sample size for the sample in 20.5 that would justify the manufacturers claim (assume a sample mean of 2050 lbs)?

20.8 In a certain country, men have a mean height of 5 ft 6 in, standard deviation 3 in, and women have a mean height of 5 ft 3 in, standard deviation 2 in. In a random sample of 100 married couples, the average height difference between husband and wife was 2 in. Does this suggest that height of partner affects the decision to propose marriage?

20.9 A survey has provided data on the television watching habits of ten-year-old children of different social classes. A sample of 220 children in social class five were found to watch for a mean period of 3.8 hours per day with a standard deviation of 0.5 hours, while a sample of 63 children in social classes one/two watched for a mean period of 3.2 hours with a standard deviation of 0.42 hours. Test, at the 1% level, whether children in social class five spend more time watching television than do those in classes one/two.

20.10 Two drugs, A and B, were tested for a certain effect on laboratory mice. Two samples, each of fifty mice, were chosen randomly, one drug administered to each, and a measure, x, of the effect, obtained. The results were as follows:

$$\text{Drug A} \quad \Sigma x = 1750 \quad \Sigma x^2 = 62,520$$
$$\text{Drug B} \quad \Sigma x = 1980 \quad \Sigma x^2 = 85,610$$

Test, at the 1% level of significance, the hypothesis that the two drugs have the same mean effect against the alternative hypothesis that drug B has a higher mean effect. You may assume that the effect on mice, of each drug, is normally distributed.

20.11 What is meant by 'standardising' a normal distribution and how is it done?

A company samples the products of two suppliers of light bulbs. The 80 bulbs in the batch from supplier A are shown to have a mean life of 503 hours with a standard deviation of 43 hours. the 110 bulbs from supplier B have a mean life of 488 hours with a standard deviation of 39 hours. Test the hypothesis that there is no difference between the mean lives of bulbs from the two suppliers at the 10%, 5% and 1% levels of significance.

20.12 The following table gives incomes after tax for the U.K. for 1962 and 1963.

Income (£)	1962		1963	
	No. in thousands	Income after Tax (£m)	No. in thousands	Income after Tax (£m)
50 – 249	5,100	1,000	4,500	900
250 – 499	6,600	2,500	6,800	2,600
500 – 749	6,200	3,800	6,000	3,800
750 – 999	4,800	4,200	4,900	4,300
1,000 – 1,999	4,100	5,100	4,600	5,800
2,000 – 3,999	300	900	400	1,000
4,000 and over	100	400	100	500
Totals	27,200	17,900	27,300	18,900

Source: National Income and Expenditure.

Test the hypothesis that the mean income has not changed between 1962 and 1963

C.I.P.F.A.

20.13 Data produced by the Family Expenditure Survey show that the percentage distribution of households by income in 1965-7 for two standard regions was as follows:

Weekly Income	% of all Households	
	Yorkshire and Humberside	Northern Ireland
Under £6	6	8
£6 but under £10	9	11
£10 but under £15	12	12
£15 but under £20	14	18
£20 but under £25	17	17
£25 but under £30	13	12
£30 but under £35	9	11
£35 but under £40	6	3
£40 but under £50	8	5
£50 or more	6	3

(a) Is there a significant difference between the mean percentage income in the two regions?

(b) Explain as carefully as you can the purpose of hypothesis testing.

C.I.P.F.A.

20.14 Suppose that a survey of a number of firms in two regions of a country shows that their investment in a given year was as follows:

	Investment (£000's)					
	24–	28–	32–	36–	40–	44–48
		Number of Firms				
Region A	8	41	77	90	58	26
Region B	19	36	47	58	27	13

Is there any evidence of a difference in average investment between the two regions?

C.I.P.F.A.

20.15 The table below gives distributions of earnings for samples of workers taken from two firms:

	Earnings (£)					
	20–	22–	24–	26–	28–	30–32
Number of workers in Firm A	11	23	42	40	28	14
Number of workers in Firm B	12	36	54	62	44	38

What evidence is there of a difference between average earnings in the two firms?

C.I.P.F.A.

20.16 Permanent Houses Completed in the U.K. (including flats, each flat being counted as one unit)

Year	For Private Owners (thousands)	For Local Housing Authorities (thousands)
1963	177.8	123.9
1964	221.3	154.7
1965	217.2	165.0
1966	208.6	176.9
1967	204.2	199.7
1968	226.0	188.0
1969	186.0	180.9
1970	174.3	177.0
1971	196.3	154.8
1972	200.6	120.4
1973	190.6	102.6

Source: Annual Abstract of Statistics.

Is there a significant difference in the average number of permanent houses completed in the two groups between 1963 and 1973?

C.I.P.F.A.

20.17 In conducting a survey of food prices, two samples of prices of a given food item were collected. Sample A came from a congested city area, and Sample B was obtained in the suburbs. The results were as follows:

	Sample A	Sample B
n	14	18
$\sum_{i=1}^{n} p_i$	12.60	14.96
$\sum_{i=1}^{n} (p_i - \bar{p})^2$	1.68	1.96

where p_i = price recorded in the ith store. Test the hypothesis that there is no difference between the mean price of the particular food item in the two areas. (Use 5% level of significance.)

C.I.P.F.A.

20.18 A machine has been operating at 60% efficiency. After adjustment, nine test runs produced the following efficiencies:

64, 59, 71, 63, 68, 61, 62, 61, 63,

Test, at the 5% level, whether efficiency has improved.

20.19 In order to test the effecitivess of a drying agent in paint, the following experiment was carried out. Each of six samples of material was cut in two halves. One half of each was covered with paint containing the agent and the other half with paint without the agent. Then all twelve samples were left to dry. The time taken to dry was as follows:

	Drying Time (hours)					
	1	2	3	4	5	6
Paint with the agent	3.4	3.8	4.2	4.1	3.5	4.7
Paint without the agent	3.6	3.8	4.3	4.3	3.6	4.6

Required: Carry out a 't'-test to determine whether the drying agent is effective, giving your reasons for choosing a one-tailed or two-tailed test. Carefully explain your conclusions. A.C.A.

20.20 Ten pairs of identical twins had the following birthweights in kilograms:

Pair	1	2	3	4	5	6	7	8	9	10
First-born	3.95	3.41	3.73	4.13	3.48	4.28	3.98	4.18	4.04	3.73
Second-born	3.93	3.35	3.72	4.18	3.44	4.15	3.89	4.20	4.00	3.72

(i) Stating any assumptions which you make, test, at the 5% significance level, the hypothesis that there is no difference in weight between first-born and second-born twins.

(ii) If the data had been collected to test the hypothesis that the first-born are heavier, what conclusion would you reach?

20.21 A group of 10 students scored the following marks in a test:

Candidate	A	B	C	D	E	F	G	H	I	J
Score	53	60	61	47	40	56	75	46	82	61

After coaching, their marks were

Candidate	A	B	C	D	E	F	G	H	I	J
Score	60	58	67	51	60	72	71	48	94	76

Did coaching improve their marks?

20.22 Define the standard error of a mean of samples of n observations and use the concept to explain briefly why one has more confidence in the mean of a large sample than in the mean of a small one.

Show why it is unlikely that a sample of 64 observations with a mean of 5.2 was drawn from a population with a mean of 5.5 and a standard deviation of 0.8. Would your answer be the same if the sample contained 16 items?
A.C.A.

20.23 Suppose samples of 10 items are to be drawn from a population with 5% of items defective. Find the 5% and 1% critical values for the alternative hypothesis that the percentage defective has increased.

20.24 The internal auditor of your company has reported to you as management accountant as follows:

'From 1000 sample postings made before metrication 210 errors were discovered whereas 250 errors were found in 1000 sample postings made after metrication. It would appear that although the staff has not changed, the postings before metrication were more accurate than they are after. Special steps should be taken to regain the accuracy previously enjoyed.'

(a) What statistical hypothesis does this statement imply?
(b) Evaluate this hypothesis and state if it should influence you as management accountant in your decision on whether to take special action. I.C.M.A.

20.25 A market research company wishing to determine whether it is more usual in the North for working men to come home to a midday meal than in the South interviews two random samples each of 500 men. In the sample from the North 330 men come home midday compared with 280 in the sample from the South. Is the difference significant? I.C.M.A.

20.26 Two groups, A and B, each consist of 100 people who have a disease. A serum is given to group A but not to group B; otherwise, the two groups are treated identically. It is found that is groups A and B, 75 and 65 people, respectively, recover from the disease. Does this result support the hypothesis that the serum helps to cure the disease? I.C.M.A.

20.27 A company selling a prestige product carries out market research surveys in two consecutive years on heads of households with income in excess of £10,000 per year. The survey results were:

	1973	1974
Sample size	1300	1000
No. possessing product	351	240

Assuming the rejection area to be outside 2.33 S.D., does the 1974 survey suggest that the product's sales are declining? I.C.M.A.

20.28 Family Expenditure Survey: Household expenditure on commodities and services 1967-8.

	North (£ per week)	South-west (£ per week)
Housing, etc.	3.77	4.86
Food	6.17	6.18
Drink and tobacco	2.46	1.87
Clothing and footwear	2.07	1.79
Durables and other goods	2.79	3.50
Transport and services	4.54	5.53
Total	21.80	23.73
Number of households in sample	1000	1000

Do households in the North spend a significantly greater proportion of their total expenditure on drink and tobacco than those in the South-west? Explain fully the theoretical basis for the test you have applied. C.I.P.F.A.

20.29 A motor-car manufacturer purchases gear assemblies from a sub-contractor who undertakes to ensure that not more than 5% of his supplies will be defective. In order to provide a check on the quality of incoming supplies a random sample of 200 assemblies is selected of which 17 are found to be defective.

(a) Does the sample evidence indicate that the sub-contractor is not maintaining the quality of his supplies at the agreed level? Use the significance levels of 0.05 and 0.01.

(b) Construct an interval estimate for the proportion of all the sub-contractors supplies that are defective. Use a confidence level of 95%.

(c) What is the minimum sample size required to estimate the proportion in part (b) of the question to within ± 0.5% at the 95% confidence level.

(d) As the result of a board decision the firm now purchases gear assemblies from a second sub-contractor. A random sample of 500 gear assemblies from his supplier reveals that 20 units are defective. Test the hypothesis that there is a significant difference in quality between the two sub-contractors' supplies. Use significance levels of 0.01 and 0.05.

20.30 In a public opinion poll of 500 people, 70% favoured a certain policy. In another district 60% of the 500 interviewed favoured the policy. Is there a significant difference of opinion?

20.31 For question 20.30, what is the minimum sample size that would make a difference between 60% and 70% significant?

Chapter Twenty One

The Chi Square Test

We will now pick up and develop a point we made in the last chapter — that in all the tests we have so far studied we have assumed that the population from which the sample was drawn was normally distributed and that the sample we drew was truly random. Even if the population was not normally distributed, we learned from the Central Limit Theorem that the sampling distribution of sample means would be normally distributed. You will realise that for much of the work we do it is this latter population (a population of sample means) from which we draw our individual sample. Thus whether we are considering a population known to be normally distributed or are considering a sample drawn from a population with unknown characteristics we have based our tests and our estimations on the Normal Distribution curve. Even when we discussed small samples and used the t test we were making an implicit assumption that, provided our population distribution does not depart too far from normal, the t test will yield valuable results. All the empirical evidence we have tends to support the view that the t test is insensitive to movements away from the normal distribution when we are testing null hypotheses of sample means.

Now, why do we make this assumption of a normally distributed population which we would find very hard to prove if we were asked? The point is, all the tests we have looked at involve the use of parameters such as the population mean, the population proportion and the variance. If we assume that the population is normally distributed we can infer a great deal from these parameters. You will remember, for example, that the variance of the sampling distribution of sample means (the standard error)^2 is taken as being equal to $\frac{\sigma^2}{n}$; or that 95% of the normally distributed population will fall within the range $\mu \pm 1.96\,\sigma$.

Suppose, however, that we do not know, and are unwilling to assume, the precise form of the population distribution. We can not now make any assumptions about the value of a parameter. Yet often a statistical test would appear to be useful. In such circumstances we would have to use a distribution free, or nonparametric, test technique. You will understand of course, that even if we could confidently make presumptions about the population distributions or parameters, we need not do so. We could always make our test using nonparametric methods. Thus for every parametric test there is a corresponding nonparametric test.

The methodology we use for nonparametric tests is identical to that which we have already used for both z and t tests. Firstly, on the basis of a null hypothesis a sampling distribution of some sample characteristic is derived.

Then having prescribed an acceptable level for a Type I error a decision criterion can be fixed. If the sample characteristic exceeds the critical value it is deemed significant at the level you have chosen; if it does not fall into the region of rejection the null hypothesis is accepted, subject always to the possibility that a Type II error may have been committed. And this is the point. For samples of the same size, nonparametric tests always leave a higher possibility of a Type II error than do parametric tests. We can always offset this, of course, by taking a larger sample size. In any case nonparametric tests make up in generality what they lack in power. The conclusions we reach do not depend on assumptions which may only apply to the situation we are studying, and so they tend to be of universal application.

The Chi Square Test

Probably the most widely used of all nonparametric tests is the Chi Square Test (pronounce it 'kye' and use the symbol χ^2), and certainly it has a wider range of applications than most other tests. But first, let us consider the nature of the Chi Square distribution.

In fact, there is not merely one distribution. Like the t distribution there are many chi square distributions — one for each degree of freedom. Three such distributions are shown in Diagram 21.1. Notice that as the degrees of freedom become fewer, the distribution becomes positively skewed more and more. Conversely, as the number of degrees of freedom is increased, the distribution becomes approximately normal.

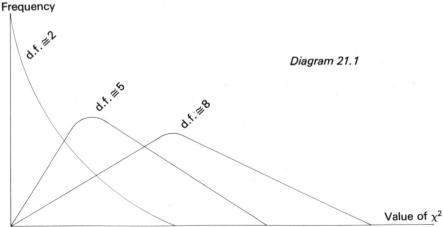

Diagram 21.1

Observed and Expected Frequencies

Suppose you have been given the task of assessing consumer preference with respect to five washing powders — Zip, Whito, Acme, Blanco and Bleacho. You supply a small quantity of each powder in boxes marked A, B, C, D and E to 1000 housewives, and ask them to test the powders and state which powder they prefer. The result of your enquiry is as follows:

Washing powder	Zip	Whito	Acme	Blanco	Bleacho
No. preferring that powder	187	221	193	204	195

Now you suspect that housewives are unable to distinguish between washing powders. Any preference that they have will be influenced not by the powder's characteristics, but by the persuasive powers of the advertisements promoting the powder, and by the attractiveness of the packet. For this reason, the housewives do not know which powder they are using (remember they are labelled A, B, C, D and E, and each powder is packed in a plain white box). Returning to statistical jargon, we have formulated the null hypothesis that housewives are unable to distinguish between the powders, and we will use the sample evidence to test the validity of this hypothesis. Of course, the appropriate alternative hypothesis is that housewives can distinguish between the powders.

Earlier in this book, we implied that the null hypothesis to be tested should always be formulated in such a fashion that we would know what to expect if it was true. Assuming that our null hypothesis is correct, then, we expect that an equal number would report that they preferred each of the powders. In other words, the results expected from our sample would be:

Washing powder	Zip	Whito	Acme	Blanco	Bleacho
No. expected to prefer that powder	200	200	200	200	200

The frequencies that we obtain in our sample we will call the *observed frequencies,* and the frequencies we expect on the basis of our null hypothesis we will call the expected frequencies. We shall now devise a significance test on whether there is a significant difference between the observed and expected frequencies.

The method we shall apply for testing the null hypothesis is to measure the total deviation of the observed from the expected frequencies, i.e. $\Sigma(O - E)$.

Product	Observed	Expected	(O − E)
Zip	187	200	−13
Whito	221	200	21
Acme	193	200	−7
Blanco	204	200	4
Bleacho	195	200	−5

As you would expect $\Sigma(O - E) = 0$, so we are faced with a similar problem to when we were attempting to measure dispersion. If $\Sigma(O - E)$ always equals zero, then we cannot use it as a measure of testing our null hypothesis. When considering the standard deviation, we got round this problem by squaring the deviations, and we can use the same method in this context; i.e. calculate $\Sigma(O - E)^2$. However, a further adjustment is necessary: we divide the $(O - E)^2$ values by E. Now why do we do this? Well, the reason is not apparent in this example, so let us consider another case.

Observed	Expected	(O−E)	(O−E)²
150	200	−50	2500
150	100	50	2500

Clearly the deviation of 50 is twice as significant on an expected value of 100 as it is on an expected value of 200, and to take this into account we can weight the (O−E)² values by dividing by E. So we have

Observed	Expected	(O−E)	(O−E)²	$\dfrac{(O-E)^2}{E}$
150	200	−50	2500	12.5
150	100	50	2500	25

which shows precisely the information required. Now if we calculate the sum of the

$$\frac{(O-E)^2}{E}$$

values, we have a weighted sum of the deviations of the observed and expected values, and we call this statistic χ^2.

$$\chi^2 = \sum \frac{(O-E)^2}{E}$$

Calculating the value of χ^2 for our washing powder example we have

Observed	Expected	(O−E)	(O−E)²	$\dfrac{(O-E)^2}{E}$
187	200	−13	169	0.845
221	200	21	441	2.205
193	200	−7	49	0.245
204	200	4	16	0.080
195	200	−5	25	0.125
		0		3.500

If our null hypothesis is true, then our expected value for χ^2 is zero, as we would expect our observed and expected values to have the same value. But as we are sampling, the actual values of χ^2 will fluctuate from sample to sample. Now χ^2 is like Z in so far as it is just another kind of standard score — so to test our null hypothesis we can use the same method as we did in the last chapter. We will compare our value for chi-square with certain critical values that we would not expect to be exceeded. However, there are two problems involved in obtaining these critical values.

When we examined sampling distributions, we learnt that as long as the sample size is sufficiently large then we can expect the sample statistic to be Normally distributed. We could use Normal Distribution Tables to obtain critical values, and compare them with the Z score of the sample statistic. Now here is where the first problem occurs — χ^2 is not Normally distributed, so the Normal Distribution Tables are of no help to us. We must obtain our critical values from χ^2 tables. Now there are many χ^2 distributions, but only one Standard Normal Distribution — so here is where our second problem arises — which χ^2 distribution should we use?

Just as with the t distribution there are many chi-square distributions — one for each degree of freedom. The distribution applicable in this case is the one with 4 degrees of freedom, and this is illustrated in Diagram 21.2. Can you see why there are four degrees of freedom in this case? Clearly, when we assigned expected values to the numbers buying each detergent, we were not completely free to assign any values that we wished. In order that we could compare expected values with observed values, the total of the expected values had to be 1000. This means that any hypothesis we adopt leaves us free to assign expected values to any four of the five brands. For example, we could assign an expected frequency of 150 to Whito, 300 to Acme, 200 to Blanco and 150 to Bleacho — then the expected frequency for Zip *must* be $1000 - (150 + 300 + 200 + 200 + 150) = 300$, otherwise we would not have a total frequency of 1000. So we have just four degrees of freedom when fixing expected values (generalising for problems like this, if we have n observed values, then we will have $(n-1)$ degrees of freedom in calculating the expected values).

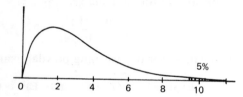

Diagram 21.2

Consulting the χ^2 tables with 4 degrees of freedom, we see that the critical values are 9.49 at the 5% level, and 13.3 at the 1% level. As our value of χ^2 is 3.5 (much less than both of the critical values), we would accept the hypothesis that housewives are unable to distinguish between the washing powders.

Errors Resulting from the Use of χ^2

Before we go any further, it is worth considering the errors that can occur if we use χ^2. A glance at Fig. 29.1 shows that the distribution is continuous, and we have used it to test discrete data. Clearly, then, this is a possible source of error. Now this error is not really serious *unless there is only one degree of freedom,* and when $v = 1$ we compensate for this error by applying *Yate's correction.* All that is involved in this correction is that we subtract ½ from the *absolute* difference between the observed and expected values. Let us examine an example to see how we apply the correction.

In a sample of 1000 people, 452 said they would vote Socialist at the next election. We wish to test the hypothesis that 50% of the electorate vote Socialist. On the basis of this hypothesis, then, we have:

	Observed	Expected
Socialist	452	500
Not Socialist	548	500

In this case, we have only one degree of freedom, so our critical values are 3.84 (at the 5% level) and 6.63 (at the 1% level).
Also, as $v = 1$ we must apply Yate's correction

Observed	Expected	(O – E)	|O – E| – ½	(|O – E| – ½)²	(|O – E| – ½)² / E
452	500	– 48	47.5	2256.25	4.5125
548	500	48	47.5	2256.25	4.5125
					$\chi^2 = 9.0250$

So we can see that the sample result is significant at the 1% level. We reject the null hypothesis and conclude that less than 50% of the electorate vote Socialist.

There is a second type of error that can arise from using the χ^2 distribution. If the expected frequency in any cell is unusually small, the use of the χ^2 test might lead to erroneous conclusions. Two general rules concerning small cell frequencies are:

i) If there are only two cells the expected frequencies in each cell should be five or more. We could, then, use the χ^2 test in the following case.

	Observed	Expected
Agreeing	643	642
Disagreeing	4	5

We could not, however, use it in this case

	Observed	Expected
Agreeing	643	644
Disagreeing	4	3

ii) If there are more than two cells, χ^2 should not be used if more than 20% of the expected frequencies are less than 5.

According to this rule we could use the χ^2 test for the examination data on the left. Only one cell out of six has a frequency less than 5, i.e. only 16.67% of the cells.

No. of students		Grade	No. of students	
Observed	Expected		Observed	Expected
18	16	A	30	32
39	37	B	110	113
8	13	C	86	87
6	4	D	23	24
82	78	E	5	2
10	15	F	5	4
–	–	U	4	1
163	163		263	263

The χ^2 test should not, however be used for data on the right, since 3 of the 7 cells have an expected frequency of less than 5, i.e. 43%.

The reasoning behind this rule can be seen if we examine the data carefully. You will see that there is a very close relationship between the observed and the expected frequencies. The maximum difference is three. It would be commonsense to assume that there is not significant different between the observed and the expected frequencies. Yet if you compute the value of χ^2, you will find it to be 14.01, greater than the critical value at the .05 level. Contrary to commonsense we would reject the null hypothesis. The dilemma can be resolved if we combine the last three classes into a single class:

Grade	Observed	Expected
E and below	14	7

Using these revised frequencies the computed value of χ^2 is 7.26, less than the critical value for 4 degrees of freedom at the .05 level which is 9.49. The null hypothesis would be accepted, and we would argue that there is no difference between the observed and the expected frequencies. This is, of course, a much more logical conclusion.

Testing Goodness of Fit to a Probability Distribution

Do you remember that in the chapter on probability distributions we investigated the pattern of V1 rockets falling on London during World War II? We divided an area of 144 square kilometres into 576 equal squares, and counted the number of bombs per square. We wanted to find if the bombs were falling at random, or were 'clustered' to a greater degree than could be ascribed to chance. If the bombs did fall randomly, then we should be able to predict the frequency distribution of bombs per square by using the Poisson distribution. The observed distribution was:

No. of bombs per square	0	1	2	3	4	5
Frequency	229	211	93	35	7	1

This distribution forms our observed frequencies, and has a mean of 0.93 bombs per square. We now formulate the null hypothesis that the bombing pattern could be predicted by the Poisson distribution with a mean 0.93. (This may not sound like a null hypothesis, but it would if we restated it slightly — there is no difference between the bombing pattern and the Poisson distribution with a mean of 0.93.) On the basis of our null hypothesis, we can calculate the expected frequencies like this:

$$576\, e^{-0.93} \left[1 + 0.93 + \frac{0.93^2}{2!} + \frac{0.93^3}{3!} + \frac{0.93^4}{4!} + \frac{0.93^5}{5!} \right]$$

$$= 576[0.3946 + 0.3670 + 0.1706 + 0.0529 + 0.0123 + 0.0023]$$
$$= 227.3 + 211.4 + 98.3 + 30.5 + 7.1 + 1.3$$

When we dealt with the Poisson distribution previously, we were content to deal with probabilities. Here we are more interested in expected values, and we can obtain these by multiplying the probabilities by 576 — the total frequency. Notice that the sum of the expected frequencies is 575.9 — implying that we expect 0.1 squares to have more than 5 bombs. It would seem sensible, then, to combine this 0.1 into the 5 bombs per square class. We can now calculate a value for χ^2 to see how well the observed data fits the expected data.

Observed	Expected	(O – E)	(O – E)²	$\frac{(O-E)^2}{E}$
229	227.3	1.7	2.89	0.0127
211	211.4	–0.4	0.16	0.0008
93	98.3	–5.3	28.09	0.2858
35	30.5	4.5	20.25	0.6639
7 } 1 }	7.1 } 1.4 }	–0.5	0.25	0.0294
				$\chi^2 = 0.9926$

We must now investigate the number of degrees of freedom in calculating the expected values. The first thing to notice is that after combining the last two classes we have five classes in total. We used up one degree of freedom in making both sets of frequencies have the same total, and we used up another degree of freedom by making both sets of frequencies have the same mean. So we must have had $5 - 2 = 3$ degrees of freedom when calculating the expected values. Generalising, for a Poisson distribution with n classes, there will be $v = n - 2$ degrees of freedom in calculating the expected frequencies.

Consulting the χ^2 table, with $v = 3$ we would reject the null hypothesis at the 5% level if χ^2 exceeds 7.81, and at the 1% level of χ^2 exceeds 11.34. Now, as our value of χ^2 is far less than these critical values, we must accept the null hypothesis and conclude that the observed frequencies fit the Poisson frequencies excellently. The differences between the observed and expected values could well be accounted for by sampling fluctuations.

Now let us see if the data would fit a binomial distribution. First, we must find the probability that a particular square is bombed. Now we know that in any binomial distribution, the mean is np. In this case the mean is 0.93, and we can take n to be 5. So

$$0.93 = 5p$$
$$p = 0.186$$

and $q = 1 - 0.186 = 0.814$. So for examples of five squares, the probability distribution would be given by $(0.814 + 0.186)^5$

$$= (0.814)^5 + 5(0.814)^4(0.186) + 10(0.814)^3(0.186)^2$$
$$+ 10(0.814)^2(0.186)^3 + 5(0.814)(0.186)^4 + (0.816)^5$$
$$= 0.3574 + 0.4083 + 0.1866 + 0.0426 + 0.0049 + 0.0002$$

Multiplying the probabilities by 576 would give the expected frequencies:

Observed	Expected	(O−E)	(O−E)²	(O−E)²/E
229	205.9	23.1	533.61	2.59
211	235.2	−24.2	585.64	2.49
93	107.5	−14.5	210.25	1.96
35	24.5	10.5	110.25	4.50
7 }	2.8 }	5.1	26.01	8.97
1 }	0.1 }			
				20.51

The null hypothesis we have used here is that the observed frequencies could be described by the binomial distribution $(0.814+0.186)^5$. To test the goodness of fit, we now need to know the degrees of freedom in calculating the expected values. As with the Poisson distribution, we used up two degrees of freedom — one to make the totals agree and one to make the means agree. Again, we have $5-2=3$ degrees of freedom, and so we can use the critical values of the last example. Applying these rules, we must reject the null hypothesis at the 1% level. The observed data could not be described by a binomial distribution.

We shall not give an example of testing goodness of fit to a normal distribution, as the method is essentially similar, but it would be as well to investigate the number of degrees of freedom involved in such tests. To fit a normal distribution to observed data, we will use up three degrees of freedom, as the distributions must have the same mean, standard deviation and total. Assuming we have n classes in our expected frequencies, then, to test goodness of fit to a normal distribution would involve $n-3$ degrees of freedom.

Contingency Tables

To illustrate a further application of χ^2 testing, let us now examine a question recently set by the A.C.A.

Using the χ^2 distribution as a test of significance, test the statement that the number of defective items produced by two machines, as shown in the following table, is independent of the machine on which they were made.

	Machine Output		Total
	Defective Articles	Effective Articles	
Machine A	25	375	400
Machine B	42	558	600
	67	993	1000

We are asked to test the statement that the number of defective items is independent of the machines on which they were made, and we adopt this as our null hypothesis. Now if the statement is not true, then the number of defectives will depend on the machine on which they were made, and the table will enable us to calculate the degree of dependence. A table constructed in this way (to indicate dependence or association) is called a *contingency table*. 'Contingency' means dependence — many of you, for example, will be familiar with the terms 'contingency planning'; i.e. plans that will be put into operation *if* certain things happen.

As usual, let us suppose that our null hypothesis is true. Given this assumption, then we are only interested in the totals in the table, i.e. that out of 1000 articles, 67 were defective, and that out of 1000 articles 400 were produced on machine A. Using only this information, we can now predict the expected number of defectives produced by the two machines. We have—

$$\frac{400}{1000} = \frac{4}{10}$$

of the total output is produced on machine A, and as 67 items are defective, then we would expect

$$67 \times \frac{4}{10} = 26.8$$

of them to have been produced by machine A. Now as machine A produces 400 items, then $400 - 26.8 = 373.2$ items can be expected to be non-defective. Likewise, it follows that as we can expect 67 defective items, $67 - 26.8 = 40.2$ of them can be expected to be produced by machine B. Finally, as we can expect B to produce 600 items, we can expect $600 - 40.2 = 559.8$ of the items to be non-defective. The expected table, then, would look like this:

	Machine Output		Total
	Defective Articles	Non-defective Articles	
Machine A	26.8	373.2	400
Machine B	40.2	559.8	600
	67.0	933.0	1000

Notice that after we have calculated just one of the values in the table, the remainder will be fixed by subtraction. So we have $v = 1$ degree of freedom in calculating the expected values for a 2×2 contingency table and we must apply Yate's correction.

Observed	Expected	$(O-E)$	$\lvert O-E \rvert - \frac{1}{2}$	$\dfrac{(\lvert O-E \rvert - \frac{1}{2})^2}{E}$
25	26.8	-1.8	1.3	0.063
375	373.2	1.8	1.3	0.005
42	40.2	1.8	1.3	0.042
558	559.8	-1.8	1.3	0.003
				$\chi^2 = 0.113$

With $v = 1$ we would reject the null hypothesis if $\chi^2 > 6.64$ (1% level). Now as our value is much less than this, there is no reason to reject the null hypothesis — the number of defective items is independent of the machine on which they were produced.

If you are going to use the above technique, then make absolutely sure that the samples are *independent*. We could not use it if the same sample is being tested under different conditions (i.e. 'before and after' investigations). It could not be used in the following case: a sample of 100 housewives were given a standard washing powder to test and report whether the powder was satisfactory or unsatisfactory. The same women were then asked to test a new super powder. The results were—

		New Powder	
		Satisfactory	Unsatisfactory
Old Powder	Satisfactory	20	10
	Unsatisfactory	50	20

Examining this table, we see that 60 women 'changed their minds', i.e. 50 changed from unsatisfactory to satisfactory, and 10 changed from satisfactory to unsatisfactory. Now suppose we formulate the null hypothesis that the type of powder had no effect on their response (i.e. they could not distinguish between them), then we would except an equal number to change their minds in either direction. So we would expect 30 to change from unsatisfactory to satisfactory and 30 to change from satisfactory to unsatisfactory. This now enables us to calculate χ^2.

Observed	Expected	(O−E)	|O−E| − ½	(|O−E| − ½)² / E
50	30	20	19.5	12.675
10	30	−20	19.5	12.675
				25.350

Using the critical values for $v = 1$, we must reject the null hypothesis at the 1% level. The difference between the observed and expected values cannot be attributed to chance and we must conclude that association between response and type of powder is established.

If we were to test for association for tables larger than a 2 × 2, then we could use a very similar analysis. Let us consider a 2 × 3 table comprising independent samples. Suppose that a random sample of men and women indicated their view on a certain proposal as follows:

	In-Favour	Opposed	Undecided	
Women	118	62	25	205
Men	84	78	37	199
	202	140	62	404

At a level of significance of (a) 0.01 and (b) 0.05, test the hypothesis that there is no difference in opinion between men and women in so far as this proposal is concerned.

(C.I.P.F.A.)

We adopt the null hypothesis that there is no association between the response and the sex of the person interviewed. On this basis we may deduce that the proportion of the sample who are female is $\frac{205}{404}$, and as 202 people are in favour of the proposal, the expected number of women in favour of the proposal is

$$\frac{205}{404} \times 202 = 102.5$$

Also, as 140 people are against the proposal, the expected number of women against the proposal is

$$\frac{205}{404} \times 140 = 71$$

We can now obtain the remaining expected values by subtraction i.e.

Expected number of undecided women $= 205 - (102.5 + 71) = 31.5$
Expected number of men in favour $= 202 - 102.5 = 99.5$
Expected number of men opposed $= 140 - 71 = 69$
Expected number of undecided men $= 62 - 31.5 = 30.5$

So the expected table looks like this:

	In Favour	Opposed	Undecided
Women	102.5	71	31.5
Men	99.5	69	30.5

Observed	Expected	$(O-E)$	$\frac{(O-E)^2}{E}$
118	102.5	15.5	2.34
62	71	-9	1.14
25	31.5	-6.5	1.34
84	99.5	-15.5	2.41
78	69	9	1.17
37	30.5	6.5	1.38
			$\chi^2 = 9.78$

In this example, we have two degrees of freedom in calculating the expected values. Consulting the table, we see that the critical values for χ^2 are 5.99 at the 5% level and 9.21 at the 1% level, hence we would reject the null hypothesis accepting the alternative hypothesis (that men and women think differently) with 99% confidence.

Summary on Degrees of Freedom

We see then, that the χ^2 test is very versatile, and it might be useful here to summarise the number of degrees of freedom.

(a) For goodness of fit tests with n classes, $v = n - 1$.

(b) For goodness of fit tests to a binomial or Poisson distribution with n classes, $v = n - 2$.
(c) For goodness of fit tests to a normal distribution with n classes, $v = n - 3$.[1]
(d) For tests of association for a contingency table with m rows and n columns $v = (n \times 1)(m - 1)$.

In conclusion, it should be noted that of all significance tests, χ^2 is probably the most widely used. However, many statisticians are highly critical of χ^2 testing, reasoning that there are other tests available which are far more suitable. We do not consider this book to be aimed at a level sufficient to discuss the criticisms in detail, but we will leave you one point to ponder. None of the null hypotheses we have considered with respect to goodness of fit can be *exactly* true, so if we increase the sample size (and hence the value of χ^2) we would ultimately reach the point when all null hypotheses would be rejected. All that the χ^2 test can tell us, then, is that the sample size is too small to reject the null hypothesis!

Exercises to Chapter 21

21.1 (a) What does the χ^2 test test?
(b) The number of rejects in six batches of equal size were:

Batch	Number of Rejects
A	270
B	308
C	290
D	312
E	300
F	320

Test the hypothesis that the difference between them is due to chance using a level of significance of 0.05. I.C.M.A.

21.2 The number of breakdowns that have occurred during the last 100 shifts is as follows:

Number of breakdowns per shift	0	1	2	3	4	5
Frequency:						
Expected number of shifts	14	27	27	18	9	5
Actual number of shifts	10	23	25	22	10	10

Show whether the manager is justified in his claim that the difference between the number of actual and expected breakdowns is due to chance. It has been customary to use a significance level of 0.05. I.C.M.A.

---

1. When calculating expected normal distribution frequencies we use up three degrees of freedom, one to make the totals agree, one to make the means agree and one to make the standard deviations agree.

21.3 A manufacturer of fashion garments for the younger age groups suspects that the market for his product has changed recently. Sales records for previous years showed that 14% of buyers were below 16 years of age, 38% were 16 to 20 years of age, 26% were 21 to 25 years and 22% were over 25. A random sample of 200 recent buyers, however, showed the following results:

Age	Under 16	16-20	21-25	Above 25
Frequency	22	62	60	56

Required:
(1) Compare the results of the sample with those expected from previous records.
(2) What null hypothesis should be tested using these data?
(3) Carry out a chi-squared (χ^2) test of significance.
(4) Express your conclusions in terms meaningful to management.

A.C.A.

21.4 The quality control department of a certain firm randomly selects 200 components and tests them for surface defects. The results were as follows:

Number of surface defects	0	1	2	3	4	5	6 and over
Number of components	90	62	31	13	3	1	0

Fit a Poisson distribution to this data, and test for goodness of fit.

21.5 Assuming that a component never contains more than five defectives, fit a binomial distribution to the data in 16.4, and test for goodness of fit.

21.6 The following experiment was performed 64 times: a coin was tossed until it came up heads and x, the number of tosses to achieve a head, was recorded. The following data were obtained:

x	Frequency
1	24
2	17
3	16
4	3
Greater than 5	4
Total	64

Are the data consistent with the hypothesis that the coin was fair?

C.I.P.F.A.

21.7 An examination of the purchase department's records for a random sample of 100 orders for a certain commodity reveals the following distribution of lead-times, that is, delays between placing an order and receiving the consignment.

Lead-times	Orders
Under 14 days	28
14 and under 21 days	52
21 days and over	20

If it is assumed that the lead-times are normally distributed with mean 16 days and standard deviation 5 days, what frequencies of orders would be expected for each category of lead-time?

Apply the chi-squared test at a 90% significance level to check whether the observed data contradict the assumptions of normality.

21.8 (a) An association between two characteristics, which is apparent from some data, is found to be of borderline statistical significance. Explain this conclusion and say what action should be taken in order to reach a clearer decision.

(b) Three types of machine A, B and C, are used by a company to produce items which are very prone to a certain type of imperfection. A random sample of 100 items from each machine showed the number of perfect and imperfect items to be as follows:

	Perfect	Imperfect
A	33	67
Machine B	39	61
C	48	52

Required: Use the chi-squared (χ^2) test to examine at the 0.05 level of significance whether there is any association between machine type and proneness to imperfection.

A.C.A.

21.9 Suppose that a random sample of men and women indicated their view on a certain proposal as follows:

	In Favour	Opposed	Undecided
Women	118	62	25
Men	84	78	37

At a level of significance of (a) 0.01, (b) 0.05, test the hypothesis that there is no difference in opinion between men and women in so far as this proposal is concerned.

C.I.P.F.A.

21.10 Two factories using materials purchased from the same supplier and closely controlled to an agreed specification produce output for a given period classified into three quality grades as follows:

	Output in Tons			
Quality grade:	A	B	C	Total
Factory:				
X	42	13	33	88
Y	20	8	25	53
Total	62	21	58	141

(a) Do these output figures show a significant difference at the 5% level?
(b) What hypothesis have you tested?

I.C.M.A.

21.11 Suppose that a random sample of men and women indicated their view on a proposal of public importance as follows:

	Opposed	Undecided	In Favour
Men	156	74	170
Women	122	50	236

At levels of significance of (a) 0.01 and (b) 0.05, test the hypothesis that there is no difference between the views of men and women on this issue.

I.C.M.A.

21.12 A certain drug was administered to 100 volunteers with common colds, and 35 reported that the drug brought some relief. The next winter, a new improved drug was given to the same group, and this time 43 reported that the drug brought some relief. Are we justified in assuming that the second drug is more likely to bring relief?

21.13 Pharmaceuticals Ltd. have developed three new respiratory drugs. They submitted the drugs to clinical trials, the results of which are given in the table below. Determine whether or not these figures indicate a real difference in response of the diseases to the drugs, and explain your conclusion. Use the chi-square (χ^2) distribution as a test of significance at the 0.01 level.

Drug	Percentage of Patients Recovered	Percentage of Patients Not Recovered
A	65	35
B	90	10
C	85	15

A.C.A.

21.14 (a) A marketing information firm has kept records on a random sample of 144 drivers who bought new cars in 1965. Drivers were classified according to age in 1965 and records kept include data on how many other new cars each driver has bought in the last ten years. This information is tabulated below.

Additional New Cars Bought	Age of Driver			Total
	20 and under 30	30 and under 40	40 and over	
0	19	5	12	36
1	23	13	12	48
2	11	18	7	36
More than 2	7	12	5	24
Total	60	48	36	

Determine whether there is any evidence that there is a relationship between propensity to buy new cars and age of drivers in U.K.

(b) How useful might you conclusion be for a car assembly firm which is investigating the problem of advertising directed at specific age groups?

I.C.A.

21.15 A market research agency has been commissioned to report on the smoking habits of men and women in a particular area. Two samples were taken, one of 200 men and the other of 200 women. The findings are shown in the following table:

	Smokers	Non-Smokers	Total
Men	110	90	200
Women	104	96	200

You are required to:
(a) test the hypothesis that difference in sex has no effect on the number of smokers in the two samples, and
(b) explain your conclusion in (a).
Use the chi-square (χ^2) distribution as a test of significance, at the 0.05 level.

21.16 The Acca Co. Ltd. tests all components built into its products to ensure that they have attained the standard of quality required. There are three checkers, *A*, *B* and *C*, who undertake this work. The table below shows the numbers of components accepted and rejected by the checkers when they tested three batches of component 1703 recently delivered to the factory. You are required to test the hypothesis that the proportions of components rejected by the three are equal and explain the significance of your conclusion. Use the chi-square (χ^2) distribution as a test of significance, at the 0.05 level of significance.

The Acca Co. Ltd. Components Quality Test

	Checker A	Checker B	Checker C	Total
Accepted	44	56	50	150
Rejected	16	24	10	50

A.C.A.

21.17 The number of students passed and failed by three examiners in a certain examination were as follows:

	Mr. X	Mr. Y	Mr. Z
Passed	50	47	56
Failed	5	14	8

Test the hypothesis that the proportions of students failed by the three examiners are equal.

C.I.P.F.A.

21.18 The following data is based on a random sample of 1000 businessmen:

	Profit Maximisers	Revenue Maximisers	Total
Studied economics	200	100	300
Did not study economics	200	500	700
Total	400	600	100

Test by *two* different statistical techniques whether the above information suggests any association between education in economics and the objectives of businessmen.

C.I.P.F.A.

21.19 Samples of students with varying educational backgrounds obtained the following results in their final degree examinations:

	Third Class	Second Class (lower)	Upper Second/First
Secondary	24	58	18
Comprehensive	36	112	52
Grammar	80	230	90
Public/Direct Grant	60	200	40

Test whether there is any association between pre-university education and the class of degree.

C.I.P.F.A.

Chapter Twenty Two

Regression and Correlation

If you have listened carefully to news bulletins, or read the business pages of a responsible newspaper you will probably have come across the phrase "economic indicator" or "business indicator". Such an indicator is an event which, although it may be important in its own right, is of even more importance in helping us to predict what is going to happen to some other variable in the future. It is commonly believed, for example, that government expenditure will affect the level of employment. What is important, of course, is knowledge of how much extra expenditure is needed to create, say, one thousand new jobs, and what is the time lag before the unemployed actually start work. Surely a man who can identify and quantify relationships such as this is worth his weight in gold to any company.

No text books can teach you to identify every relationship which may be important in a particular set of circumstances, but you should have a good idea of those relationships which are relevant to your own work. You will know, for example, that next months sales figures are influenced by this month's advertising, or that the major influence on your departmental budget is the behaviour of labour costs. What you have to learn is how a knowledge of such relationships can be used to enable you to make predictions (or estimates). After all, in forecasting sales figures or preparing a departmental budget you *are* being asked to make predictions.

Bivariate Distributions

At some time, every businessman must have faced the problem as to whether or not he should advertise. Some believe that their sales depend on the amount of advertising they undertake; others hold equally firmly to the view that they key to increasing their sales lies in providing service, in offering a wide selection of goods, or in giving an unconditional guarantee of satisfaction. The only way of settling this sort of argument for any individual firm is to examine the facts.

Ron's Department Store has recently hired a consultant to help boost its sales and thereby the morale of its management (and hopefully its employees). The consultant, a marketing specialist, has reasoned that sales are a function of advertising expenditure. Thus he has decided to try to relate sales to the amount spent on advertising. The data from the company's records are produced for him by the accountant and he is presented with the following information.

	Advertising X (£000)	Sales Y (£ mill)
1966	100	9
1967	105	8
1968	90	5
1969	80	2
1970	80	4
1971	85	6
1972	87	4
1973	92	7
1974	90	6
1975	95	7
1976	93	5
1977	85	5
1978	85	4
1979	70	3
1980	85	3

Now this series is rather different from others we have considered. Whereas in the series we have so far studied each item has had a single value, in this series there are two items associated with each value. We call distributions like this *Bivariate Distributions*.

We cannot, of course, draw any meaningful conclusions from the data as its stands. Possibly it would help if we were to present the information in the form of a diagram − in this case the scatter diagram.

Before we get too involved with the scatter diagram itself, let us consider the variables which are the subject of analysis. Firstly, what is the nature of the data which comprises the two variables? Can it be treated as an enumeration of the population; or must we treat it as a sample? The answer may seem very simple, but there are serious implications underlying it. We shall be calculating measures which contain constants. If we treat the data as a population we will treat the constants we calculate, α and β (alpha and beta), as population parameters, absolutely accurate and unchanging. If the data, however, is treated as a sample, the constants we calculate will be a and b, which are no more than estimates of α and β. As such they might be wrong and will certainly be subject to some sampling error.

Secondly which is the dependent and which the independent variable? In terms of the analysis we shall assume one variable to be a function of the other. But in terms of the actual data, the analysis itself cannot determine which is which. This determination can be made only on the basis of your own knowledge of the subject matter. In so far as Ron's Department Store is concerned there is little doubt that most people would agree that advertising expenditure is the independent variable and sales the dependent. We shall proceed to analyse on this assumption, even if one or two of you do disagree.

The Scatter Diagram

The purpose of the scatter diagram, as you know, is to illustrate diagrammatically any relationship that may exist between the dependent and independent variables. To the extent that it succeeds it can help the analyst in three ways:

1. It indicates generally whether or not there appears to be a relationship between the two variables.
2. If there is a relationship it may indicate whether it is linear or non linear.
3. It the relationship is linear, the scatter diagram will show whether it is negative or positive.

If you will look at Diagram 22.1 you will see that the data does provide some evidence that the higher the level of advertising, the higher the level of sales achieved. Certainly the same level of advertising is associated occasionally with differing levels of sales, just as the same level of sales is associated with differing levels of advertising. But as you can see, the general trend of the points on this diagram is an upward slope from left to right. It is indicated that there is a relationship, that it is linear and that it is a positive relationship in the sense that as one variable rises so does the other.

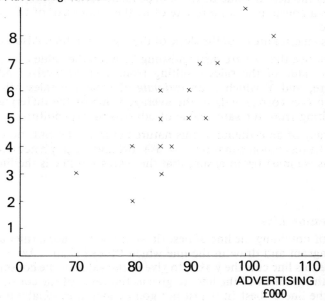

Diagram 22.1

The scatter diagram, then, is the graph of a bivariate distribution and is used to obtain an indication of whether association exists between the two variables. Having discovered that in this case an association is indicated, we shall now try to discover the nature of the relationship and the degree of association between the variables. If we can do that we can use our knowledge to estimate what the value of sales will be for any given level of advertising.

Regression

Firstly we will attempt to discover the nature of the relationship by calculating an equation enabling us to estimate the value of one variable given that we know the value of the other. The variable we are trying to predict is the dependent variable (the one plotted coventionally on the y axis) while the variable we are using as a basis for prediction is the independent variable (plotted on the x axis). In this book we will confine ourselves to an analysis of the simple linear case; simple implying two variables only such that y depends on the value of x and not on another variable; linear implying that the relationship between x and y is a straight line relationship. In this simple case the equation which best fits the data is an equation of the form

$$\hat{Y} = a + bX$$

where $\hat{Y}$ is an estimate of the average value of Y corresponding to a given value of X

X is the actual value of the independent variable

a is a constant — an estimate of α, the y intercept of the regression line

b is an estimate of β, the slope of the regression line. Also a constant.

You will notice that we are distinguishing between the value of Y, which is the *actual* value of the sales resulting from a given level of advertising expenditure, and $\hat{Y}$ which is an *estimate* of what the sales will be. Our estimate in fact corresponds to the average value of the differing level of sales resulting from the same value of advertising expenditure.

The accuracy of an estimate of this nature naturally depends on the extent to which the regression equation $\hat{Y} = a + bX$ and its graph actually fits the data. Thus we must try to ensure that the regression line is the line of best fit.

The Regression Line

One way of obtaining the line of best fit would be to place a ruler across the scatter diagram and draw in the line which "looks right". We could then find where this line cuts the y axis, to give us the value of the constant a, and calculate the gradient of the line, to give us the value of the constant b. But guessing the line of best fit is just not good enough, especially if decisions are to be taken on the basis of our estimates. What we need is some criterion by which we can determine what we mean by "best fit".

One criterion we could adopt would be to minimise the magnitude of the error involved, that is, the differences between the value of the points on the diagram (Y) and the value of the points on the regression line we have drawn (Ŷ). We could, that is, minimise the value of $\Sigma (Y - \hat{Y})$. Problems arise with this criterion however, In Diagram 22.2a three errors are shown, one positive and two negative. They offset each other and the net error is zero. In this case, by minimising the error we have obtained what seems to be quite a good fit. Look now at Diagram 22.2b.

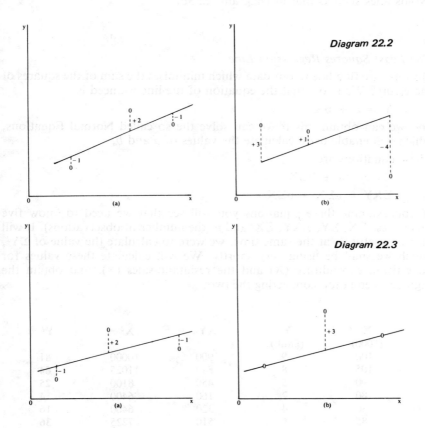

Diagram 22.2

Diagram 22.3

This regression line, too, reduces the error to zero, yet quite clearly in this case the fit is bad. Reducing the error to a minimum then, is no criterion for positioning the line of best fit. It may give a good fit, but it is just as likely to give a bad one.

What, then, if we were to minimise the absolute error, ignoring the sign? Once again this will not guarantee a line of best fit. In Diagram 22.3, the line in diagram (a) is clearly a better fit than the line in diagram (b), yet line b minimises the absolute error. Statisticians reject both of these criteria as unsatisfactory and have defined the line of best fit as "that line which minimises the sum of the squares of the errors — the least squares regression line."

The use of this criterion has several advantages. Firstly, by squaring the errors we remove all negative signs and so avoid negative errors cancelling out the positive ones; secondly, squaring tends to emphasise the larger errors and the minimising criterion means that we are trying to avoid the larger errors; thirdly, all points are taken into account so the criterion avoids lines such as that in Diagram 22.3b.

The Least Squares Regression Line

How can we fit a line to our data which minimises the sum of the squares of the errors? We know that the equation of the line we need is

$$\hat{Y} = a + bX$$

and we can obtain this if we can solve the so called Normal Equations, which will enable us to calculate the values of a and b.

These equations are:

$$\Sigma Y = na + b\Sigma X$$
$$\Sigma XY = a\Sigma X + b\Sigma X^2$$

If you examine these equations you will see that we need to know five quantities, ΣX, ΣY, ΣXY, ΣX^2 and n (the number of observations). It will also be useful if, at the same time, we were to calculate the value of ΣY^2, which we shall be using very shortly. We will calculate these values for advertising expenditure (X) and the resultant sales (Y), and obtain the regression equation connecting the two.

X (£000)	Y (£mill)	XY	X^2	Y^2
100	9	900	10000	81
105	8	840	11025	64
90	5	450	8100	25
80	2	160	6400	4
80	4	320	6400	16
85	6	510	7225	36
87	4	348	7569	16
92	7	644	8464	49
90	6	540	8100	36
95	7	665	9025	49
93	5	465	8649	25
85	5	425	7225	25
85	4	340	7225	16
70	3	210	4900	9
85	3	255	7225	9
1322	78	7072	117532	460

We can now insert these figures in the Normal equations and calculate the value of a and b.

$$78 = 15a + 1322b$$
$$7072 = 1322a + 117532b$$

Now these equation look fearsome, but they can be solved as normal simultaneous equations. If we multiply the first equation by 1322 — the coefficient of a in the second equation we get

$$103116 = 19830a + 1747684b$$

Now multiply the second equation by 15, i.e. the coefficient of a in the first equation and we get

$$106080 = 19830a + 1762980b$$

We now have a pair of simultaneous equations in which the coefficient of the term a is the same. If we subtract one of these equations from the other we will have an equation not containing the term a.

$$\begin{array}{rcl}106080 & = & 19830a + 1762980b \\ 103116 & = & 19830a + 1747684b \\ \hline 2964 & = & 15296b \\ b & = & 0.193776\end{array}$$

Now, given the value of b, we can insert this value in the first equation and so find the value of a.

$$\begin{array}{rcl}78 & = & 15a + (0.193776 \times 1322) \\ 78 & = & 15a + 256.1721 \\ -178.1721 & = & 15a \\ a & = & -11.8781\end{array}$$

We have calculated the values of both a and b and can insert them in the regression equation $\hat{Y} = a + bX$.

Thus the regression equation connecting sales and advertising expenditure is

$$\hat{Y} = -11.8781 + 0.19378X$$

or more simply $\hat{Y} = -11.9 + 0.19X$

You will agree that, even with electronic calculators, solving simultaneous equations with figures of this magnitude is cumbersome and productive of errors. We can however solve the normal equations using only symbols and so obtain expressions which enable us to obtain the value of a and b more directly and this is infinitely preferable.

The statistics necessary to find a and b are the variance of X, and what we call the covariance of X and Y. The variance of X we will symbolise as S^2_x, and as you know this is calculated as

$$S^2_x = \frac{\Sigma x^2}{n} - \left(\frac{\Sigma x}{n}\right)^2$$

The covariance is symbolised as S^2_{xy} and it is calculated as

$$S^2_{xy} = \frac{\Sigma XY}{n} - \frac{\Sigma X \Sigma Y}{n.n}$$

The expression from which you can calculate the value of b is

$$b = \frac{\text{Covariance}}{\text{Variance}} = \frac{S^2_{xy}}{S^2_x}$$

and the expression for a is

$$a = \frac{1}{n}(\Sigma Y - b\Sigma X)$$

We have already calculated all the totals necessary for these expressions. Substituting we find

$$S^2_{xy} = \frac{\Sigma XY}{n} - \frac{\Sigma X \Sigma Y}{n.n} = \frac{7072}{15} - \frac{1322 \times 78}{15 \times 15} = 13.17333$$

$$S^2_x = \frac{\Sigma X^2}{n} - \left(\frac{\Sigma X}{n}\right)^2 = \frac{117532}{15} - \left(\frac{1322}{15}\right)^2 = 67.98222$$

Hence $\quad b = \dfrac{13.17333}{67.98222} = 0.193776$

and $\quad b = \dfrac{1}{15}(78 - 1322 \times 0.193776) = -11.8781$

As you can see, the values of a and b are exactly the same as when calculated from the Normal Equations, but with the advent of electronic calculators, a great deal of tedious work is taken out of obtaining the regression equation.

How well does this equation fit the original data? Let us calculate what our estimate of the value of Y would be for the various values of X, and then compare them with the observed values.

Using the regression equation

$$\hat{Y} = -11.8781 + 0.1938X$$

we find that

For a value of X	Observed Y	Estimated $\hat{Y}$
100	9	7.5019
105	8	8.4709
90	5	5.5639
80	2	3.6259
80	4	3.6259
85	6	4.5949
87	4	4.9825
92	7	5.9515
90	6	5.5639
95	7	6.5329
93	5	6.1453
85	5	4.5949
85	4	4.5949
70	3	1.6879
85	3	4.5949

Well, as you can see the estimation of the value of Y is not perfect, but on the whole, we think you will agree, it is pretty good.

The equation we have calculated is known as the equation of the regression line of y on x because we are estimating the value of Y given the value of X. Thus if we had spent £75000 on advertising we would estimate a sales figure by putting X = 75 (thousand) in the regression equation, i.e.

$$\hat{Y} = -11.8781 + (75 \times 0.1938) = 2.6569 = £2656900$$

What does this prediction mean? Obviously it cannot mean that every time we spend £75000 on advertising we would have sales of exactly £2656900. Conditions could never be quite the same on each occasion, and the fact that we spent some money last month would, of itself influence the effect of spending more this month, since advertising has a cumulative effect. In fact, actual sales would fluctuate, even if advertising expenditure did not. The figure we estimate for sales is, in fact, an average figure.

We calculated the regression equation from observed values of X ranging from £70000 to £105000 and used that equation to estimate the level of sales resulting from advertising expenditure of £75,000. When we take a value of the independent variable within the range of the observed values we call the process interpolation (placing between). We might have asked, however, for the estimated value of sales from advertising expenditure of £120000.

$$\hat{Y} = -11.8781 + (0.1938 \times 120) = £1137790$$

Here we used a value of X outside the range of observed values. We call this process extrapolation (placing outside). Both estimates are subject to error, but an extrapolated estimate is less reliable than an interpolated estimate. Can you see why? Within the observed range of X we know how the data behaves, and how well the straight line fits. Outside the observed range we do not know how the data behaves, and the straight line may no longer be a good fit for these values of X. This is illustrated in Diagram 22.4 and we can see that great caution is needed when making extrapolated estimates.

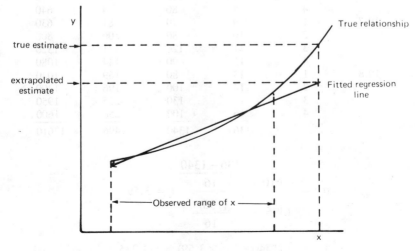

Diagram 22.4

Linear Regression and the time series

We saw in an earlier chapter that the trend of many time series is linear and we learned to calculate that trend using a moving average. Now although a time series is not strictly a bivariate series, we can obtain the linear trend by using regression analysis.

Suppose we are examining the figures of quarterly profit made by a firm. If the data appears to demand a linear trend we can assume that profits depend on time and convert time into an independent variable by numbering successive quarters 1,2,3,4,5 and so on.

Profits have varied over the last four years quarter by quarter, but have been generally rising. The figures, in thousands of pounds are:

Year	Quarter			
	1	2	3	4
19-5	40	60	90	70
19-6	50	70	110	80
19-7	70	80	120	90
19-8	80	100	130	100

If we number the quarters from 1 to 16 and use this as the independent variable, we can convert the time series into a bivariate distribution.

Year	Quarter	X	Y	X^2	XY
19-5	1	1	40	1	40
	2	2	60	4	120
	3	3	90	9	270
	4	4	70	16	280
19-6	1	5	50	25	250
	2	6	70	36	420
	3	7	110	49	770
	4	8	80	64	640
19-7	1	9	70	81	630
	2	10	80	100	800
	3	11	120	121	1320
	4	12	90	144	1080
19-8	1	13	80	169	1040
	2	14	100	196	1400
	3	15	130	225	1950
	4	16	100	256	1600
		136	1340	1496	12610

$$b = \frac{12610 - \frac{136 \times 1340}{16}}{1496 - \frac{(136)^2}{16}} = 3.59$$

$$a = \frac{1}{16}(1340 - 136 \times 3.59) = 53.235$$

Thus the equation of our trend line for this series is:

$$\hat{Y} = 53.235 + 3.59X$$

We can obtain trend values by inserting into this equation $X = 1,2,3...16$. Having thus obtained the trend we could use it to calculate the deviations from the trend and to calculate the average seasonal variation. The original data and the trend are plotted in diagram 22.5. As you know, only two points are needed to plot a straight line, and we have chosen $X = 1$ and $X = 10$.

When $X = 1$, $\hat{Y} = 56.825$ and when $X = 10$, $\hat{Y} = 89.135$.

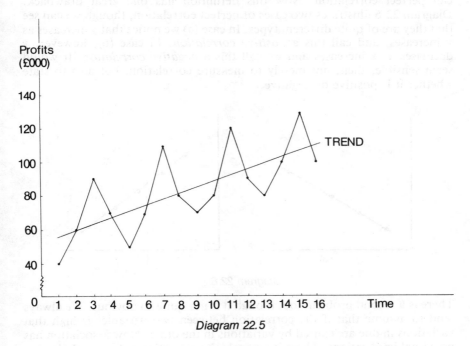

Diagram 22.5

So far in this chapter we have been concentrating on the nature of the relationship between two variables but we must also pay attention to the degree of relationship. We will now turn our attention to measuring the extent to which two variables are related when that relationship is less than perfect.

Correlation
Our analysis of regression has shown us that it is possible to obtain an expression showing the relationship between two variables such that for any value of X it is possible for us to estimate the value of Y associated with it. Sometimes the estimate we make is consistently good and there seems to be a very close relationship between the two variables. At other times the estimate is very poor indeed and we are led to think that the relationship is weak. Naturally we would like to know more about this relationship and in

particular we would like to be able to measure its strength. Statisticians have designed a measure, the coefficient of correlation, which measures the relationship between the variables by asking the question, "how much of the variation in the value of Y can be explained by variations in the value of X"

Suppose we take an extreme case. Having calculated the regression equation we find that the regression line is such that every point in the scatter diagram falls exactly on the line. The whole of the variation in Y is explained by variations in the value of X. We have here a case of what we call perfect correlation. Now this definition has one great drawback. Diagram 22.6 illustrates two cases of perfect correlation, though we can see that they are of quite different types. In case (a) we notice that y increases as x increases, and call this a *positive correlation*. In case (b), however, y decreases as x increases and we call this a *negative correlation*. It would seem sensible, then, not merely to measure correlation, but also to state whether it is positive or negative.

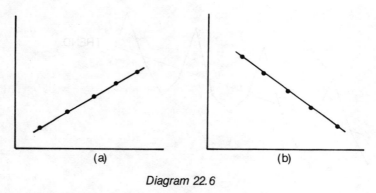

Diagram 22.6

There is a second problem too, though it is not quite so obvious. We always tend to assume that if the correlation between two variables is high that variations in one are caused by variations in the other. Now association has no casual implications. There is a very high association between admissions into mental institutions and the number of T.V. licences issued. But it would be a bold man who maintained that we have proved that watching television causes insanity. We have here a case of spurious or nonsense correlation and we must always beware of this. In using the correlation coefficient we must be sure in our own mind that association between two variables is likely, or at least possible.

Bearing in mind this warning how will we go about measuring correlation?

Concepts of Variation

So important is the idea of variation in the analysis of correlation that it would be as well at it much more closely. Firstly we will draw a scatter diagram of our data and superimpose on the diagram a line representing the mean value of Y ($\bar{Y}$), (Diagram 22.7a). This line will, of course be a

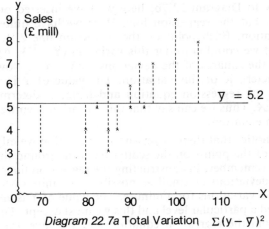

Diagram 22.7a Total Variation $\Sigma(y - \bar{y})^2$

horizontal straight line since the mean value of Y is the same for all values of X. As you can see there is a great deal of variation of the values of Y about $\bar{Y}$, and we will measure the total variation by totalling the squares of the individual deviations. Can you see why we must square the deviations? Obviously if we have calculated the value of $\bar{Y}$ correctly, the sum of the deviations must be equal to zero. Thus, as a measure of total variation we have

Sum of the squares of the deviations from $Y = \Sigma (Y - \bar{Y})^2$

You will have realised already that if we take the average of the deviations squared we will have the variance of Y.

$$S_y^2 = \frac{\Sigma(Y - \bar{Y})^2}{n} = \frac{\Sigma Y^2}{n} - \left(\frac{\Sigma Y}{n}\right)^2$$

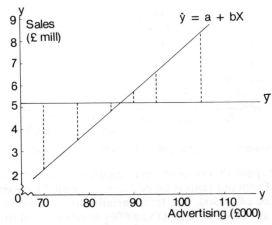

Diagram 22.7b Explained Variation $\Sigma(\hat{y} - \bar{y})^2$

If you turn now to Diagram 22.7b, here we have inserted on the scatter diagram not $\bar{Y}$, but the regression line. Here we have quite a different concept of variation. Each point on the regression line differs from the value of $\bar{Y}$, and we could consider this variation $(\hat{Y} - \bar{Y})$. Again we will take the sum of the squares of the deviations, $\Sigma(\hat{Y} - \bar{Y})^2$. Now there is an important characteristic of this variation. The value of $\hat{Y}$ is in each case calculated from the regression equation and hence is determined by the value we give to X. Thus we can say that this variation is entirely explained by the regression equation.

Finally you will notice that there is yet another variation we might consider – the variation of the points on the scatter diagram around the regression line. As you will remember, in constructing the regression line we made the square of these deviations as small as possible. We minimised $\Sigma(Y - \hat{Y})^2$. Think about these variations for a minute. The value of Y is an established fact; sales reached a particular level and this we must accept. The value of $\hat{Y}$ we calculated from the regression equation and we have argued that the variation of $\hat{Y}$ from $\bar{Y}$ is fully explained by that equation. But there remains a part of the variation of Y (equal to $Y - \hat{Y}$) which is unaccounted for. It is a part of the total variation which is not explained by the regression equation. It is, then, an unexplained variation. This variation is shown in Diagram 22.7c. Summarising then, we have:

$$\text{Total variation squared} = \Sigma(Y - \bar{Y})^2$$
$$\text{Explained variation squared} = \Sigma(\hat{Y} - \bar{Y})^2$$
$$\text{Unexplained variation squared} = \Sigma(Y - \hat{Y})^2$$

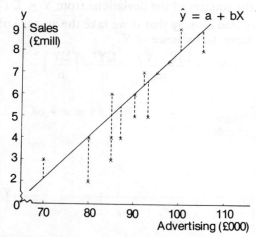

Diagram 22.7c Unexplained Variations $\Sigma(y - \hat{y})^2$

Suppose we had spent £85000 on advertising and as a result actual sales had been £6000000. From our regression equation we would have estimated that sales (Y) would be £4594900. The total variation is the actual value of sales (Y) less the average value of sales (£5200000) which is equal to £800000. Of this total variation an amount equal to $\hat{Y} - \bar{Y}$ is explained by the regression equation equal to £4594900 − £5200000 i.e. £−605100. We would have expected sales to be £605100 below the average, whereas in fact they were £800000 above the average. There is then a variation of £1405100 which we have not explained.

What we have done in fact is to divide total variation into two parts − that part which we can explain, and that part which remains unexplained.

Total Variation − Explained Variation + Unexplained Variation

Let us now calculate these three variations from the data for Ron's Department Store.

			Total		Unexplained		Explained	
X	Y	$\hat{Y}$	$Y - \bar{Y}$	$(Y - \bar{Y})^2$	$Y - \hat{Y}$	$(Y - \hat{Y})^2$	$\hat{Y} - \bar{Y}$	$(\hat{Y} - \bar{Y})^2$
100	9	7.5019	3.8	14.44	1.4981	2.2443	2.3019	5.299
105	8	8.4709	2.8	7.84	−0.4709	0.2217	3.2709	10.699
90	5	5.5639	−0.2	0.04	−0.5639	0.3180	0.3639	0.132
80	2	3.6259	−3.2	10.24	−1.6259	2.6436	−1.5741	2.478
80	4	3.6259	−1.2	1.44	0.3741	0.1400	−1.5741	2.478
85	6	4.5949	0.8	0.64	1.4051	1.9743	−0.6051	0.366
87	4	4.9825	−1.2	1.44	−0.9825	0.9653	−0.2175	0.047
92	7	5.9515	1.8	3.24	1.0485	1.0994	0.7515	0.565
90	6	5.5639	0.8	0.64	0.4361	0.1902	0.3639	0.132
95	7	6.5329	1.8	3.24	0.4671	0.2182	1.3329	1.777
93	5	6.1453	−0.2	0.04	−1.1453	1.3117	0.9453	0.894
85	5	4.5949	−0.2	0.04	0.4051	0.1641	−0.6051	0.366
85	4	4.5949	−1.2	1.44	−0.5949	0.3540	−0.6051	0.366
70	3	1.6879	−2.2	4.84	1.3121	1.7216	−3.5121	12.335
85	3	4.5949	−2.2	4.84	−1.5949	2.5437	−0.6051	0.366
$\bar{Y} = 5.2$				54.40		16.1101		38.3

Rounding errors cause the sum of explained plus unexplained variation to differ from total variation by 0.01.

Measuring the Relationship

Armed with our knowledge of variation we can now go ahead to devise a measure of the degree of relationship between X and Y. Suppose that the whole of the variation Y were explained by the regression equation. We could then argue that the relationship was perfect − that we have perfect correlation. But if only a part of the variation was explained, the correlation between X and Y would be less than perfect. As a first approach then we could use as a measure of the relationship between the two variables that proportion of the variation in the values of the dependent variable which is explained. Thus, we have calculated the total variation squared to be 54.4 and the explained variations squared to be 38.3. As a first measure we have:

$$\frac{\Sigma(\hat{Y} - \bar{Y})^2}{\Sigma(Y - \bar{Y})^2} = 0.704$$

We call this measure the Coefficient of Determination, symbolised as r^2. It is not, perhaps, as well known as the Coeffieicnt of Correlation but it is, in fact, far more meaningful. The value of the coefficient of determination cannot, of course, be greater than 1 since we cannot explain a greater proportion of total variation than the whole; nor can it be less than zero since we cannot have less than no variation explained.

Since we measured variation by taking the sum of the squares of the individual variations, it would seem logical to take the square root of the coefficient of determination. This we define as the coefficient of correlation.

$$r = \sqrt{\frac{\text{Explained Variation}}{\text{Total Variation}}} = \sqrt{\frac{\Sigma(\hat{Y} - \bar{Y})^2}{\Sigma(Y - \bar{Y})^2}} = \sqrt{\frac{38.3}{54.4}} = 0.839$$

Thus the coefficient is an abstract measure of the relationship between variables based on a scale ranging between $+1$ and -1.[1] Whether r is positive or negative depends on the nature of the relationship between X and Y. If the sign of b in the regression equation is $+$, as X increases so does Y, and we have positive correlation. On the other hand, if the sign of b is negative, you will find that as X increases the value of Y decreases and we have negative correlation.

Now there is nothing wrong in calculating the Coefficient of Correlation in this way, but it is, you must agree, a most cumbersome way of doing it. Not only have you to calculate the regression equation first, you have then to use that equation to obtain the values of $\hat{Y}$, obtain $\Sigma(\hat{Y} - \bar{Y})^2$ and then express this as a proportion of total variation squared. Like all such methods, it has its uses, and explains precisely what we are doing. But what we need is a method of calculating the Coefficient of Correlation directly without calculating the regression equation first.

Pearson's Product Moment Correlation Coefficient

We can calculate the value of the Coefficient of Correlation far more simply if we use the formula

$$r = \frac{\text{Covariance (XY)}}{\sqrt{\sigma_x^2 \times \sigma_y^2}} = \frac{S_{xy}^2}{\sqrt{S_x^2 \times S_y^2}}$$

As you will remember, the covariance (XY) is

$$\frac{\Sigma XY}{n} - \frac{\Sigma X . \Sigma Y}{n.n}$$

The variance of X is

$$\frac{\Sigma X^2}{n} - \left(\frac{\Sigma X}{n}\right)^2$$

The variance of Y is

$$\frac{\Sigma Y^2}{n} - \left(\frac{\Sigma Y}{n}\right)^2$$

You will notice the similarity between this formula and the formula we used earlier to calculate the value of b in the regression equation. In fact the numerator is the same and one term in the denominator is the same.

1. Remember that $\sqrt{1} = \pm 1$.

Let us now confirm that, in fact, this formula will yield the same Coefficient of Correlation as our previous calculations.

$$S_{xy}^2 = \frac{\Sigma XY}{n} - \frac{\Sigma X . \Sigma Y}{n.n} = \frac{7072}{15} - \frac{1322 \times 78}{15 \times 15} = 13.17333$$

$$S_x^2 = \frac{\Sigma X^2}{n} - \left(\frac{\Sigma X}{n}\right)^2 = \frac{117532}{15} - \left(\frac{1322}{15}\right)^2 = 67.98222$$

$$S_y^2 = \frac{\Sigma Y^2}{n} - \left(\frac{\Sigma Y}{n}\right)^2 = \frac{460}{15} - \left(\frac{78}{15}\right)^2 = 3.62667$$

$$r = \frac{13.17333}{\sqrt{67.98222 \times 3.62667}} = 0.839$$

The interpretation of r

One disturbing feature of many students' work is that although they can calculate the value of the coefficient of correlation in a perfectly efficient manner they completely ignore any effort to interpret what their results mean. What is the point of calculating a statistic if you cannot interpret it? We have calculated that r is +0.839 for the data we were working with. Thus it is clear that a strong positive correlation exists. But what does +0.839 mean? If we had another bivariate distribution for which r = +0.42, clearly our first coefficient implies a stronger correlation than the second. But is it twice as strong? In fact it is far more meaningful to consider the Coefficient of Determination. Do you remember that this measures the proportion of total variation that can be explained by the regression equation? When r = 0.839, r^2 = 0.704 and we can conclude that 70.4% of the variations in Y can be explained by the regression equation, leaving less than 30% to be explained by other factors. But if we consider a coefficient of correlation of 0.42, we have a coefficient of determination of 0.42^2 = 0.1764, which implies that only 17.64% of the variations in Y are explained by the regression equation. Thus a coefficient of correlation of 0.839 is four times as strong as one of 0.42.

We would like, however, to issue a few words of warning about the interpretation of even a high coefficient. Even though analysis indicates that correlation exists we are not justified in assuming that there is therefore a cause and effect relationship. We must never fall into the trap of assuming that cause and effect exists when it is nonsense to do so. Sometimes a high correlation is obviously nonsensical. There is, as noted earlier, a high positive correlation between the annual issue of television licences and the annual admissions into mental institutions. It would be ludicrous to suggest that cause and effect exists here. The only logical conclusion that one can draw is that, quite by chance, both statistics were increasing at the same rate. In this case it is quite obvious that cause and effect is not proved by a high correlation coefficient — but sometimes things are not quite so clear cut. The high correlation between infant mortality and the extent of overcrowding that was found in Bethnal Green between the First and Second World Wars does seem to suggest that overcrowding causes high infant mortality. In fact, however, both are probably a result of low income levels.

If, then, we can never use correlation analysis to prove a cause and effect relationship you may wonder if it is worth study. Well, we think it is. Correlation can give added weight to a relationship that theory suggests exists. We could use it, for example, to verify the economic theory that consumption depends on income. Again, correlation can sometimes suggest that causal relationships exist in areas that were not previously suspected — a technique much used in medical research. Here correlation can often point to promising avenues of investigation.

Finally we would like to point out that a low correlation coefficient does not necessarily mean a low degree of association. After all, the coefficient we have calculated measures the strength of a linear relationship. The relationship may be very high, but curvilinear, and our coefficient would not indicate this.

Significance Test for r

Let us now return to Ron's problem. We have calculated a regression equation which indicates the nature of the relationship between sales and advertising expenditure and our correlation coefficient seems to be fairly high. But our results are obtained from a sample and we must now face up to the question as to whether the evidence we have justifies the conclusion that there is some correlation. After all, we know that conclusions based on a small sample can be very wide of the truth. The obvious way of testing our conclusions is to undertake a significance test on our correlation coefficient.

We will assume that our sample has been drawn from a population with a zero correlation coefficient.

Thus
$$H_0: r = 0$$
$$H_1: r \neq 0$$

If H_0 is true we would expect r to be distributed about zero with a standard error

$$S_r = \sqrt{\frac{1-r^2}{n-2}} = \sqrt{\frac{1-0.704}{15-2}} = \sqrt{0.022769} = 0.15089$$

This estimate is made with 13 degrees of freedom.

$$\text{The test statistic } t = \frac{0.839-0}{0.15089} = 5.56$$

For a 2 sided test with 13 degrees of freedom the critical value for t is 2.160 (at the 5%) level and 3.012 (at the 1% level).

We will accept H_0 and conclude that our correlation coefficient is highly significant. We must still remember, however, that we have not proved a cause and effect relationship and that in making the test we assume that our sample was drawn from a normally distributed population.

Rank Correlation

Positive correlation implies that large values of x are associated with large values of y and small values of x with small values of y. The converse is true, of course, of negative correlation. A useful method of assessing whether the data show any correlation would be to *rank* them, i.e. list the values of the variables as 1st, 2nd, 3rd and so on, the lowest rank being assigned to the lowest value of the variable.

Example

The following data refers to Gross Investment as a percentage of National Income and the percentage growth in National Income in nine selected countries. We are trying to assess if the level of investment is associated with the growth in National Income. In order to make an initial judgement the data has been ranked and the difference in the ranks of the two series listed in the column headed D.

	Gross Investment (%) x	National Income Growth (%) y	x Ranked	y	D	D^2
Belgium	19.2	3.4	4	2	2	4
Denmark	18.8	3.6	3	4	−1	1
France	19.5	4.9	5	6½	−1½	2.25
Germany	25.9	7.1	7	9	−2	4
Italy	21.8	5.6	6	8	−2	4
Netherlands	26.0	4.9	8	6½	1½	2.25
Norway	28.1	3.8	9	5	4	16
United Kingdom	16.4	2.6	1	1	0	0
United States	18.4	3.5	2	3	−1	1
						34.5

You will notice that in ranking the growth of National Income two countries have a growth rate of 4.9%. We cannot rank them 6th and 7th, or even equal 7th. Since the two countries are equal and between them occupy both the 6th and 7th ranks we rank them equal at $6½ = \frac{(6+7)}{2}$. Now, looking at the ranks it does seem that there is some positive correlation here. While only the United Kingdom has the same rank in both series, only in the case of Norway is there any marked difference in the ranks. Generally countries with a higher level of gross investment tend to have the higher growth rate.

We can in fact get a very good approximation to the correlation coefficient using ranks rather than the original data. Moreover the calculations are easy as the formula simplifies to

$$P = 1 - \frac{6\Sigma D^2}{n(n^2-1)}$$

where D is the difference in ranks and n the number of observations. This is known as *Spearman's Rank Correlation Coefficient* and in this case $P = 1 - \frac{6 \times 34.5}{9(81-1)} = 1 - \frac{207}{720} = 0.7125$. Thus there is a positive correlation, but since $P^2 = 0.5076$, only about 50% of the variation of y is explained. Correlation is not strong. Our knowledge of economics leads us to think that possibly we are examining the wrong variables. It may be net investment rather than gross that affects growth. This too could account for the difference in ranks in the case of Norway. Possibly much of the high level of investment here is replacement of worn out equipment, or possibly it is investment which does not lead directly to growth.

One great advantage of Rank Correlation is that it can be used to rank *attributes* which cannot be given a numerical value. Thus we could assess the consistency with which different panels of judges assess the contestants in beauty contests such as Miss World. In the same way we could assess whether different methods of selecting applicants for employment are likely to lead to the same results. Do we for example end up with the same ranking from a personal interview as we would from a written test or a psychological test? Thus we can use rank correlation in many cases where Pearsons' correlation coefficient is not applicable.

Finally, notice that in the table $\Sigma D = 0$. This must be so, (do you see why?) and provides a useful check on our initial calculations.

Exercises to Chapter 22

22.1 (a) What is meant when two series of recorded figures such as expenditure on advertising and volume of sales are said to be linearly correlated?
(b) If, in such a case, a regression line can be established what practical use does it have?
(c) With the aid of scatter diagrams distinguish between positive and negative correlation.

22.2 Describe the construction of a scatter diagram and a line of best fit. Illustrate by means of scatter diagrams the following three types of relationship between two variables:
(a) strong positive relationship;
(b) weak negative relationship;
(c) absence of relationship.   A.C.A.

22.3 *Advertising and Related Sales of Two Companies*

Company X		Company Y	
Sales	Advertising	Sales	Advertising
(£000)	(£000)	(£000)	(£000)
190	9.0	439	51.4
100	14.5	632	58.8
279	18.3	856	82.2
241	11.4	896	86.8
176	20.8	626	73.6
314	23.1	717	78.8
223	29.6	749	67.4
120	25.1	605	64.1
156	34.6	468	66.4
73	38.2	513	58.9
104	18.7	803	74.6
71	30.9	484	54.5
184	38.6	726	71.9
101	42.6	771	81.0
253	36.9	664	72.1

(a) What is meant by (i) *positive correlation,* (ii) *zero correlation?*
(b) Describe briefly the limitations of scatter diagrams.
(c) Draw *one* scatter diagram on which both firms' results are illustrated.
(d) Interpret the results of the diagram.

22.4 (a) Calculate the coefficient of correlation for the following data:

Domestic Manufacturing Ltd. Payroll and Value of Product × Output

Month	Total Payroll (£000)	Total Value of Product × Output (£000)
Jan.	5	8
Feb.	6	9
March	7	9
April	9	11
May	8	13

(b) What is the purpose of finding the correlation coefficient in respect of the above data, and what does the value of the coefficient indicate?

A.C.A.

22.5 *Television Receiving Licences Current Monochrome and Colour*

	Monochrome (millions)	Colour (millions)
Jan.	15.2	1.4
Feb.	15.1	1.5
March	15.0	1.6
April	14.9	1.7
May	14.9	1.9
June	14.8	2.0
July	14.8	2.1
Aug.	14.7	2.2
Sept.	14.7	2.3
Oct.	14.5	2.5
Nov.	14.4	2.6
Dec.	14.2	2.8

Sources: Post Office; Department of Trade and Industry.

(a) Prepare a scatter diagram based on the data given above, mark in the mean point and briefly comment on what is revealed.
(b) Explain the difference between the product moment coefficient of correlation and the rank correlation coefficient, mentioning any advantages either might have over the other.
(c) What is the association between correlation and regression?

22.6 Explain what is meant by the term 'product moment correlation coefficient'.

Age (in years)	18	20	21	27	23	34	24	42	38	44
Length of training required (in days)	8	5	6	8	7	11	8	10	6	8

The length of training required by operatives to be able to carry out a routine task is shown in the above table along with the age of the trainees. By calculating the product moment correlation coefficient determine whether the length of training required correlates with age.

22.7 (a) Explain what is meant by 'product moment correlation coefficients' and 'rank correlation coefficients' and give examples where each one has advantages over the other.

(b)
$$\begin{array}{lcccccccccc} x & 24 & 26 & 21 & 23 & 22 & 19 & 18 & 17 & 15 & 16 \\ y & 35 & 33 & 31 & 29 & 25 & 24 & 23 & 21 & 20 & 19 \end{array}$$

Calculate the product moment correlation coefficient between the variables x and y.

22.8 (a) Measurements on two variables are found to yield a correlation coefficient of 0.35. Give two reasons why it cannot necessarily be inferred that the two variables are causally related.

(b) The following table gives the mean daily temperature and the amount of electricity consumed on 8 consecutive Mondays during a year.

Temperature (°F)	37	32	35	40	40	44	42	48
Electricity consumption (megawatt hours)	3.7	3.8	3.7	3.6	3.7	3.4	3.4	3.3

Required: Code the data about the assumed means of 40 and 3.5 so that x = temperature $-$ 40; y = electricity consumption $-$ 3.5, and calculate the coefficient of correlation between temperature and electricity consumption. Comment on your result.

22.9 The following data taken from a manufacturing company's budget relates to volume of sales and the corresponding expenses. Obtain the product moment correlation coefficient and comment, in the light of your result, on the relationship between volume of sales and the total expenses.

Volume of sales (thousands or units)	5	6	7	8	9	10
Total expenses (£000)	74	77	82	86	92	95

22.10 It is often suggested that low unemployment and a low rate of wage inflation cannot coexist. Examine the evidence below and discuss whether it supports the above contention, producing where necessary relevant statistical measures.

Year	Unemployment %	Change in Wages %
1	1.6	5.0
2	2.2	3.2
3	2.3	2.7
4	1.7	2.1
5	1.6	4.1
6	2.1	2.7
7	2.6	2.9
8	1.7	4.6
9	1.5	3.5
10	1.6	4.4

Does the recent experience agree with your discussion above?

11	2.5	4.0
12	2.5	7.7
13	2.5	5.7
14	2.7	9.5

Source: Key Statistics 1900-70, L.C.E.S.

C.I.P.F.A.

22.11 Describe with examples what the coefficient of correlation measures and state the maximum and minimum values that 'r' can take.

An applicant for a certain post has to take five intelligence tests; he obtains the following results

Intelligence Test (x)	Score Obtained (y)
1	+2
2	+2
3	−3
4	+1
5	+3

Calculate the value of the product moment correlation coefficient 'r'. Comment on the view that his score increases as he progresses from one test to the next.

22.12 (a) Explain the value of a scatter diagram in investigating the association between two sets of figures. Illustrate the scatter diagrams associated with different values of the correlation coefficient.

(b) The following table records annual rainfall and sunshine in a British coastal resort:

Period	1	2	3	4	5	6	7	8	9	10	11
Rainfall (in)	15.1	15.8	14.9	16.6	12.6	17.4	16.1	13.7	15.5	17.8	13.2
Sunshine (hr)	1692	1634	1835	1741	1876	1561	1921	1942	1822	1542	1874

Calculate the rank correlation coefficient between rainfall and sunshine.

22.13 (i) Ten students enter for two papers in the final examination of a professional body. The marks obtained by each student are:

		Paper 1	Paper 2
Total marks awarded for paper		100	10
Marks obtained by student	A	58	7
	B	56	5
	C	54	8
	D	65	10
	E	58	5
	F	60	7
	G	59	6
	H	51	2
	I	53	4
	J	56	6

Calculate the product moment correlation coefficient r between the two sets of marks and give a clear explanation of the meaning of your result.

(ii) Without doing any further calculations show the effect, if any, on r, of increasing all the marks by 15%.

22.14

	Finished Output – (thousand tons)	Expenditure on Electricity (£)
Jan.	20	106
Feb.	22	138
March	25	158
April	26	172
May	21	120
June	23	142
July	28	184
Aug.	20	102
Sept.	25	164
Oct.	29	192

Express the relationship between the finished output and the expenditure on electricity during the period January to October in the form of the product moment coefficient of correlation and comment on the meaning of the figure obtained.

22.15 (a) Explain what are meant by the terms 'correlation coefficients' and 'scatter diagrams' and comment on their use.

(b) Two instructors assess the apprentices in a course they are running and give them the following placings

Apprentice	A	B	C	D	E	F	G	H	I	J	K	L
1st Instructor	1	2=	2=	4	5	6=	6=	8	9	10	11=	11=
2nd Instructor	1=	3	5	1=	4	6	10	7	8	11	9	12

Calculate the correlation between the two sets of placings.

Using the correlation you have just calculated comment on the agreement between the two instructors.

22.16 The personnel department of a large company is investigating the possibility of assessing the suitability of applicants by using psychological tests instead of normal interview procedures. A comparative test of seven applicants was carried out using both methods. The results were as follows:

Applicant	Ranking by Interview Procedure	Ranking by Psychological Tests
A	4	5
B	1	2
C	7	7
D	6	4
E	2	1
F	3	3
G	5	6

You are required to:
(a) calculate the rank coefficient of correlation;
(b) interpret the result established.

C.I.P.F.A.

22.17

Month	Broadcast Receiving Licences Current at end of Period Monochrome Television (thousands)	Cinema Admissions Weekly Averages (millions)
April	13,667	2.89
May	13,549	2.74
June	13,427	2.10
July	13,341	2.86
Aug.	13,197	3.47
Sept.	13,025	3.15
Oct.	12,779	2.70
Nov.	12,514	2.42
Dec.	12,286	2.06
Jan.	12,051	3.21
Feb.	11,923	2.95
March	11,766	2.89

Sources: Post Office; Department of Trade and Industry.

(a) Using a ranking method, calculate the appropriate statistic to give some assessment of the relationship which exists between TV licences issued and attendance at cinemas. Comment on your results.

(b) Distinguish between negative correlation and perfect positive correlation with the help of scatter diagrams. What is *spurious correlation*?

22.18 (i) Describe, with examples, what is meant by a scatter diagram.

(ii) Describe, with examples, what the coefficient of rank correlation measures. Give the range of values that this coefficient can take.

Two personnel managers each interview eight applicants for a vacant post. After the interviews each manager awards each applicant a mark, as given below.

	Candidate							
	A	B	C	D	E	F	G	H
Manager X	12	17	14	7	10	12	6	3
Manager Y	16	15	18	5	13	10	7	4

Convert the marks into ranks and then calculate and interpret the coefficient of rank correlation.

22.19 (a) Explain what is meant by scatter diagram and coefficient of rank correlation.

(b) Two personnel managers each interview the same five candidates for a vacant post. After the interviews the managers rank the candidates according to suitability for the post.

	Candidate				
	A	B	C	D	E
Manager X	2	1	5	3	4
Manager Y	3	5	4	1	2

Obtain and interpret a coefficient of rank correlation for the two sets of rankings.

22.20 (a) Explain the connection between correlation and regression and briefly indicate how the latter may be measured.

(b) The table given below shows the ranking obtained by ten typists who were first given speed tests and then subjected to speed checks as they worked in normal conditions. Using the rank coefficient of correlation determine whether the speed tests are a reliable guide to the speed likely to be attained by a typist working in normal conditions.

Typist	Ranking In Tests	In Normal Conditions
A	1	3
B	2	4
C	3	6
D	4	1
E	5	2
F	6	8
G	7	9
H	8	10
I	9	7
J	10	5

(c) State which method of measuring correlation would give a more accurate result for (b) than the rank coefficient. What informtion would you have needed to have been able to use this more accurate method?

22.21 A group of redundant employees accepted for a re-training scheme are given an aptitude test, and six months after completion of their training period they are ranked according to the quality of their output. The following are the rankings:

Employee	Aptitude Test	Work Quality
A	2	1
B	3	2
C	1	3
D	6	4
E	7	5
F	4	6
G	8	7
H	5	8
I	12	9
J	11	10
K	9	11
L	10	12

You are required to use an appropriate statistical technique to determine whether the aptitude test is a reliable predictor of a worker's ability to perform quality work.   A.I.A.

22.22 The table below shows the rate of growth of national income and the level of gross investment as a percentage of national income in a number of countries.

Country	Gross Investment (%)	National Income Growth Rate (%)
Belgium	19.2	3.4
Denmark	18.8	3.6
France	19.5	4.9
Germany	25.9	7.1
Italy	21.8	5.6
Netherlands	26.0	4.9
Norway	28.1	3.8
United Kingdom	16.4	2.6
United States of America	18.4	3.5

Source: O.E.D. Statistics.

(a) Plot the data on a graph.
(b) Rank each country by:
 (i) level of investment and
 (ii) rate of growth.
 Calculate Spearman's coefficient of rank correlation.
(c) Interpret your answer in (b).
(d) If you were the Chancellor of the Exchequer in Norway, what comments would you have to make on your country's growth rate related to the level of investment?

22.23 (a) Explain the meaning of the term 'regression line' and describe one method of obtaining such a line.
(b) A record of maintenance cost is kept on each of several nearly identical automatic machines. These data are to be compared with such machines' age to determine whether there is any association; and if there is, how strong the association is. Use the following data to decide, by finding a correlation coefficient, the answer to these points.

Maintenance Costs (£)	Age (yrs)
120	6
50	2
180	7
60	5
110	3
20	1
90	6

22.24 A company has found that the number of private cars with current licences can be used as a basis for estimating its production requirements. The relevant figures for the past nine years taken from the Annual Abstract of Statistics are as follows:

Year	In thousands
1	8,247
2	8,917
3	9,513
4	10,303
5	10,816
6	11,228
7	11,515
8	12,059
9	12,717

Using the method of least squares, you are required to:

(a) estimate the number of private cars with current licences in year 11.

(b) provide an estimate for year 11 but assuming that because of other factors, e.g. labour unrest, only 60% of the normal annual increase will be realised in year 11.

Answers should be submitted in millions correct to one decimal place of a million. I.C.M.A.

22.25 The rate of growth of national income per head of population and the percentage of national income taken by the government in taxation in a number of countries is shown in the table:

Country	Taxation as Percentage of National Income	Growth Rate per Head (%)
	x	y
United Kingdom	35.5	1.8
France	39.4	4.2
West Germany	37.5	3.5
Netherlands	40.0	4.3
Norway	41.0	3.0
Sweden	46.3	3.2
Switzerland	25.4	2.2
U.S.A.	28.3	2.8

Source: O.E.C.D. Statistics.

Plot the ponts on a scatter diagram and, using the 3-point method, draw the *line of regression* of y on x. Comment on your result.

Explain briefly how you would calculate the *line of regression* of x on y.

22.26 *General Knowledge Test of 100 Questions Given to 20 Students of Different Ages*

Student	Age Years	Months	Number of Questions Answered Correctly
A	16	8	39
B	16	2	40
C	17	9	45
D	17	1	47
E	18	6	46
F	18	7	67
G	19	10	45
H	20	4	53
I	19	8	49
J	20	7	46
K	20	10	54
L	20	8	52
M	22	0	57
N	22	2	53
O	21	11	61
P	21	5	56
Q	19	11	48
R	21	8	56
S	18	11	52
T	21	10	64

(a) What is meant by a regression line?
(b) Draw a scatter diagram to illustrate the above results.
(c) Sketch in a line of best fit.
(d) Estimate, using this line of best fit, the number of questions likely to be answered (correctly) by a 17-year-old student.

22.27 (a) Write a short note on regression lines.
(b) Explain the difference between 'the regression line of y on x' and 'the regression line of x on y'.
(c) Calculate the product moment correlation between the following sets of figures:

Intelligence Test Scores	Income at Age 23 (in £100's)
121	16
84	6
105	19
93	18
126	18
109	17
97	22
112	13
116	14
86	9
97	20
103	21

22.28 The following table gives the annual turnover and annual expenditure on advertising of fifteen mail order firms.

Annual Turnover (£100,000) (Y)	Annual Expenditure on Advertising (£1000) (X)
4.5	22
5.1	28
5.3	31
5.4	31
5.9	35
6.0	43
6.5	43
6.6	48
6.6	43
6.8	49
6.9	56
7.0	52
7.0	57
7.5	58
8.7	61

Plot these figures on a scatter diagram and calculate a regression line by use of the three-point method. Draw this regression line on the scatter diagram. Describe how this line can be used to estimate the likely turnover for a given expenditure on advertising. Comment on the accuracy of such an estimate.

22.29 Sales of groggets for each of the past five years have been as follows:

Year ended 31 May 1972	12,500
Year ended 31 May 1973	15,480
Year ended 31 May 1974	18,640
Year ended 31 May 1975	21,480
Year ended 31 May 1976	24,400

The seasonal pattern of sales is represented by the quarterly indices 70, 90, 160, 80. You are required to use the least-squares method to determine the trend line of the company's sales, project the trend line into the following year and calculate the expected sales for each quarter of that year.   A.I.A.

22.30 (a) From the following information draw a scatter diagram and by the method of least squares draw the regression line of best fit.

Volume of sales (in thousands of units)	5	6	7	8	9	10
Total expenses (£000)	74	77	82	86	92	95

(b) What will be the expected total expenses when the volume of sales is 7500 units?

(c) If the selling price per unit is £11, at what volume of sales will the total income from sales equal the total expenses?   I.C.M.A.

22.31 The following data has been collected over eight periods:

Period	Units of Output	Total Cost (£)
1	10,000	32,000
2	20,000	39,000
3	40,000	58,000
4	25,000	44,000
5	30,000	52,000
6	40,000	61,000
7	50,000	70,000
8	45,000	64,000

Draw a scatter diagram and by the method of least squares draw a straight line which best fits the data.
Give the equation of the line and estimate the cost likely to be incurred at the output levels of 26,000 units and 48,750 units. I.C.M.A.

22.32 A machine will run at different speeds but the higher the speed the sooner a certain part has to be replaced. Trial observation gives the following data:

Speed (revolutions per minute)	Life of Drill-head (hours)
18	162
20	154
20	171
21	165
23	128
26	138
26	140
28	129
31	125
32	106
32	97
40	95
41	103
42	109
43	69

Plot the figures on a scatter diagram. Determine the equation to the regression line. Plot the line on the scatter diagram and estimate the life of the drill-head if the machine operates at 30 revolutions per minute.
I.C.M.A.

22.33 *Personal Saving*

Quarter/Year	A (£ millions)	B (£ millions)	C (£ millions)	D (£ millions)
I	673	684	918	923
II	569	582	555	662
III	570	533	549	658
IV	391	412	477	647

Source: Economic Trends.
Fit a straight line trend to the above data, and find the seasonal components. Compare the actual figures for year D with those predicted using this trend and seasonal components approach. Consider the possible reasons for the divergences, paying particular attention to the quarter with the largest divergence. C.I.P.F.A.

22.34 *Annual Percentage Change in Manufacturing in a Number of Countries*

Country	Annual Percentage Change	
	Total Output	Output per Man-hour
Germany	10.0	5.6
Italy	8.9	6.7
Austria	6.8	5.3
France	6.5	5.9
Netherlands	6.1	4.6
Belgium	3.8	3.5
U.S.A.	3.7	3.2
Canada	3.5	3.0
U.K.	3.4	2.6
Denmark	3.3	2.3
Sweden	3.3	3.2
Ireland	2.8	2.1

Source: C. T. Saunders: *Journal of the Royal Statistical Society, A.*

Calculate the regression equation for total output on output per man-hour. Estimate the percentage change in total output corresponding to a 4% change in output per man-hour. How would you estimate the percentage change in output per man-hour corresponding to a 4% change in total output?   C.I.P.F.A.

22.35 The advertising expenditure and product sales of a company for eight consecutive months are as follows:

Month	1	2	3	4	5	6	7	8
Advertising expenditure (£000)	26	52	18	47	51	36	18	26
Sales (hundreds)	102	142	135	115	156	171	127	110

The company has developed a relationship predicting the sales in each month from the amount of advertising expenditure in that and the previous months as:

$$s_i = 85 + 0.6a_i + 0.9a_{i-1}$$

where s_i are the predicted sales in month i, in hundreds, and a_i is the advertising expenditure in month i, in thousands of pounds.

Required:
(1) Calculate the estimated sales in months 2 to 8 using the formula and comment on the results.
(2) What is:
  (i) the maximum discrepancy in any month between actual sales and expected sales?
  (ii) the maximum percentage discrepancy, when the discrepancy is expressed as a percentage of actual sales?
(3) What would be the effect on sales if advertising expenditure were reduced by £5000 per month?   A.C.A.

Chapter Twenty Three

Statistical Sources

There is little doubt that the second half of the twentieth century is the age when the data bank has come into its own. Data is collected on a large scale, not only by the government, but also by firms specialising in making market surveys and taking public opinion polls and by industrial firms and international organisations such as the United Nations or European Economic Community. This data is stored and made available to investigators through data retrieval systems. At one extreme we may be seeking data from a highly sophisticated computer system; at the other we will be painstakingly searching through the files in a filing cabinet.

It is small wonder that in the last half century the nature of the problem facing the statistical investigator has changed. As recently as 1920 the main problem was that data on many important problems was just not available, and the statistician had to find some means of organising his own survey to obtain primary data. Today the problem is often one of choice — of knowing where data is stored and selecting that which is most suited to immediate needs. There is still a great deal of primary data collected of course, particularly when we are concerned with marketing our product or with assessing public opinion which, at best, is fickle. Yet modern sophisticated methods of sampling still have to be based on secondary data. Even so simple a task as assessing which way the electorate will vote will require not only the register of electors, but also information on the age structure and the social structure of the areas to be sampled. The survey itself is in fact very often only the tip of the iceberg. Where then are we to look for the data which may be essential to us?

Before we start hunting through the mass of published information we must realise that much of the data we need may already be held in the records of our own firm or trade organisation. A survey into market trends could well begin with an analysis of salesmens' performance and their opinion of how the market will develop analysed by both product and area. We may not get all we require, but it will give use a good start. In the same way a great deal of useful information is held in wages books, petty cash records, ledgers, personnel records and even in the published final accounts of the firm. One of the greatest advantages of using such internal data is that it is still possible to ask questions of those who originally recorded the data, to clarify definitions, to discuss problems and possibly even to extend the data. It is just not possible to obtain this supplementary information when we are using data produced by an outside body.

At the same time we must remember that internal data is, by its very nature, narrowly based. It will relate to a single firm, or at best to a single industry. If we look at the sales analysis of a firm of seedsmen we will certainly get a very clear picture of sales trends for that firm. We may be able to extend the picture and argue that these trends are typical of the seed industry as a whole. What we cannot do is to pretend that the figures we have tell us anything about other industry or about British industry in general.

Only when we have exhausted such internal sources should we turn to published data. No matter how narrow or specific our enquiry, however, we will still have to turn to secondary data because, if we are to put our results into perspective, there is still a mass of background information that we need; and often this data can be obtained only from published sources. Unemployment within our own industry becomes more meaningful when we can put it into the context of the unemployment rate within the region or within the United Kingdom; the rate of inflation means little until we compare it with the rate of increase in salaries and the rate of inflation in other countries. Much of the data we need will already have been produced by independent survey bodies such as Gallup Poll, or by research units within firms or the universities, but by far the most important source of such secondary statistics is the Central Statistical Office, responsible for publishing data collected by the wide range of government departments. Whether you want to know the average July temperature at London Airport, the number of children in primary schools last year, or the Balance of Payments in 1972, you will find it tucked away among the government statistics published by the Central Statistical Office.[1]

The Annual Abstract of Statistics

In the United Kingdom possibly the best source of this official data is the Annual Abstract of Statistics published by the Central Statistical Office. It collates and summarises information relating to the United Kingdom produced by government departments (such as the Department of Education and Science) and obtained from national surveys (such as the census of distribution or the census of population). As you might expect it tends to give information for the United Kingdom as a whole rather than individual towns or industries. If we need more detail we may have to refer to far more comprehensive reports issued by the appropriate government departments. Population statistics, for example, comprise 29 tables in the Annual Abstract, but the Office of Population Census alone publishes comprehensive and detailed reports on each individual county, together with a Preliminary Report, and a general report. Additionally, the Registrar General's Annual Review and Quarterly Reviews supplement the Census reports.

---

1. Students are urged to obtain a copy of the Stationary Office publication — "Government Statistics — a brief guide to sources" published by and issued free by The Press and Information Services, Central Statistical Office, Great George Street, London SW1P 3AQ.

The Annual Abstract is precisely what its name implies — an abstract, but it is a good starting point for any investigation. Not only does its fifteen sections cover such diverse topics as weather, output, finance, education and transport, but in most cases information is given on a comparative basis for at least ten years so that any marked trend can be spotted easily. For more recent years most data is given monthly. Additionally, the Central Statistical office publish the Monthly Digest of Statistics, bringing the information in selected fields up to date month by month.

Population and Vital Statistics

Not every statistician will agree with us but we tend to think that of all the data produced by the government, that relating to population is the most useful. General knowledge of the population and its age-sex distribution is of course necessary for efficient administration, but think of what else is needed. We need age distribution to plan schools for the next generation, to assess the effect of changes in (say) family allowances, or to plan a policy for pensions; we need a knowledge of the causes of death to allocate funds for medical research; information on housing is the first step in planning a housing policy; questions on educational qualifications help the formulation of an educational policy — and so on. Information on all these topics, and many more, can be found in the reports on the Census of Population and the Annual Reports of the Registrar General. There is little doubt that the Census of Population is one of the most useful sources of data, and certainly, along with its necessary accompaniment — the compulsory registration of births, deaths and marriages — it is the most useful single source of information relating to vital statistics.

The census consists of a complete enumeration of the persons in a country on a given night together with questions on age, sex, marital status, occupation and the industry in which they work. In addition the modern census collects information on social conditions such as the housing of the population or the educational standards reached. In the United Kingdom the first census was taken in 1801 and apart from 1941, one has been taken every 10 years since that date. The government now have, in fact, the powers to take a census every five years if they so wish, and one such was taken in 1966.

Obviously, every census must ask certain questions which are never changed — such as name, sex, age etc., but the opportunity is usually taken to ask additional questions to throw light on social problems that rear their head from time to time. In 1921, for example, questions were asked relating to widows and orphans in view of the manpower losses between 1914 and 1918; in 1961 a 10% sample was asked to answer additional questions relating to educational attainment; and in 1966 questions were asked relating to the mobility of the population, on car ownership and garage space, and on the journey to work.

While such statistics are essential, in many developing countries a complete census is not possible, probably because of the distances involved, the lack of administrative machinery for collection and analysis of data, possibly because of the literacy rate, or the knowledge that the population would not willingly give the information required. In such cases, a limited census or even a sample survey may well be the only way to obtain the information about the population and its living conditions.

It would be foolish to pretend, however, that even in the United Kingdom the census information is completely accurate. There are sources of error at every stage. The distribution and collection of forms is in the hands of part-time enumerators who are each responsible for an area. We can never be absolutely sure that every house has received a form or that every form is returned. It is not likely, of course, but is at least possible that an enumerator may not know where a house is, and may not bother to find out, particularly if the weather is very bad, as it was on census day 1951. A much more probable source of error, however, lies in completing the questionnaire. Some people may misunderstand the question, others may only be able to guess the answer; others will believe that the answer they give is correct when, in fact, it is not. Educational standards in the U.K. vary tremendously and the quality of the completed questionnaires is bound to vary — some mistakes will inevitably occur. Admittedly some mistakes will be picked up by the enumerator on the doorstep; others will be spotted during the process of tabulation. But some will remain. How can any enumerator know that John Smith is not in fact married to the woman living in his house if he claims that he is married?

The census office itself is a fruitful source of error — such as errors arising in transferring information from the forms to punched cards or magnetic tape. That errors do exist has been shown by special post-census sample surveys designed to check accuracy, and by such techniques as comparing the death certificates of people who die shortly after the census with information given in their completed questionnaires. Yet errors are relatively few and the census office is able to claim a remarkably high degree of accuracy for most of the data obtained from the census.

Information of this nature is supplemented annually by the reports of the Registrar General in respect of births, deaths, marriages, divorces, etc., while figures for immigration and emigration can be brought up to date from Home Office statistics. In the U.K. such figures are highly reliable since there are legal pressures to ensure that registration takes place. It is not possible to bury a corpse without a death certificate, or to obtain a passport wihout a birth certificate, doctors and hospitals notify the Registrar of all births to ensure that parents ultimately register their children. Not all countries, however, have these same pressures and in some countries the births of female children are regularly not notified, while in others deaths are not registered. Sometimes of course the discrepency becomes obvious when the rate of male and female births (which stays remarkably stable) begins to change, or when the death rate differs markedly from a neighbouring country with very similar social conditions. Nor must we think that the information required is confined to the mere fact of the birth or marriage. The Population (Statistics) Acts require that informants also give such information as the age of parents, their occupation, the duration of the marriage and so on. So much information is available, in fact, that the analysis of the data has given rise to a whole new branch of statistics — demography, the analysis of vital statistics.

Economic Statistics

Since the time of the Domesday Book in the eleventh century governments have been interested in data on how the economy is progressing. In the very early days, it was, of course, merely a question of listing manpower for military purposes and personal wealth for taxation purposes. It is interesting to remember that in many underdeveloped parts of the world people are still averse to giving information for fear that it might be used to assess them for taxes. By the sixteenth century interest had developed in statistics of overseas trade since the philosophy of Mercantilism demanded a balance of payments surplus as a criterion of success. But the twentieth century is, after all, the century of economic planning, and for successful planning a mass of wide-ranging information is needed. Today government departments produce statistical data on almost every aspect of economic life — on prices and incomes, on employment and productivity, on production and distribution, on trade and finance. Once again you will find information on all these in the Annual Abstract, but let us take one or two of the major fields and look behind the statistical tables.

(a) Price Indexes

It is commonplace that whenever a body of workers ask for an increase in wages, one of the points that they make strongly is that the increase is designed to compensate for the rise in the level of prices, and they usually quote in evidence changes in the Index of Retail Prices. In Britain this index, introduced in 1947 to replace the old Cost of Living Index, is probably the best method of measuring the changes in the standard of living that we have. Constructed monthly and published in the Department of Employment Gazette it is used not only as a measure of inflationary trends, but also as the foundation of incomes policy.

If you think back to chapter twenty three you will recall that there are three main elements in the construction of a price index — the selection of commodities to be included, the determination of weight for each commodity or group of commodities, and obtaining the appropriate prices at the relevant point of time.

The present index originated in 1953 with the Household Expenditure Survey conducted by the then Ministry of Labour. This was an effort to discover the precise pattern of household expenditure in the United Kingdom. A sample of 20,000 households was selected and each person over 16 in the sample was asked to keep a detailed record of what he spent over a three-week period. The survey produced a 65% response, some 13,000 forms being returned. In the event it was found that low-income families, and particularly those living mainly on state pensions, had an expenditure pattern so different from the majority of households that they could not be regarded as typical and so were ignored. Thus the usable sample was cut down to 11,600 household budgets. It was felt that this sample reflected the spending habits of over 90% of the public and was sufficiently accurate to form the basis of a new index of retail prices. The weights to be used were determined by the percentage of income spent on individual commodities, and the survey also indicated the commodities that should be included in the index. Thus with a total weight of 1000, the weight for food was 350 and alcoholic drink 71, indicating that the average consumer spent 35% of his income on food and 7.1% of his income on alcoholic drink. On this basis a new index was introduced based on 17th January 1956 = 100. This index remained in existence for six years only, until January 1962.

Now as you know, if an index is to be really useful, the base should not be changed too frequently. Why, then, was this index dropped after only six years? It was soon realised that as wages and prices rose in the post-war world, people's consumption patterns changed fairly rapidly. Three cases in particular are outstanding: we were spending a far higher proportion of our income on such things as houses and motor-vehicles, but we were spending a much lower proportion on food. Hence it was decided that the weights used in the construction of the index should be changed year by year. This annual revision was based on a smaller Family Expenditure Survey covering a sample of 5000 households each year. These surveys are continuous throughout the year and each member of the household is required to keep a detailed account of his spending for a fortnight. It is not, of course, possible to use a single year's survey as a basis for revising the weights. Some items such as cars, carpets and washing machines are bought only infrequently, and this is likely to cause a large sampling error. So the weights are determined by the average of the previous three years' expenditure patterns disclosed by the survey. A study of the way the weights have changed over the years tells us a great deal about the way our expenditure patterns have changed.

Group	Weights used in construction of I.R.P.						
	1914	1952	1956	1962	1966	1970	1974
Food	60	399	350	319	298	255	253
Alcoholic drink		78	71	64	67	66	70
Tobacco		90	80	79	77	64	43
Housing	16	72	87	102	113	119	124
Fuel and light	8	66	55	62	64	61	52
Duradle household goods		62	66	64	57	60	64
Clothing and footwear	12	98	106	98	91	86	91
Transport and vehicles			68	92	116	126	135
Services		91	58	56	56	55	54
Meals consumed away from home						43	51
Miscellaneous		44	59	64	61	65	63
Others	4						
	100	1000	1000	1000	1000	1000	1000

You can see how over the years the percentage of income spent on food has fallen from 60% in 1914 to 25% in 1974. Since few people were starving in the U.K. in 1974 this is indicative of the rising standard of living. We can see also, signs of the revolution in power. Coal, the major fuel in 1914, absorbed 8% of income, whereas by 1974 coal and electricity absorb only 5.2%. The percentage of income spent on tobacco has also fallen rapidly, but most important of all in indicating the quality of life is the fact that if we take the four basic necessities of life — food, clothing, housing and fuel and light, we find that in 1914 90% of income was spent on these; by 1974 we spent only 52%. A study of these weights reveals a great deal about society.

Having solved the problem of obtaining a satisfactory weighting, and having made sure that the commodities covered by the index are those actually bought by the man in the street, we are still left with crucial problem of what prices to use. Everyone knows that even in a small town or village, prices vary widely from shop to shop. Hypermarkets and supermarkets sell goods at prices which smaller shops cannot possibly hope to match. Prices differ according to the brand of the commodity we buy and, of course, prices differ from area to area, reflecting partly the degree of competition in the area and partly the costs of distribution. Every motorist knows that even in his own area there may be as much as 6p difference in the price of a gallon of petrol. If he travels long distances he will also see that as he moves away from the vicinity of the oil refineries, the *general* level of prices tends to rise. The result is that while some people are paying x pence for a gallon of petrol, others may have to pay as much as ten pence more. Which of these prices are we to use when we are constructing a retail price index? The Department of Employment uses an interesting technique to try to ensure that the prices they use are representative.

On the Tuesday nearest to the 15th of the month prices are collected by visiting several shops selling the same type of good, care being taken that the types of shop visited are those that handle the bulk of household spending. This is done in several areas of different types ranging from London and the cities to small townships with a population under 5000. Thus, both different types of retail outlets and different geographical areas are represented.

Price relatives are now calculated for each item in each town. Let us take bread as an example. The price relatives for a particular city where five shop have been visited are 182.3, 181.9, 183.1, 182.5 and 182.4. These price relatives are combined with those for other cities to give a simple unweighted average. Thus we have a price relative for cities. The same analysis is now carried out for the other population groups, and the resultant price relatives averaged to give a price relative for bread for the country as a whole. In the same way we get price relatives for every commodity covered by the index.

So far, weighting has not mattered, but now we begin to combine the items to obtain an index for each of the major classes, and the question of weights becomes important. Although we know that in 1974 the class 'food' was given a weighting of 253, in order to obtain the index of food prices, this total weight has to be broken down into its constituent parts. We might, for example, apply weights of 29 to milk and cheese products, 31 to bread and cereals and so on. Once this has been done it is a simple matter to obtain an index for food, and having done the same for all other groups, to obtain an overall index of retail prices.

Can you see now why so many individuals cannot believe what the Retail Price Index tells us? As individuals we buy only a selection, and often only a small selection of the items included in the index. The items we buy may have gone up in price more than others that we do not buy, and it seems to us that the index understates the price rises. Or we may buy from a grocer whose prices have been steady when others are charging more. The index will then seem to overstate the rise in prices. Like many statistical measures, the Index of Retail Prices is a good and useful measure when applied to large numbers. It is not so useful when applied to the individual.

(b) Manpower, Production and Distribution

If you ask any economist for his 'short list' of aims for economic policy it is almost certain that among the items listed would be 'to promote economic growth'. At the risk of annoying our economist friends we will interpret this to mean simply that we are trying to increase the volume of output. Equally surely we would guess that another aim would be 'to eliminate unemployment', and a third to 'raise the standard of living'. So long as such things as these are accepted as aims of economic policy, it is little wonder that modern governments devote a great deal of time to the collection of statistics of manpower, production and distribution.

Manpower

It might seem strange, but in spite of the twentieth century being the age of manpower planning, it was not until 1945 that we had any reliable detailed information from the Ministry of Labour[2] about the size and distribution of the working population. Before 1939 there was highly reliable data relating to unemployment, but so far as the working population was concerned, all we had were estimates based on the Census of Population, the Census of Production and the Ministry of Labour's annual estimate. The war changed all that. It is impossible for a country to wage total war without detailed knowledge of the manpower available, and of industry's need for manpower. So the U.K. emerged from the war in 1945 with more information about manpower than ever before — but as you will appreciate it is not possible to rely on pre-war and post-war comparisons.

Just as the wartime planning necessitated the collection of manpower statistics, so post-war planning demanded, not only knowledge of the size of the working population, but also its age and sex structure, its occupational skills, etc. Collection of such statistics was facilitated in 1948 by the passing of the National Insurance Act and the introduction of the Standard Industrial Classifications. The former required every gainfully employed person to register and in theory we should have obtained a complete census of all those in employment. Unfortunately, there are some omissions resulting from the failure of some individuals to register —

2. Now the Department of Employment and Productivity.

particularly married women and the self-employed. The Standard Industrial Classification was designed to provide a standard pattern of classification by industry which could be used by all industrial statistics, and so facilitate comparison. Unfortunately from this point of view, there were major revisions in the classification in 1958 and 1968, and there have been many minor changes since. The effect of the major changes was to reclassify some industries as distributive and service industries which had previously been classed as manufacturing. You can see the effect this had in the following table, which classifies manpower according to two different Standard Industrial Classifications.

Change in the Composition of Employment (000's)
June 30th 1959

	1948	1958
Males		
Manufacturing	6,271	5,738
Distribution	1,578	1,689
Other	7,459	7,881
Females		
Manufacturing	2,898	2,739
Distribution	1,422	1,520
Other	3,569	3,630
Total	23,197	23,197

Source: Treasury Bulletin for Industry 1962.

The more recent changes have not been as drastic as this. They have involved largely the subdivision of industry into two or more component parts, as with the extracting in 1974 of 'Handtools and Implements' from the more general industry 'Cutlery Spoons Forks etc'.

	Thousand Employees		
	1972	1973	1974
Hand Tools and Implements	—	—	21
Cutlery Spoons Forks etc.	35	35	15

Unless you are very careful you can be left with the impression that there has been an unpheaval in certain industries at the time the classification was changed. However minor such changes may seem, however desirable they may be, they have one unfortunate effect — many of the series are not comparable over more than a few years.

One of the simplest ways of collecting data relating to employed persons was to use the quarterly exchange of insurance cards, since those who do hand in full insurance cards at the end of a quarter form a random sample of the working population. In fact, for many years, the annual figures published were based on cards handed in during the second quarter, supplemented by returns from employers having five or more workers. This gave a sample estimated to cover 75% of the working population. Now that the United Kingdom has abolished insurance cards, it seems that a major source, apart from the employers' annual return, will be found in the records of the Department of Health and Social Security or the Inland Revenue.

We would not wish to leave you with the impression that manpower statistics are concerned only with totals. Statistics are also produced relating to unemployment, unemployment rates, temporary lay-offs, days lost through strike action, average salaries and wages in the different industrial groups, holiday entitlement, etc. Moreover, most of the data is brought up to date monthly in the Department of Employment Gazette, and quarterly in the publications on Incomes, Prices, Employment and Production. Information today is far removed from the guesses and estimates of the 1930's.

Production

Without doubt one of the most important sources for data on industrial production is the Census of Production, the results of which are published by the Business Statistics Office. The full census is a massive document, but you will find a first-class summary of the relevant information in the Annual Abstract of Statistics. Although the expression 'census' is used, we must not imagine that it covers every firm in the United Kingdom. Firstly, there is a rather narrow definition of production. The census covers extractive, manufacturing and building industries. So important sectors of the economy such as agriculture and fisheries, distribution and the service industries are excluded. Secondly, even within this limited field, not every firm is included. The small firm is only required to give the nature of the trade and the size of its labour force. However, in some industries where a significant proportion of total output is produced by small firms, the full census requires such firms to make a simplified return.

In spite of its limited scope, the Census of Production provides us with a great deal of information. It shows us the way in which national industrial output is divided between different industries, and the study of this data over time brings out the trend and relative importance of individual industries. Central planning would be difficult without data like this, especially in respect of the distribution of labour and the allocation of capital resources. It is particularly important for the information it gives relating to changes in stocks of raw materials and finished goods, to capital expenditure on plant and machinery, and to building work undertaken. Without information such as this, national income statistics would be much less reliable, and the Index of Industrial Production subject to major errors.

Naturally, much of the information contained in the census report can be obtained elsewhere. The Department of Energy publishes much information on the coal industry and all forms of fuel production; The British Steel Corporation produces data on steel production; the building industry is covered by reports from the Department of the Environment. If you are interested in the details of a particular industry, say the tonnage of herrings landed at Hull, you would turn to the very comprehensive reports of the government departments, extracts from which appear in the Annual Abstract of Statistics. But the only place all this information is brought together is in the report on the Census of Production. Without this we would find it very difficult to estimate the total value of industrial production in the country.

In the United Kingdom, the first Census of Production was undertaken in 1907, and between the wars four more were taken. From 1948 it was intended to take a full census every year, but in fact the cost both to the state and to industry was heavy, and in the event a full census has been taken only every five years, while in the intervening years we have undertaken a sample survey, or a limited survey in which fewer questions are asked.

The census suffers from the normal defects of any effort to provide comprehensive and accurate information. It is expensive and it takes a considerable time to analyse and publish, and can only be undertaken at long intervals. To provide up-to-date information between censuses, the Department of Trade and Industry publish quarterly a series of sample statistics relating to different aspects of industry such as changes in stocks, capital expenditure and industrial buildings. The Department also sells statistical surveys of individual industries.

Distribution

Although Napoleon called England a nation of shopkeepers, it was not until 1951 that any official enquiry attempted to determine how many retail establishments there are in the United Kingdom, what kind of shops (multiples, independents) they were, what they sold or what was their turnover. The effort to obtain such information well illustrates the problems we face when we are attempting to take a census in a completely new field. Before any information could be obtained the government had to build up a sampling frame. This was done from May to October 1950, enumerators all over the country listing names and addresses of traders who fell within the scope of the census, 'as far as could be judged from the outside of the premises'. They were not allowed to enter the shop and talk to the owner or his employees, but were asked to distinguish between shops, stalls, depots and other kinds of premises. So thorough was this initial enumeration that the shops omitted must be negligible.

Having established a sampling frame the next step was to draft the questionnaire. This was the first time such a census had been undertaken, and a careful examination of objectives was necessary. It was decided that the census should provide information about

(a) the number and size of retail and wholesale outlets;
(b) the value of services provided by the distributive sector;
(c) the relative efficiency of distribution in different areas, i.e. the number of shops per head of the population.

Although the first census was a complete survey obtaining a 91% response rate to the questionnaire, a committee on the Census of Distribution and Production (the Vernon Smith Committee) recommended in 1954 that sample surveys be undertaken between the dates of the full census. It was decided to take a 12% sample and a most interesting thing about it was the way it was selected.

Firstly, every distributive outlet with a turnover in excess of £100,000 was to be included. This covered all the multiple stores, department stores and the majority of the co-operative societies. But once again there was no up-to-date sampling frame for the smaller independent retailers, so as in 1950 an enumeration had to be made in the relevant sample areas. These areas were chosen as follows:

(1) Certain areas existed in which it was believed that important changes had taken place since 1950, e.g. the new towns and central London. In these areas a complete enumeration of distributive outlets was undertaken and a 20% sample selected.

(2) In Greater London a sample of electoral wards stratified by size and distinguishing between mainly residential and mainly commercial was selected, and every shop in the selected wards was included in the survey.

(3) In the large towns (over 100,000 population) a sample of streets stratified by the number of shops there had been in 1950 was taken.

(4) Other towns were sampled by taking a cross-section of local authority areas stratified by population by sales in 1950, and by population changes since 1950.

(5) In rural districts a regional sample stratified by population density or population changes was selected.

The survey was undertaken by post with a very energetic follow-up. In the end the response rates were a remarkable 96% for larger traders, 89% for independent traders, and even 75% for street traders, hawkers, etc. It was estimated that the error was no more than 0.5% of the overall totals.

The 1961 Census of Distribution covered wholesale establishments which had been omitted from the sample survey, and to minimise cost only a 5% sample of establishments were given the full questionnaire. The remainder were asked only for details of employments and turnover together with some descriptive information. Thus we now obtain at five-year intervals details of the number, type and size of retail and wholesale firms, their turnover, stocks, capital expenditure, and hire purchase debt.

You may cast some doubt on the value of a census that takes nearly three years to publish its findings, and it is worth asking the reason for the delay. The first reason is the sheer magnitude of the operation — almost half a million retailers are involved, and it is no exaggeration to say that the returns take well over a year to come in. Even if a high proportion of firms are willing to co-operate the follow-up takes a considerable period of time. Time is absorbed too by a vast volume of correspondence from retailers asking questions about the completion of the forms.

Secondly, in the editing of the forms, many errors and omissions are found, and the correction of these errors involves both greater correspondence and delay. Finally, the analysis and printing of results, although computers are now used, is a tedious process.

Nevertheless, the importance of the analysis cannot be overestimated. Not only would National Income Statistics be unreliable without reasonably accurate figures of the contribution of the distributive trades to the Gross Domestic Product, but also the information is vital for the planning of, say, city shopping centres, or for the provision of shopping facilities in new towns. Moreover, the census returns form a new starting point for monthly and quarterly estimates which are vital in day-to-day planning.

National Income Statistics

All of the data that we have discussed in the last few paragraphs is still sectional data, no matter how widely the net is spread. It is concerned with only one aspect of the economy. What we must now do is turn our attention to the performance of the economy as a whole. This is seen only when we bring together the various parts of the economy in the statistics of National Income. We find these statistics in the book *National Income and Expenditure* published by Her Majesty's Stationery Office. This is not the place to go into details of what the expression 'National Income' means. However, we should note that all production is intended to satisfy consumer wants, and the more that is produced the greater can our consumption be. Thus what the National Income Statistics are measuring is basically the way our standard of living is changing.

Now any economist will tell you that National Income can be measured in three ways — by aggregating expenditure, by aggregating income, or by aggregating output. In theory, whichever method we use should given the same totals, but in practice there is always a difference which appears in the tables as 'residual error'. This has at times been as much as £700 million. Obviously we have not reached perfection!

To the extent that the data we have is accurate, overall planning is facilitated, but we must admit that despite a quarter of a century's experience in producing these figures, and despite many improvements that have taken place, several of the more important aggregates are little more than approximations. It is for this reason that each successive edition of *National Income and Expenditure* (called the 'Blue Book') amends the figures for the previous years.

There are differences in opinion as to what should be included in National Income calculations, and this makes international comparisons very difficult. For example in the peasant economies of the underdeveloped countries food grown by the farmers and eaten themselves is a very important part of the National Income and can be justifiably included. But in the United Kingdom such consumption is not included in the total of National Income. In fact in this country we include only such goods and services as are exchanged for money; and this cuts out a host of services which contribute to our welfare. If your wife or mother washes your shirt it does not affect the National Income statistics, but if you send it to a laundry the National Income is affected. Thus any change in social habits whereby people begin to pay for things they previously did for themselves will raise the National Income — even though the goods and services available have not increased.

One further point to be borne in mind is that the National Income is necessarily measured in money terms — and the value of money changes. Thus we may find that the National Income has risen by 10% since the last year, but if inflation is at the rate of 15% can we really say we are better off?

In spite of these criticisms, however, the annual Blue Book on National Income and Expenditure contains a mass of information vital to the planners in society and to the students of society.

Beginning with the National Income measured by the expenditure method, the Blue Book reconciles this figure with the National Income measured by the income of the various factors of production (wages, profits, interest and rent). This is far more than an academic exercise. The first table indicates the changing pattern of people's consumption. Here we can see for example, the changing expenditure on health, defence or education, the switch of expenditure to the service industries and the rise of the consumer durable industry. The second part of the table indicates the distribution of income in society; the proportion of income earned by the performance of work as compared with the proportion received from rents. Here we can trace the increasing share of the National Income going to the 'working class'. The third reconciliation indicates the Gross National Product by industry. Thus we can assess the contribution of each industrial group to the National Income. Other tables in the Blue Book take many of these points and in turn analyse them in much more detail.

This type of data is obtained from three main sources — the statistics assembled by the Inland Revenue, the Census of Production and Distribution, and the accounts of central and local government. But these are supplemented by a wide range of other sources. Unfortunately the coverage is far from complete, and in many cases the accuracy is impaired by the fact that the data used has been compiled for purposes other than National Income estimates.

The data we have discussed in this chapter is very important, but please remember that it is only a fraction of the data published each year. Much of this is a result of the day-to-day work of the government departments, some is an offspring of special surveys, and a great deal more is the result of the work of the statistical departments of banks, insurance companies, finance houses and the like. Overlooking and co-ordinating the work of the government departments is the Central Statistical Office producing the Annual Abstract of Statistics, the most important of all sources, and the starting point for a knowledge of what is available.

Exercises to Chapter 23

23.1 Outline the procedures which are employed when a Census of Population is undertaken in the United Kingdom, or in a country of your choice.

I.C.S.A.

23.2 Which statistics, published in the United Kingdom, are available relating to population? Mention the main publications in which they are to be found and write a summary of the details given in each.

23.3   Write an account of the contents of the Population and Vital Statistics section of the Monthly Digest of Statistics.

23.4   Describe where and in what form figures concerning population of the United Kingdom can be found. Say what use could be made of this information by an organisation of your own choice, e.g. an insurance company, a chain of supermarkets, a local authority or any other with which you are familiar.

23.5   Describe the principal official sources of data relating to (a) population and (b) employment.

23.6   (a) Explain the purpose and use of a Census of Distribution and describe briefly the information collected in the Census of Distribution (Great Britain) 1971.
(b) Describe the statistical information you would expect to find in the publication *Economic Trends*. A.C.A.

23.7   Describe the statistical information which is obtained from the Census of Production.

23.8   Write an essay on the construction and use of *two* official index numbers published by the government.

23.9   Explain how and by whom the General Index of Retail Prices is prepared, where it is published and what purpose it serves.

23.10 Either
(a) Outline the structure of The Family Expenditure Survey and explain its purpose. How reliable are its findings? Indicate some of the changes in the 1973 survey which affects the comparison with earlier years.
Or
(b) Describe The General Household Survey, commenting briefly on some of its findings. How useful is the survey to government?

23.11 (a) Describe the construction of the Official Retail Prices Index used in the United Kingdom, and
(b) explain how it could be used by:
  (i) businessmen,
  (ii) consumers, and
  (iii) trade unions. A.C.A.

23.12 What methods are available to indicate the general change in retail price levels in a community? Support your answer by referring to the construction of statistical indicators in the United Kingdom or a country of your choice. I.C.S.A.

23.13 Describe the Department of Employment Index of Retail Prices. (Your answer should show knowledge of the purpose of the index, the groups of items included, the method of collecting prices, the method of determining weights and the calculation of the index number itself.) Discuss the importance of such an index.

23.14 Discuss the statistical material available on National Income and Expenditure in the United Kingdom.

23.15 What are the main contents of the monthly employment statistics prepared by the Department of Employment and Productivity?

23.16 As a trade union official, you have been given the task of submitting a well-documented wage claim on behalf of your members.
    (a) What kind of problems would you expect to encounter in making your case?
    (b) What sources of economic statistics would you consult?
    (c) What kind of information would you seek from those sources to justify your claim?

23.17 (a) Write a summary of the contents of either the Department of Employment Gazette or the Monthly Digest.
    (b) Name two tables that regularly appear in the publication you have selected in (a) and briefly summarise the details given in them.

23.18 Describe TWO of the main summary sources of statistics published regularly by government departments.

23.19 (a) Give details of the official statistical publications that should be found in a statistics department of a large business.
    (b) Describe in detail the construction and use of one of the main indicators found in one of these publications.

23.20 Write briefly on four of the following sources of economic statistics. You should comment on the breadth and detail of the information in the publications you choose.
    (a) Economic Trends;
    (b) Annual Abstract of Statistics;
    (c) The Family Expenditure Survey;
    (d) The Monthly Digest of Statistics;
    (e) National Income and Expenditure (the *Blue* Book);
    (f) The Department of Employment Gazette.

23.21 Describe the statistical information provided in THREE of the following official publications:
    (i) Annual Abstract of Statistics;
    (ii) Business Monitors;
    (iii) Department of Employment Gazette;
    (iv) Family Expenditure Survey Report.   A.C.A.

23.22 (a) If your employer wished to gain information on the following, suggest which publications would provide this:
    (i) Wages rates and average earnings.
    (ii) Retail and wholesale prices.
    (iii) Industrial production.
    (iv) Regional and national unemployment.
    (v) Value and volume of monthly and annual imports and exports of goods.
    (b) Give examples of the purpose for which he may wish to use this information.

23.23 Describe the construction and use of two of the following Economic and Business Indicators:
(i) General Index of Retail Prices.
(ii) Index of Industrial Production.
(iii) Index Numbers of Wholesale Prices.
(iv) Unemployment and Employment Statistics.
(v) National Income Statistics.

23.24 (a) What information is available in the Monthly Digest of Statistics concerning any *three* of the following:
(i) Population and vital statistics.
(ii) External trade.
(iii) Prices.
(iv) Social Services?
(b) Explain briefly the purpose and the derivation of the Index of Industrial Production.

23.25 (a) State *two* main official publications which give statistics for the United Kingdom relative to each of the following:
(i) Index of retail prices.
(ii) Short time worked.
(iii) Sales of vehicles.
(iv) Imports.
(v) Personal incomes.
(vi) Temperatures, rainfall and sunshine.
(b) State *two* main montly publications for which the compilers are:
(i) Central Statistical Office;
(ii) Department of Employment.
Describe briefly the contents of any *one* of these publications.

23.26 State with reasons FOUR important economic indicators which are used in charting the economy.

Answers to Questions

CHAPTER 1

1.12 Taking classes as 650-659, 660-669, etc.
Frequencies are 2 5 6 14 26 18 13 10 3 3 = 100

1.14 Frequencies are
0 5 13 7 9 27 19 19 10 7 4 = 120

1.16 Suggested classes 10-19, 20-29 etc.
Frequencies are 2 6 10 10 14 10 6 6 2 = 66

CHAPTER 2

2.4 Taking classes 110-119, 120-129 etc.
Frequencies are 1 3 7 14 8 5 2 = 40

2.6 Taking classes as 10 but under 15, 15 but under 20 etc.
Frequencies are 3 4 4 3 3 2 3 = 22

CHAPTER 4

4.2 (b) 18.588 ± 0.056, i.e. ± 0.3013%
(c) 341 ± 9, i.e. ± 2.64%.

4.4 Receipts = £4,900,000 ± 87,500
Cost = £4,350,000 ± 55,000
So Profit = £550,000 ± 142,500

4.6 (a) Deposits = 10,969 ± 40.7 (million $)
(b) Cash = 1,360 ± 10.7 (million $)

4.8 (a) (i) 5720 ± 0.96%, (ii) 5480 ± 1.004%
(b) (i) 0.865, (ii) 395

4.10 (b) (i) ± 43,000 tones
(ii) Assuming series discrete, max. error = + 85,914 tons, but if continuous, max. error is 86,000 tons
(c) (i) 135.92 ± 2.77 tons
(ii) 29,250lb ± 11.43%

4.12 (a) £176,000 ± £32,802.5
(b) ± 18.64%.

4.14 (i) EFTA, £1276.6m, EEC £1753.9m
(ii) EFTA, £1270m, EEC £1750m
(iii) 0.52%

4.16 (b) (i) Output = 2000 ± 65.5
Unit Profit = $£1 \pm 0.1$
Weekly Profit = $£2000 \pm 272.05$
(ii) 50 ± 6.5

CHAPTER 5

5.2 211

5.4 Median = 2, mean = 2.48

5.6 Assume final group is '£30 and under £40'
If assumed mean is 17.5, and if $c = 5$,
$\Sigma f = 376$, $\Sigma fd = -224.5$ } hundreds
$\bar{x} = £1451.50$

5.8 Assuming employment starts at 18, and retirement age is 65, $\bar{x} = 43.61$ years

5.10 Assume final group is 70-79
If assumed mean is 25, and if $c = 10$, $\Sigma f = 756$, $\Sigma fd = 452$.
$\bar{x} = 30.98$ years

5.12 Frequencies are: 7, 16, 12, 11, 10, 4, $\Sigma f = 60$
If assumed mean is 549.5, and if $c = 20$, then $\Sigma fd = 13$.
$\bar{x}$ 553.83 miles

5.14 Factory A: 27.23 years, Factory B: 37.09 years

5.16 £24 – £36: 5,085,000
£32 – £48: 4,860,000
£24 – £48: 8,235,000

5.18 Median = 84p
Mode = 76.9p
Taking an assumed mean of 75, and $c = 10$, $\Sigma f = 395$, $\Sigma fd = 453$. $\bar{x} = 86.47$p

5.20 Frequencies are: 3, 13, 23, 25, 22, 9, 4, 1, $\Sigma f = 100$
Mode = 17 years 8½ months

5.22 Taking an assumed mean of 6.5, $\Sigma fd = 212$. $\bar{x} = 7.26$ hours
Median = 7.41 hours

5.24 Taking an assumed mean of 250, and $c = 100$, $\Sigma f = 460$, $\Sigma fd = -6$. $\bar{x} = £248.7$
Total value of orders = £114,402
Mode = £206.48

5.26 Mean = £1162.95   5.28 54,583,214
      Mode = £859.14
      Median = £1002.47

CHAPTER 6

6.4 Trend 1347.5, 1382.5, 1440, 1532.5, 1608.75, 1658.75, 1715.5, 1800.0, 1878.75, 1926.25, 1997.5.

6.6 Seven Day moving average:
1137.1, 1145.7, 1148.6, 1148.6, 1148.6, 1150, 1148.6, 1155.7, 1168.6, 1174.3, 1172.9, 1168.6, 1171.4, 1172.9, 1181.4, 1202.9, 1207.1, 1205.7.

6.8 Trend
2.55, 2.725, 2.95, 3.21, 3.475, 3.64, 3.675, 3.61, 3.41, 3.16, 2.9, 2.61, 2.425, 2.36.

CHAPTER 7

7.2 Trend: 14.875, 16.0, 16.75, 16.875, 17.5, 18.625, 19.375, 19.5, 19.625, 20.75, 22.5, 23.625.
Actual/Trend: 1.0756, 0.5, 0.8358, 1.7185, 0.9142, 0.4832, 0.929, 1.7435, 0.8662, 0.4337, 0.8444, 1.7777.

7.4 Trend 350.75, 350.625, 347.75, 341.625, 334.125, 325.875, 313.125, 299.125, 290.125.
Deviations -0.75, 44.375, -14.75, -21.625, 1.875, 34.125, -5.125, -20.125, -15.125.
Seasonal Variation: -11, -22, $-5\frac{1}{2}$, $+38\frac{1}{2}$.

7.6 Trend 505.625, 515.875, 515.25, 510, 507.625, 507.625, 514.625, 523.875.
Deviations -39.625, 10.125, 25.75, 31, -67.625, 2.375, 23.375, 20.125
Seasonal Variation $+24$, $+25$, -54.5, $+5.5$
Deseasonalised Series 456, 495, 520.5, 520.5, 517, 516, 494.5, 504.5, 514, 519, 547.5, 525.5.

7.8 Trend 180.125, 183.75, 205.125, 238, 251.625, 261.875, 259.25, 252.25, 254.125, 249, 245.625, 245.625
Deviations -46.125, 29.25, -18.125, -40, 59.375, 37.125, -49.25, 4.75, -23.125, 74, -44.625, -20.625
Seasonal Variation -34, -16, 0, $+50$.

7.10 Trend 13.125, 13.5, 13.875, 14.375, 15.125, 15.75, 16.375, 16.875
Actual/Trend 1.219, 0.6666, 1.009, 1.113, 1.239, 0.7619, 1.0381, 1.0666
Seasonal Variation 1.09, 1.1718, 0.7144, 1.0238
Deasonalised Data 13, 13, 14, 13, 14, 15, 14.5, 17, 17, 17, 17, 18

7.13 With $\alpha = 0.5$, forecasts are 13, 13.5, 14.75, 11.88, 12.94, 14.47, 15.73, 13.87, 15.43, 16.71, 18.35, 15.68.

399

CHAPTER 8

8.6 Simple index = 200. Weighted Index = 194.4

8.8
Materials:	100	98.8	99.0	97.8	98.6	98.2
	98.8	101.4	101.5	103.1	106.5	112.6
Output :	100	99.6	99.8	99.3	99.7	100.3
	100.9	102.9	103.3	104.1	104.8	105.6
Food :	100	100.7	101.3	100.4	101.5	103.2
	103.2	105.1	105.2	105.4	106.3	107.9

8.10 $x = 121$ Index = 166.9

8.12 Price index 1964 = $\dfrac{5623.198}{60.42} = 93.1$

 1965 = $\dfrac{5914.008}{60.42} = 97.9$.

8.14 Base weighting (Laspeyre's) = $\dfrac{9556}{7274} \times 100 = 131.4$

 Current weighting (Paasche's) = $\dfrac{9700}{7318} \times 100 = 132.5$

8.16
Group 1	100	106	113	122	128
Group 2	100	112	119	129	136
Group 3	100	102	106	107	110

8.18
	Laspeyre's	Paasche's	Fisher's
Price Index	106.6	108.2	107.4
Volume Index	106.9	108.5	107.7

8.20 $\Sigma p_0 q_0 = 833$ $\Sigma p_0 q_1 = 805$ $\Sigma p_0 q_2 = 797$ $\Sigma p_0 q_3 = 865$
Index of production 1967 = 96.6, 1968 = 95.7 1969 = 103.8

8.22 $\Sigma wi = 122{,}927.5$
(b) Index of Production = 122.9
(c) Rises to 125.6
(d) 81.

8.24 1955 value = $\Sigma p_0 q_0 = 6{,}800{,}650$
$\Sigma p_0 q_n = 5{,}792{,}619$
Volume Index (Laspeyres) = $\dfrac{5792619}{6800650} \times 100 = 85.2$

CHAPTER 9

9.2 Range = 67; $\bar{x}$ = 48; Median = 48; Q_1 = 36; Q_3 = 63;
Quartile Deviation = 27; Mean Deviation = 14.08.

9.4 Median = £79.5

Under £88	= 45
£63 but under £75	= 14
Over £96	= 11

Quartile Deviation = £23

9.6 Q_1 = £4093.20 Q_2 = £7749.15 Q_3 = £11,001
Q.D. = £6970.80

9.8

	Median	S.Q.R.
Week 1	22.5 cwt	9.5 cwt
Week 2	25 cwt	7.5 cwt

9.10 $\bar{x}$ = £91.28 σ = £28.72
using mid points £29.95, £49.95 etc.

9.12 $\bar{x}$ = 136.45p σ = 33.9p

9.14 σ = £1065.

9.16 Taking upper limit as 110
$\bar{x}$ = 62.72 thousand σ = 13.52 thousand
Reasonable rule might be to keep constituencies within the range $\bar{x} \pm \sigma$

9.18 $\bar{x}$ = £260 σ = £23.9 C.V. = 9.19%.

9.20 Taking upper limit as 80 years

	$\bar{x}$	σ	c.v.
Males	32.42	21.6	66.6%
Females	35.77	23.14	64.7%

9.22 Taking lower limit as 11 and upper limit as 80
Range = 69; $\bar{x}$ = 45.76; σ = 13.48; c.v. = 29.46%

9.24 $\bar{x}$ = £1200 σ = £390.73 Median = £1112.22
Coefficient of Skewness = 0.674.

CHAPTER 10

10.2 (i) 4.75% (ii) 99.242%

10.4 a) 3,085 b) 668 c) 6915 d) 8413

10.6 16.1372 ozs

10.8 0.023, 0.006, 121 gms

10.10 NB this distribution is discrete, so if we wish to find percentage less than 229 we require area to left of 228.5. Likewise, percentage greater than 280 given by area to right of 280.5.
i) 22.66% ii) 10.65%.

10.12　a) 294　　b) 6　　c) 32

10.14　Taking mid point of first group to be 35, and of last group to be 9, $\bar{x}$ = 23.74, σ = 6.127
Taking range as $\bar{x} \pm 30$, range = 5.36 to 42.12%.

10.16　1) 30.8% 2) 53.3% 3) 7.68 weeks 4) 15.9%

10.18　a) 11.51% b) 57.62% c) 13.27% d) 2.28% e) 95.44%

CHAPTER 11

11.2　$4/11$

11.4　b) i) 0.6591 ii) 0.2866
　　　c) i) 0.08 ii) 0.52 iii) 0.2

11.6　i) $1/3$ ii) $1/36$ iii) $91/216$ iv a) $7/15$ b) $1/15$

11.8　a) $1/12$ b) $1/2$ c) $1/4$ d) $1/2$

11.10　a) 0.03 b) 0.33

11.12　a) 0.8938 b) 0.1025

11.14　i) 0.5838 ii) 0.05

11.16　0.52　0.6425

11.18　i) $5/12, 1/4, 1/3$
　　　　ii) $11/80$
　　　　iii) $5/24$
　　　　iv) $2/5$
　　　　v) $9/10$
　　　　vi) $20/33, 8/33, 5/33$

11.20　b) 0.6513, 44

11.22　i) 0.0165 ii) 0.0179

11.24　a) $64/132600$ b) $16/5525$

11.26　P (article has both defects) = 0.03.

CHAPTER 12

12.2　EMV(A) = £5,000 EMV(B) = £32,500, EMV(C) = £85,000 so buy C.

12.4　EMV (with forecast) = £124,943. Value of information = £39,943, so buy information.

12.6　Maximax: stock 16. Maximin: stock 10. minimax: stock 12.

12.8　Newagent never makes a loss, decision is repeatable, so use EMV.

12.10　If p = probability that expenditure is increased,
$$400{,}000P + 100{,}000(1-p) - 50{,}000 > 129{,}687$$
$$p > 0.2656.$$

12.12　Maximax: buy Allied
　　　　Maximin: buy Guided, if system replaced do not expand

12.14　£0.3672m.

CHAPTER 13

13.2　$100\left(\left(\frac{1}{2}\right)^{10} + {}^{10}C_1\left(\frac{1}{2}\right)^9\left(\frac{1}{2}\right) + {}^{10}C_2\left(\frac{1}{2}\right)^8\left(\frac{1}{2}\right)^2 + {}^{10}C_3\left(\frac{1}{2}\right)^7\left(\frac{1}{2}\right)^3\right) = 17$

13.4　0.8208. $\bar{x}$ = 0.26 p = 0.052

13.6   a) (i) 0.0894 (ii) 0.0338 b) (i) 0.0902 b) (ii) 0.0404

13.8   a) 0.1336 b) i) 0.0746 (ii) 0.0758

13.10 0.3526

13.12 0.0144
$$e^{-2.5}\left(1 + 2.5 + \frac{(2.5)^2}{2!} + \frac{(2.5)^3}{3!} + \ldots\right)$$

13.14 88%.

13.16
No. of defectives	0	1	2	3	4	5
No. of samples	18.3	31	26.3	14.8	6.3	2.1

13.18 $\bar{x} = 25$, $\sigma = 4.937$, Proportion = 0.01659

13.20 i) 6.059% ii) 0.81% iii) 87.95%.

CHAPTER 16

16.2   a) 25.14% b) 37.07% c) 53.28%

16.4   90% limits either $100 + 3.84$ or $100 - 3.84$
       95% limits either $100 + 4.935$ or $100 - 4.935$
       99% limits either $100 + 6.99$ or $100 - 6.99$

16.6   95% limits: 2 ± 0.00392
       99% limits: 2 ± 0.005152
       Sample mean (2.005) met infrequently, so machine probably needs resetting.

16.8   4.55% (2.275% have too large a clearance and 2.275% will not fit).

16.10 95% limits for sample means = 3 ± 0.707
       99% limits for sample means = 3 ± 0.929
       Limits calculated on assumption that men and women pair randomly. Actual mean difference outside 99% limits – this casts doubts on assumption.

16.12 95% limits = 0.07 ± 0.0224, 99% limits = 0.07 ± 0.0294.

16.14 95% limits = 0.3 ± 0.02842, sample proportion well within limits, so proportion buying Blotto's Gin could well be 0.3.

CHAPTER 17

17.1   Warning Limits = 25.1 ± 0.396
       Action Limits = 25.1 ± 0.624
       Minimum tolerances ± 1.395
       A.W.C.L. = 23.771 to 26.229.

17.5
	An	R	W	A	C
	0.4299	1.328	0.377	0.594	0.734

17.6 Warning Limits = 25.02 ± 0.0226
Action Limits = 25.02 ± 0.356
Firm cannot meet specifications.

17.8 Action Limits = 17.748 to 27.252
A.W.C.L. = 13.872 to 26.128
Reset machine to 20.

17.10 0.505.

17.12 Range for washers = ± 0.00027, so accept order.

17.14 $\bar{\bar{x}} = 755.2$, $\bar{R} = 17.2$, min tolerance = ± 22.84
Warning Limits = 755.2 ± 6.48 Action Limits = 755.2 ± 10.22
A.W.C.L. = 737.62 to 762.38
Samples, 2, 5, 11, 12, 20 outside warning limits

17.16 $\bar{x} = 136$, σ 7.563
Warning Limits = 136 ± 14.82 Action Limits = 136 ± 23.37

17.18 Proportion rejected = 0.042, standard error = 0.0142
Warning Limits = 2.8 to 14
Action Limits = 0 to 17.18
It appears that proportion rejected has increased.

17.20 Warning Limits = 1.2 to 10.8
Action Limits = 0 to 13.5.

CHAPTER 18

18.2 £18.50 ± £1.176

18.4 $\bar{x} = -£2.80$, $\sigma = £10.45$
95% Interval estimate = $-£2.80 \pm 2.0482$
Least favourable = $-£48,482$; most likely = $-£28,000$
Most favourable = $-£7518$.

18.6 i) 99.8% interval estimate = £218 ± 6.57 ii) £245.10
N.B. answers assume finite population adjustment made, as the estimate refers to the 365 days in the current year.

18.8 95% interval estimate = 0.55 ± 0.0976. n > 537
N.B. second part of question is one-sided.

18.10 ii) 0.395 iii) 0.395 ± 0.0677, 0.395 ± 0.08905
iv) 2,295 v) taking ± standard deviations as covering entire normal distribution, n = 7320.

18.12 $\hat{\mu} = 19.7$ $\hat{\sigma} = 0.46$

18.14 25

18.16   17 and 26.

18.18   n ≥ 168

18.20   n ≥ 55.

CHAPTER 20

20.2   (i) Reject if number of defectives greater than 30
       (ii) Reject if number of defectives less than 9 or greater than 31.

20.4   For two sided test, reject if sample means outside range 4.99804 to 5.00196cm. For one sided test, reject if sample mean greater than 5.001645.

20.6   Test statistic t = 1.41. Sample does not substantiate claim.

20.8   Test statistic of Z = 2.773. This suggest choice of marriage partner is affected by height.

20.10 Test statistic of Z = 2.5. Difference is significant at 1% level.

20.12
	Mean	Variance
1962	£658.09	£221,020.9
1963	£692.31	£241,120.7

This question could be approached in two ways
1) That the data as forming two samples from a population of incomes earned over a long time period (past and future)
   test statistic Z = 0.263. Differences not significant.
2) Treat the data as forming two populations, in which case significance test would not be appropriate. However, using the methods of chapter 4 we could examine errors involved and conclude that
   Maximum value for mean income in 1962 = £672.28
   Minimum value for mean income in 1963 = £678.05
so we can conclude with certainty that mean income has increased.

20.14 Test statistic Z = 3.12, difference significant at 1% level.

20.16
	Mean	Variance
Private	200.26	271.3
L.A.	158.54	871.04

1) If data treated as two samples, test statistic t = 3.9, difference significant at 1% level.
2) If data considered as two populations
   Minimum mean for private = 200,214
   Maximum mean for L.A.   = 158,586
   Average for private sector is greater.

20.18 Test statistic t = 2.85, difference significant at 5% but not at 1% level.

20.20 Assuming that the samples are paired, test statistic t = 2.004. Differences not significant.

20.22 For samples of 64, test statistic Z = 3. Significant at 1% level.
For samples of 16, test statistics t = 1.452. Not significant.

20.24 Test statistic Z = 2.125 Significant at 5% but not at 1% level.

20.26 Test statistic Z = 1.543. Difference not significant.

20.28 Test statistic Z = 2.595. Difference significant at 1% level.

20.30 Test statistic Z = 3.315. Significant at 1% level.

CHAPTER 21

21.2 $\chi^2 = 7.884$, $v = 5$, not significant

21.4 $\bar{x} = 0.9$. Expected values: 81.3, 73.2, 32.9, 12.6
$v = 2$, $\chi^2 = 4.292$, not significant

21.6 Expected values: 32, 16, 8, 8
$v = 3$, $\chi^2 = 10.188$
Significant at 5% but not at 1% level

21.8 $v = 2$, $\chi^2 = 4.751$, no association

21.10 $v = 2$, $\chi^2 = 1.502$, no association

21.12 Insufficient information given — need to know the number who 'changed their minds'

21.14 $v = 6$, $\chi^2 = 14.741$, association established at 5% level

21.16 $v = 2$, $\chi^2 = 3.379$, no difference in proportion

21.18 $v = 1$, $\chi^2 = 125.4$, most highly significant
Using difference between proportions, $Z = 11.2$

CHAPTER 22

22.4 $\Sigma x = 35$, $\Sigma y = 50$, $\Sigma x^2 = 225$, $\Sigma y^2 = 516$, $\Sigma xy = 360$, $r = 0.79$

22.8 Taking $x = (x - 40)$ $y = (y - 3.5)$
$\Sigma x = -2$, $\Sigma y = 0.6$, $\Sigma x^2 = 182$, $\Sigma y^2 = 0.28$, $\Sigma xy = -6.2$, $r = -.93$

22.10 $\Sigma x = 18.9$, $\Sigma y = 35.2$, $\Sigma x^2 = 37.01$, $\Sigma y^2 = 132.22$, $\Sigma xy = 64.7$, $r = -0.56$
Including additional years
$\Sigma x = 29.1$, $\Sigma y = 62.1$, $\Sigma x^2 = 63.05$, $\Sigma y^2 = 330.25$, $\Sigma xy = 133.85$, $r = 0.4026$.

22.12 $\Sigma d^2 = 374$, $p = -0.7$

22.14 $\Sigma x = 39$, $\Sigma y = 478$, $\Sigma x^2 = 245$, $\Sigma y^2 = 31812$, $\Sigma xy = 2764$, $r = 0.986$

22.16 $\Sigma d^2 = 8$, $p = 0.857$

22.18 $\Sigma d^2 = 12.5$, $p = 0.85$

20.20 $\Sigma d^2 = 76$, $p = 0.5394$

22.22 $\Sigma d^2 = 34.5$, $p = 0.7125$

22.24 $\Sigma x = 45$, $\Sigma y = 95{,}315$, $\Sigma x^2 = 285$, $\Sigma xy = 508{,}810$.
$y = 7904.3 + 537.25x$
(a) 13.8 (mil), (b) 13.6 (mil)

22.30 $\Sigma x = 45$, $\Sigma y = 506$, $\Sigma x^2 = 355$, $\Sigma xy = 3872$, $y\ 4.4x + 51.33$ (b) 84.33, (c) 7.78

22.32 $x = x - 20$, $y = y - 100$
$\Sigma x = 143$, $\Sigma y = 391$, $\Sigma x^2 = 2413$, $\Sigma xy = 484$, $y = 217 - 3.09x$.
When $x = 30$, $y = 124.3$

22.34 $\Sigma x = 48$, $\Sigma y = 62.1$, $\Sigma x^2 = 218.5$, $\Sigma xy = 286.59$.
$y = 1.44x - 0.59$. When $x = 4$, $y = 5.17\%$.

22.36   (1) Sales: 139.6, 142.6, 129.4, 157.9, 152.5, 128.2, 116.8 (hundreds)
(2) 1850, 10.82%
(3) decline of 75,000.

Tables

1. Common Logarithms
2. Normal Distribution
3. Values of e^{-x}
4. Student's t Critical Points
5. χ^2 Critical Values

COMMON LOGARITHMS $\log_{10} x$

x	0	1	2	3	4	5	6	7	8	9	Δ_m +	1 2 3	4 5 6 ADD	7 8 9
10	·0000	0043	0086	0128	0170	0212					42	4 8 13	17 21 25	29 34 38
						0212	0253	0294	0334	0374	40	4 8 12	16 20 24	28 32 36
11	·0414	0453	0492	0531	0569	0607					39	4 8 12	16 19 23	27 31 35
						0607	0645	0682	0719	0755	37	4 7 11	15 19 22	26 30 33
12	·0792	0828	0864	0899	0934	0969					35	4 7 11	14 18 21	25 28 32
						0969	1004	1038	1072	1106	34	3 7 10	14 17 20	24 27 31
13	·1139	1173	1206	1239	1271	1303					33	3 7 10	13 16 20	23 26 30
						1303	1335	1367	1399	1430	32	3 6 10	13 16 19	22 26 29
14	·1461	1492	1523	1553	1584	1614	1644	1673	1703	1732	30	3 6 9	12 15 18	21 24 27
15	·1761	1790	1818	1847	1875	1903	1931	1959	1987	2014	28	3 6 8	11 14 17	20 22 25
16	·2041	2068	2095	2122	2148	2175	2201	2227	2253	2279	26	3 5 8	10 13 16	18 21 23
17	·2304	2330	2355	2380	2405	2430	2455	2480	2504	2529	25	2 5 7	10 12 15	17 20 22
18	·2553	2577	2601	2625	2648	2672	2695	2718	2742	2765	24	2 5 7	10 12 14	17 19 22
19	·2788	2810	2833	2856	2878	2900	2923	2945	2967	2989	22	2 4 7	9 11 13	15 18 20
20	·3010	3032	3054	3075	3096	3118	3139	3160	3181	3201	21	2 4 6	8 11 13	15 17 19
21	·3222	3243	3263	3284	3304	3324	3345	3365	3385	3404	20	2 4 6	8 10 12	14 16 18
22	·3424	3444	3464	3483	3502	3522	3541	3560	3579	3598	19	2 4 6	8 10 11	13 15 17
23	·3617	3636	3655	3674	3692	3711	3729	3747	3766	3784	18	2 4 5	7 9 11	13 14 16
24	·3802	3820	3838	3856	3874	3892	3909	3927	3945	3962	18	2 4 5	7 9 11	13 14 16
25	·3979	3997	4014	4031	4048	4065	4082	4099	4116	4133	17	2 3 5	7 9 10	12 14 15
26	·4150	4166	4183	4200	4216	4232	4249	4265	4281	4298	16	2 3 5	6 8 10	11 13 14
27	·4314	4330	4346	4362	4378	4393	4409	4425	4440	4456	16	2 3 5	6 8 10	11 13 14
28	·4472	4487	4502	4518	4533	4548	4564	4579	4594	4609	15	2 3 5	6 8 9	11 12 14
29	·4624	4639	4654	4669	4683	4698	4713	4728	4742	4757	15	1 3 4	6 7 9	10 12 13
30	·4771	4786	4800	4814	4829	4843	4857	4871	4886	4900	14	1 3 4	6 7 8	10 11 13
31	·4914	4928	4942	4955	4969	4983	4997	5011	5024	5038	14	1 3 4	6 7 8	10 11 13
32	·5051	5065	5079	5092	5105	5119	5132	5145	5159	5172	13	1 3 4	5 7 8	9 10 12
33	·5185	5198	5211	5224	5237	5250	5263	5276	5289	5302	13	1 3 4	5 6 8	9 10 12
34	·5315	5328	5340	5353	5366	5378	5391	5403	5416	5428	13	1 3 4	5 6 8	9 10 12
35	·5441	5453	5465	5478	5490	5502	5514	5527	5539	5551	12	1 2 4	5 6 7	8 10 11
36	·5563	5575	5587	5599	5611	5623	5635	5647	5658	5670	12	1 2 4	5 6 7	8 10 11
37	·5682	5694	5705	5717	5729	5740	5752	5763	5775	5786	12	1 2 4	5 6 7	8 10 11
38	·5798	5809	5821	5832	5843	5855	5866	5877	5888	5899	11	1 2 3	4 6 7	8 9 10
39	·5911	5922	5933	5944	5955	5966	5977	5988	5999	6010	11	1 2 3	4 6 7	8 9 10
40	·6021	6031	6042	6053	6064	6075	6085	6096	6107	6117	11	1 2 3	4 5 7	8 9 10
41	·6128	6138	6149	6160	6170	6180	6191	6201	6212	6222	10	1 2 3	4 5 6	7 8 9
42	·6232	6243	6253	6263	6274	6284	6294	6304	6314	6325	10	1 2 3	4 5 6	7 8 9
43	·6335	6345	6355	6365	6375	6385	6395	6405	6415	6425	10	1 2 3	4 5 6	7 8 9
44	·6435	6444	6454	6464	6474	6484	6493	6503	6513	6522	10	1 2 3	4 5 6	7 8 9
45	·6532	6542	6551	6561	6571	6580	6590	6599	6609	6618	10	1 2 3	4 5 6	7 8 9
46	·6628	6637	6646	6656	6665	6675	6684	6693	6702	6712	9	1 2 3	4 5 5	6 7 8
47	·6721	6730	6739	6749	6758	6767	6776	6785	6794	6803	9	1 2 3	4 5 5	6 7 8
48	·6812	6821	6830	6839	6848	6857	6866	6875	6884	6893	9	1 2 3	4 4 5	6 7 8
49	·6902	6911	6920	6928	6937	6946	6955	6964	6972	6981	9	1 2 3	4 4 5	6 7 8

COMMON LOGARITHMS $\log_{10} x$

x	0	1	2	3	4	5	6	7	8	9	$\Delta_m +$	1	2	3	4	5	6	7	8	9
															ADD					
50	·6990	6998	7007	7016	7024	7033	7042	7050	7059	7067	9	1	2	3	4	4	5	6	7	8
51	·7076	7084	7093	7101	7110	7118	7126	7135	7143	7152	8	1	2	2	3	4	5	6	6	7
52	·7160	7168	7177	7185	7193	7202	7210	7218	7226	7235	8	1	2	2	3	4	5	6	6	7
53	·7243	7251	7259	7267	7275	7284	7292	7300	7308	7316	8	1	2	2	3	4	5	6	6	7
54	·7324	7332	7340	7348	7356	7364	7372	7380	7388	7396	8	1	2	2	3	4	5	6	6	7
55	·7404	7412	7419	7427	7435	7443	7451	7459	7466	7474	8	1	2	2	3	4	5	6	6	7
56	·7482	7490	7497	7505	7513	7520	7528	7536	7543	7551	8	1	2	2	3	4	5	6	6	7
57	·7559	7566	7574	7582	7589	7597	7604	7612	7619	7627	8	1	2	2	3	4	5	6	6	7
58	·7634	7642	7649	7657	7664	7672	7679	7686	7694	7701	8	1	2	2	3	4	5	6	6	7
59	·7709	7716	7723	7731	7738	7745	7752	7760	7767	7774	7	1	1	2	3	4	4	5	6	6
60	·7782	7789	7796	7803	7810	7818	7825	7832	7839	7846	7	1	1	2	3	4	4	5	6	6
61	·7853	7860	7868	7875	7882	7889	7896	7903	7910	7917	7	1	1	2	3	4	4	5	6	6
62	·7924	7931	7938	7945	7952	7959	7966	7973	7980	7987	7	1	1	2	3	3	4	5	6	6
63	·7993	8000	8007	8014	8021	8028	8035	8041	8048	8055	7	1	1	2	3	3	4	5	6	6
64	·8062	8069	8075	8082	8089	8096	8102	8109	8116	8122	7	1	1	2	3	3	4	5	6	6
65	·8129	8136	8142	8149	8156	8162	8169	8176	8182	8189	7	1	1	2	3	3	4	5	6	6
66	·8195	8202	8209	8215	8222	8228	8235	8241	8248	8254	7	1	1	2	3	3	4	5	6	6
67	·8261	8267	8274	8280	8287	8293	8299	8306	8312	8319	6	1	1	2	2	3	4	4	5	5
68	·8325	8331	8338	8344	8351	8357	8363	8370	8376	8382	6	1	1	2	2	3	4	4	5	5
69	·8388	8395	8401	8407	8414	8420	8426	8432	8439	8445	6	1	1	2	2	3	4	4	5	5
70	·8451	8457	8463	8470	8476	8482	8488	8494	8500	8506	6	1	1	2	2	3	4	4	5	5
71	·8513	8519	8525	8531	8537	8543	8549	8555	8561	8567	6	1	1	2	2	3	4	4	5	5
72	·8573	8579	8585	8591	8597	8603	8609	8615	8621	8627	6	1	1	2	2	3	4	4	5	5
73	·8633	8639	8645	8651	8657	8663	8669	8675	8681	8686	6	1	1	2	2	3	4	4	5	5
74	·8692	8698	8704	8710	8716	8722	8727	8733	8739	8745	6	1	1	2	2	3	4	4	5	5
75	·8751	8756	8762	8768	8774	8779	8785	8791	8797	8802	6	1	1	2	2	3	4	4	5	5
76	·8808	8814	8820	8825	8831	8837	8842	8848	8854	8859	6	1	1	2	2	3	4	4	5	5
77	·8865	8871	8876	8882	8887	8893	8899	8904	8910	8915	6	1	1	2	2	3	4	4	5	5
78	·8921	8927	8932	8938	8943	8949	8954	8960	8965	8971	6	1	1	2	2	3	4	4	5	5
79	·8976	8982	8987	8993	8998	9004	9009	9015	9020	9025	6	1	1	2	2	3	4	4	5	5
80	·9031	9036	9042	9047	9053	9058	9063	9069	9074	9079	5	1	1	2	2	3	3	4	4	5
81	·9085	9090	9096	9101	9106	9112	9117	9122	9128	9133	5	1	1	2	2	3	3	4	4	5
82	·9138	9143	9149	9154	9159	9165	9170	9175	9180	9186	5	1	1	2	2	3	3	4	4	5
83	·9191	9196	9201	9206	9212	9217	9222	9227	9232	9238	5	1	1	2	2	3	3	4	4	5
84	·9243	9248	9253	9258	9263	9269	9274	9279	9284	9289	5	1	1	2	2	3	3	4	4	5
85	·9294	9299	9304	9309	9315	9320	9325	9330	9335	9340	5	1	1	2	2	3	3	4	4	5
86	·9345	9350	9355	9360	9365	9370	9375	9380	9385	9390	5	1	1	2	2	3	3	4	4	5
87	·9395	9400	9405	9410	9415	9420	9425	9430	9435	9440	5	0	1	1	2	2	3	3	4	4
88	·9445	9450	9455	9460	9465	9469	9474	9479	9484	9489	5	0	1	1	2	2	3	3	4	4
89	·9494	9499	9504	9509	9513	9518	9523	9528	9533	9538	5	0	1	1	2	2	3	3	4	4
90	·9542	9547	9552	9557	9562	9566	9571	9576	9581	9586	5	0	1	1	2	2	3	3	4	4
91	·9590	9595	9600	9605	9609	9614	9619	9624	9628	9633	5	0	1	1	2	2	3	3	4	4
92	·9638	9643	9647	9652	9657	9661	9666	9671	9675	9680	5	0	1	1	2	2	3	3	4	4
93	·9685	9689	9694	9699	9703	9708	9713	9717	9722	9727	5	0	1	1	2	2	3	3	4	4
94	·9731	9736	9741	9745	9750	9754	9759	9763	9768	9773	5	0	1	1	2	2	3	3	4	4
95	·9777	9782	9786	9791	9795	9800	9805	9809	9814	9818	5	0	1	1	2	2	3	3	4	4
96	·9823	9827	9832	9836	9841	9845	9850	9854	9859	9863	4	0	1	1	2	2	2	3	3	4
97	·9868	9872	9877	9881	9886	9890	9894	9899	9903	9908	4	0	1	1	2	2	2	3	3	4
98	·9912	9917	9921	9926	9930	9934	9939	9943	9948	9952	4	0	1	1	2	2	2	3	3	4
99	·9956	9961	9965	9969	9974	9978	9983	9987	9991	9996	4	0	1	1	2	2	2	3	3	4

Areas in the right-hand tail of the normal distribution

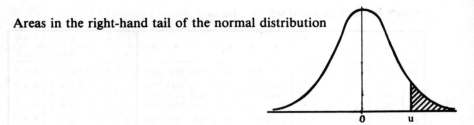

z	.00	.01	.02	.03	.04	.05	.06	.07	.08	.09
0.0	.5000	.4960	.4920	.4880	.4840	.4801	.4761	.4721	.4681	.4641
0.1	.4602	.4562	.4522	.4483	.4443	.4404	.4364	.4325	.4286	.4247
0.2	.4207	.4168	.4129	.4090	.4052	.4013	.3974	.3936	.3897	.3859
0.3	.3821	.3783	.3745	.3707	.3669	.3632	.3594	.3557	.3520	.3483
0.4	.3446	.3409	.3372	.3336	.3300	.3264	.3228	.3192	.3156	.3121
0.5	.3085	.3050	.3015	.2981	.2946	.2912	.2877	.2843	.2810	.2776
0.6	.2743	.2709	.2676	.2643	.2611	.2578	.2546	.2514	.2483	.2451
0.7	.2420	.2389	.2358	.2327	.2296	.2266	.2236	.2206	.2177	.2148
0.8	.2119	.2090	.2061	.2033	.2005	.1977	.1949	.1922	.1894	.1867
0.9	.1841	.1814	.1788	.1762	.1736	.1711	.1685	.1660	.1635	.1611
1.0	.1587	.1562	.1539	.1515	.1492	.1469	.1446	.1423	.1401	.1379
1.1	.1357	.1335	.1314	.1292	.1271	.1251	.1230	.1210	.1190	.1170
1.2	.1151	.1131	.1112	.1093	.1075	.1056	.1038	.1020	.1003	.0985
1.3	.0968	.0951	.0934	.0918	.0901	.0885	.0869	.0853	.0838	.0823
1.4	.0808	.0793	.0778	.0764	.0749	.0735	.0721	.0708	.0694	.0681
1.5	.0668	.0655	.0643	.0630	.0618	.0606	.0594	.0582	.0571	.0559
1.6	.0548	.0537	.0526	.0516	.0505	.0495	.0485	.0475	.0465	.0455
1.7	.0446	.0436	.0427	.0418	.0409	.0401	.0392	.0384	.0375	.0367
1.8	.0359	.0351	.0344	.0336	.0329	.0322	.0314	.0307	.0301	.0294
1.9	.0287	.0281	.0274	.0268	.0262	.0256	.0250	.0244	.0239	.0233
2.0	.02275	.02222	.02169	.02118	.02068	.02018	.01970	.01923	.01876	.01831
2.1	.01786	.01743	.01700	.01659	.01618	.01578	.01539	.01500	.01463	.01426
2.2	.01390	.01355	.01321	.01287	.01255	.01222	.01191	.01160	.01130	.01101
2.3	.01072	.01044	.01017	.00990	.00964	.00939	.00914	.00889	.00866	.00842
2.4	.00820	.00798	.00776	.00755	.00734	.00714	.00695	.00676	.00657	.00639
2.5	.00621	.00604	.00587	.00570	.00554	.00539	.00523	.00508	.00494	.00480
2.6	.00466	.00453	.00440	.00427	.00415	.00402	.00391	.00379	.00368	.00357
2.7	.00347	.00336	.00326	.00317	.00307	.00298	.00289	.00280	.00272	.00264
2.8	.00256	.00248	.00240	.00233	.00226	.00219	.00212	.00205	.00199	.00193
2.9	.00187	.00181	.00175	.00169	.00164	.00159	.00154	.00149	.00144	.00139
3.0	.00135									
3.1	.00097									
3.2	.00069									
3.3	.00048									
3.4	.00034									
3.5	.00023									
3.6	.00016									
3.7	.00011									
3.8	.00007									
3.9	.00005									
4.0	.00003									

Values of e^{-x} (for use with Poisson Distribution)

x	.00	.01	.02	.03	.04	.05	.06	.07	.08	.09
0.0	1.0000	.9900	.9802	.9704	.9608	.9512	.9418	.9324	.9231	.9139
0.1	0.9048	.8958	.8869	.8781	.8694	.8607	.8521	.8437	.8353	.8270
.2	.8187	.8106	.8025	.7945	.7866	.7788	.7711	.7634	.7558	.7483
.3	.7408	.7334	.7261	.7189	.7118	.7047	.6977	.6907	.6839	.6771
.4	.6703	.6637	.6570	.6505	.6440	.6376	.6313	.6250	.6188	.6126
.5	.6065	.6005	.5945	.5886	.5827	.5769	.5712	.5655	.5599	.5543
.6	.5488	.5434	.5379	.5326	.5273	.5220	.5169	.5117	.5066	.5016
.7	.4966	.4916	.4868	.4819	.4771	.4724	.4677	.4630	.4584	.4538
.8	.4493	.4449	.4404	.4360	.4317	.4274	.4232	.4190	.4148	.4107
.9	.4066	.4025	.3985	.3946	.3906	.3867	.3829	.3791	.3753	.3716
1.0	0.3679	.3642	.3606	.3570	.3535	.3499	.3465	.3430	.3396	.3362
1.1	.3329	.3296	.3263	.3230	.3198	.3166	.3135	.3104	.3073	.3042
.2	.3012	.2982	.2952	.2923	.2894	.2865	.2837	.2808	.2780	.2753
.3	.2725	.2698	.2671	.2645	.2618	.2592	.2567	.2541	.2516	.2491
.4	.2466	.2441	.2417	.2393	.2369	.2346	.2322	.2299	.2276	.2254
.5	.2231	.2209	.2187	.2165	.2144	.2122	.2101	.2080	.2060	.2039
.6	.2019	.1999	.1979	.1959	.1940	.1920	.1901	.1882	.1864	.1845
.7	.1827	.1809	.1791	.1773	.1755	.1738	.1720	.1703	.1686	.1670
.8	.1653	.1637	.1620	.1604	.1588	.1572	.1557	.1541	.1526	.1511
.9	.1496	.1481	.1466	.1451	.1437	.1423	.1409	.1395	.1381	.1367
2.0	0.1353	.1340	.1327	.1313	.1300	.1287	.1275	.1262	.1249	.1237
2.1	0.1225	.1212	.1200	.1188	.1177	.1165	.1153	.1142	.1130	.1119
.2	.1108	.1097	.1086	.1075	.1065	.1054	.1044	.1033	.1023	.1013
.3	.1003	.0993	.0983	.0973	.0963	.0954	.0944	.0935	.0925	.0916
.4	.0907	.0898	.0889	.0880	.0872	.0863	.0854	.0846	.0837	.0829
.5	.0821	.0813	.0805	.0797	.0789	.0781	.0773	.0765	.0758	.0750
.6	.0743	.0735	.0728	.0721	.0714	.0707	.0699	.0693	.0686	.0679
.7	.0672	.0665	.0659	.0652	.0646	.0639	.0633	.0627	.0620	.0614
.8	.0608	.0602	.0596	.0590	.0584	.0578	.0573	.0567	.0561	.0556
.9	.0550	.0545	.0539	.0534	.0529	.0523	.0518	.0513	.0508	.0503
3.0	0.0498	.0493	.0488	.0483	.0478	.0474	.0469	.0464	.0460	.0455
3.1	.0450	.0446	.0442	.0437	.0433	.0429	.0424	.0420	.0416	.0412
.2	.0408	.0404	.0400	.0396	.0392	.0388	.0384	.0380	.0376	.0373
.3	.0369	.0365	.0362	.0358	.0354	.0351	.0347	.0344	.0340	.0337
.4	.0334	.0330	.0327	.0324	.0321	.0317	.0314	.0311	.0308	.0305
.5	.0302	.0299	.0296	.0293	.0290	.0287	.0284	.0282	.0279	.0276
.6	.0273	.0271	.0268	.0265	.0260	.0257	.0257	.0255	.0252	.0250
.7	.0247	.0245	.0242	.0240	.0238	.0235	.0233	.0231	.0228	.0226
.8	.0224	.0221	.0219	.0217	.0215	.0213	.0211	.0209	.0207	.0204
.9	.0202	.0200	.0198	.0196	.0194	.0193	.0191	.0189	.0187	.0185
4.0	0.0183									
x	.00	.01	.02	.03	.04	.05	.06	.07	.08	.09

Student's t Critical Points

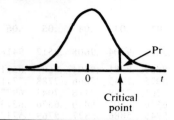

Pr\d.f.	.25	.10	.05	.025	.010	.005	.001
1	1.000	3.078	6.314	12.706	31.821	63.657	318.31
2	.816	1.886	2.920	4.303	6.965	9.925	22.326
3	.765	1.638	2.353	3.182	4.541	5.841	10.213
4	.741	1.533	2.132	2.776	3.747	4.604	7.173
5	.727	1.476	2.015	2.571	3.365	4.032	5.893
6	.718	1.440	1.943	2.447	3.143	3.707	5.208
7	.711	1.415	1.895	2.365	2.998	3.499	4.785
8	.706	1.397	1.860	2.306	2.896	3.355	4.501
9	.703	1.383	1.833	2.262	2.821	3.250	4.297
10	.700	1.372	1.812	2.228	2.764	3.169	4.144
11	.697	1.363	1.796	2.201	2.718	3.106	4.025
12	.695	1.356	1.782	2.179	2.681	3.055	3.930
13	.694	1.350	1.771	2.160	2.650	3.012	3.852
14	.692	1.345	1.761	2.145	2.624	2.977	3.787
15	.691	1.341	1.753	2.131	2.602	2.947	3.733
16	.690	1.337	1.746	2.120	2.583	2.921	3.686
17	.689	1.333	1.740	2.110	2.567	2.898	3.646
18	.688	1.330	1.734	2.101	2.552	2.878	3.610
19	.688	1.328	1.729	2.093	2.539	2.861	3.579
20	.687	1.325	1.725	2.086	2.528	2.845	3.552
21	.686	1.323	1.721	2.080	2.518	2.831	3.527
22	.686	1.321	1.717	2.074	2.508	2.819	3.505
23	.685	1.319	1.714	2.069	2.500	2.807	3.485
24	.685	1.318	1.711	2.064	2.492	2.797	3.467
25	.684	1.316	1.708	2.060	2.485	2.787	3.450
26	.684	1.315	1.706	2.056	2.479	2.779	3.435
27	.684	1.314	1.703	2.052	2.473	2.771	3.421
28	.683	1.313	1.701	2.048	2.467	2.763	3.408
29	.683	1.311	1.699	2.045	2.462	2.756	3.396
30	.683	1.310	1.697	2.042	2.457	2.750	3.385
40	.681	1.303	1.684	2.021	2.423	2.704	3.307
60	.679	1.296	1.671	2.000	2.390	2.660	3.232
120	.677	1.289	1.658	1.980	2.358	2.617	3.160
∞	.674	1.282	1.645	1.960	2.326	2.576	3.090

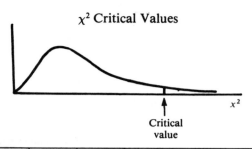

χ^2 Critical Values

Pr d.f.	.250	.100	.050	.025	.010	.005	.001
1	1.32	2.71	3.84	5.02	6.63	7.88	10.8
2	2.77	4.61	5.99	7.38	9.21	10.6	13.8
3	4.11	6.25	7.81	9.35	11.3	12.8	16.3
4	5.39	7.78	9.49	11.1	13.3	14.9	18.5
5	6.63	9.24	11.1	12.8	15.1	16.7	20.5
6	7.84	10.6	12.6	14.4	16.8	18.5	22.5
7	9.04	12.0	14.1	16.0	18.5	20.3	24.3
8	10.2	13.4	15.5	17.5	20.1	22.0	26.1
9	11.4	14.7	16.9	19.0	21.7	23.6	27.9
10	12.5	16.0	18.3	20.5	23.2	25.2	29.6
11	13.7	17.3	19.7	21.9	24.7	26.8	31.3
12	14.8	18.5	21.0	23.3	26.2	28.3	32.9
13	16.0	19.8	22.4	24.7	27.7	29.8	34.5
14	17.1	21.1	23.7	26.1	29.1	31.3	36.1
15	18.2	22.3	25.0	27.5	30.6	32.8	37.7
16	19.4	23.5	26.3	28.8	32.0	34.3	39.3
17	20.5	24.8	27.6	30.2	33.4	35.7	40.8
18	21.6	26.0	28.9	31.5	34.8	37.2	42.3
19	22.7	27.2	30.1	32.9	36.2	38.6	43.8
20	23.8	28.4	31.4	34.2	37.6	40.0	45.3
21	24.9	29.6	32.7	35.5	38.9	41.4	46.8
22	26.0	30.8	33.9	36.8	40.3	42.8	48.3
23	27.1	32.0	35.2	38.1	41.6	44.2	49.7
24	28.2	33.2	36.4	39.4	43.0	45.6	51.2
25	29.3	34.4	37.7	40.6	44.3	46.9	52.6
26	30.4	35.6	38.9	41.9	45.6	48.3	54.1
27	31.5	36.7	40.1	43.2	47.0	49.6	55.5
28	32.6	37.9	41.3	44.5	48.3	51.0	56.9
29	33.7	39.1	42.6	45.7	49.6	52.3	58.3
30	34.8	40.3	43.8	47.0	50.9	53.7	59.7
40	45.6	51.8	55.8	59.3	63.7	66.8	73.4
50	56.3	63.2	67.5	71.4	76.2	79.5	86.7
60	67.0	74.4	79.1	83.3	88.4	92.0	99.6
70	77.6	85.5	90.5	95.0	100	104	112
80	88.1	96.6	102	107	112	116	125
90	98.6	108	113	118	124	128	137
100	109	118	124	130	136	140	149

Index

Action Limits 269
Additional Information (value of) 203-205
Addition, general law of 176, 178
Alternative hypothesis 305
Annual Abstract of Statistics 380
Approximation Chapter 4
A priori probability 174
Arithmetic mean 72-80
A.W.C.L. 270-271
a value 118, 121-122

Bar Chart 19-22
  component 21-22
  compound 21-22
  negative figures 21
Base period 131-132
Bias in sample 227-228
Binomial Distribution 212-217, 337
  coefficients 215-216
  mean of 217
  normal approximation to 217-219
  standard deviation of 217
Bivariate distribution 347-348

Census of distribution 389-391
  of population 381-383
  of production 388-389
Central limit theorem 252-254
Central Tendency Chapter 5
  arithmetic mean 72-80
  geometric mean 87-88
  harmonic mean 86-87
  median 80-83
  mode 83-86
Chain base index 132
Chi-Square distribution Chapter 21
  errors in the use of 334-336
Circular diagram 22-23
Class interval 9-10
Cluster sampling 232
Collectively exhaustive events 172
Combinations 215-216
Conditional probability 183-185
Confidence limits 261-264
  for mean 261-262
  for proportions 264
Contingency tables 338-341
Control charts 271-272

Correlation Chapter 22
  coefficient of 362
  interpretation of 363-364
  negative 362
  Rank coefficient of 365-366
  significance of 364
Covariance 353
Critical values 311
Cumulative frequency curve 28-29
Curvilinear relationship 355, 364

Data, continuous 7
  discrete 8
  primary 2
  secondary 2
Decision criteria Chapter 12
Degrees of freedom 289-290, 341
Department of Employment Gazette 388
Dependent variable 350
Determination, coefficient of 361
Deviation 73, 145
Diagrams Chapter 2, 3
Difference between means 313-5, 318-20
  between proportions 320-323
Dispersion Chapter 9
  coefficient of variation 149
  mean absolute deviation 145-146
  quartile coefficient of variation 150
  quartile deviation 144
  range 142-143
  relative dispersion 149-150
  semi-interquartile range 143
  standard deviation 146-148
  variance 146
Distribution of sample means 250-252
  of sums and differences 255-256

Empirical probability 174-175
Equally likely events 172-173
Error Chapter 4
  absolute 62
  biassed 61-62
  calculations involving 62-65
  relative 62
  unbiassed 61-62
Estimation Chapter 18
  of population mean 280-282
  of variance 285-286

Expected Monetary Value (EMV) 201-202
Exponential Smoothing 113-122
Extrapolation

Factorial notation 220
Family Expenditure Survey 384
Fisher's Index 132
Forecasting 110-112
Fractional Defective Quality
  Control 274-275
Freedom, degrees of 289-290, 341
Frequency 3
  cumulative 28-29
  curve 29
  distribution 6
  expected 332
  observed 331-
  polygon 25-26

Geometric mean 87-88
Goodness of fit 336-338
Graphs Chapter 3

Harmonic mean 86-87
Histogram 23-28
Household Expenditure Survey 383
Hypothesis 300
Hypothesis Testing Chapter 19

Ideograph 51-52
Independent events 177
Independent variable 350
Index numbers Chapter 8
  base period 131-132
  chain base 132
  defined 126
  deflating a series 133-134
  Fisher 132
  Laspeyre 127-128
  price relative 129
  simple aggregative 128
  volume index 130-131
  weighting 128
Index of Retail Prices 383-386
Interpolation 88

Laspeyre index 127-128
Least squares line 352-355
Linear relationship 349-350
Linear trend 356-357
Logarithmic graph 40-46
Lorenz curve 46-49

Manpower statistics 386-388
maximax criterion 197-198
maximin criterion 198
Mean arithmetic 72-80
  estimation of population 280-282

Mean geometric 87-88
  harmonic 86-87
Mean range 273
Median 80-83
Mesokurtic 162
Midpoint 7-9
minimax criterion 198-199
Misuse of statistics 51-54
Mode 83-86
Moving average 101-103
Multiplication law 177
Multi-stage sampling 231-232
Mutually exclusive events 172

National Income statistics 391-393
Non-parametric tests 330
Normal distribution Chapter 10
Normal equations 352
Normal tables, use of 162
Null hypothesis 304

Ogive 28-29
One-sided test 306-307

Paasche Index 130
Paired samples 318
Parameters 330
Pearson's correlation coefficient 362
Percentage strata graph 35
Performance chart 300-301, 303
Personal interview 238-240
Pictogram 51-52
Pie diagram 22-23
  weakeness of 23
Pilot survey 244
Poisson distribution 219-223, 336
  conditions necessary for 221-
  e 220
Population 5
Population mean, estimate of 280-282
  variance estimate of 285-288
Population statistics 381-383
Postal enquiry 236-238
Price relative 129
Primary data 2
Probability Chapter 11
  a priori 124
  conditional 183-185
  empirical 174-5
  laws of 175-178
  measurement of 173
  subjective 175
Probability distributions Chapter 13
  binomial 212-217
  nature of 211-212
  Pascal's triangle 214-215
  Poisson distribution 219-223
Proportion, estimating of 282-283

Quality control Chapter 17
Quasi random sampling 229
Questionnaire 1-2, 240-243
Quota sampling 232-234

Random numbers 228
   sampling 227-228
   selection 227
Range 142-143
Rank correlation 365-366
Ratio scale graph 40-46
regret matrix 198-200
Regression Chapter 22
Regression line 350-352
Relative dispersion 149-150
Residuals 99, 109-110
Response rate 237-238
Retail Price Index 383-386
rollback principle 206-208

Sample 6
Sample means, distribution of 250-252
Sample proportions (tests of) 282-283
Sample size 283-285
Sample surveys Chapter 15
Sampling design Chapter 14
   cluster 232
   error 233, 235-236
   frame 228-229
   multi-stage 231-232
   quasi random 229
   quota 232-234
   stratified 229-231
   systematic 229
   theory Chapter 16
Scatter diagram 50-51, 349
Seasonal variation 99, 100
   elimination of 108-109
Secondary data 2
Semi-interquartile range 144
Significance, statistical Chapter 20
   difference between means 313-315, 318-320
   difference between proportions 320-323
   level of 309
   of sample means 311-313, 315-317
   tests of Chapter 20
Simple aggregative index 128
Skewness coefficient of 150

Spearman's coefficient of correlation 365-366
Standard deviation 146-148
Standard error of mean 251
   of proportion 260
Standard Industrial Classification 387
Standard Scores 161-163
States of nature 197
Strata graph 33-35
Strategies 197
Subjective probability 175

Table of a_n 273
Tabulation 3-5
   principles of 5
t-distribution 287-289
Test of sample means 280-282
Time Series Chapter 6 & 7, 356-8
   residuals 99, 100
   seasonal Variation 99, 100, 106-108
   trend 99, 100
Tree diagrams 181-183
Trend-calculation of 101-103
   centreing 103
   linear 356-357
Two-sided tests 307-308
Type I error 298-300
Type II error 301-303

Variable 2
   dependent 350
   independent 350
Variable sampling fraction 230
Variance 146
   estimate of population 285-286
Variation, coefficient of 149
   explained and unexplained 360
   quartile coefficient of 150
Vital statistics 381-383
Volume index 130-131

Warning limits 269
Weighted average 74-76

Yate's correction 334

Z chart 49-50
Z scores 161-163